Unique Radio Innovation for the 21st Century

Damith C. Ranasinghe · Quan Z. Sheng ·
Sherali Zeadally
Editors

Unique Radio Innovation for the 21st Century

Building Scalable and Global RFID Networks

 Springer

Editors
Dr. Damith C. Ranasinghe
University of Adelaide
School of Electrical & Electronic
Engineering
Auto-ID Lab
Adelaide SA 5005
Australia
damith@eleceng.adelaide.edu.au

Dr. Quan Z. Sheng
University of Adelaide
School of Computer Science
Adelaide SA 5005
Australia
qsheng@cs.adelaide.edu.au

Dr. Sherali Zeadally
University of the District of Columbia
Dept. Computer Science & Information
Technology
Connecticut Ave. NW., 4200
Washington, District of Columbia 20008
USA
szeadally@udc.edu

ISBN 978-3-642-03461-9   e-ISBN 978-3-642-03462-6
DOI 10.1007/978-3-642-03462-6
Springer Heidelberg Dordrecht London New York

Library of Congress Control Number: 2010931050

Cover design: KuenkelLopka GmbH, Heidelberg

Printed on acid-free paper

Springer is part of Springer Science+Business Media (www.springer.com)

Preface

Over the years, RFID has gone through many transformations, from traditionally being known as a contactless access card technology to connecting information and physical flows of objects (i.e., "Networked RFID"). Recent advances in communication technologies, middleware, and Web services are enabling the development and deployment of RFID technology which continues to evolve and has now become part of the so-called "Internet of Things". Indeed, this future Internet will provide an environment where everyday physical objects such as buildings, commodities, and in general all "things" are readable, recognizable, addressable, and even controllable using services via the Web. The capability of connecting and integrating information from both the physical world and the virtual one not only affects the way how we live, but also creates tremendous new RFID-enabled business opportunities such as the support of independent living of elderly persons, efficient supply chains, anti-counterfeiting and improved environmental monitoring. Many organisations are planning or have already exploited RFID in their main operations to take advantage of the potential for greater automation, efficient business processes, and inventory visibility.

RFID data management, scalable information systems, business process reengineering, and evaluating investments are emerging as significant technical challenges to applications underpinned by new developments in RFID technology. In this book, we present contributions from world leading experts on the latest developments and state-of-the-art results in the RFID field to address these challenges. The book offers a comprehensive and systematic description of technologies, architectures, and methodologies that are required to develop secure, scalable, and reliable RFID networks.

The book consists of five major parts. The first part presents a view into the future evolution of this technology and what might be possible in the near horizon while also taking a systematic look at privacy and security issues that cannot be ignored in the euphoria. The following four parts cover topics on RFID data management, global RFID information architectures and systems, innovative RFID-enabled applications and, business related issues.

This book serves well as a valuable reference point for researchers, educators, and engineers who are working in RFID data management, RFID application development, as well as graduate students who wish to understand, learn, and discover opportunities in this emerging research and development area. Furthermore, it is also of general interest to anyone deploying RFID applications, particularly those working on the design, development, and deployment of large-scale or open-loop networked RFID systems. The comprehensive coverage of topics related to information systems will also be of benefit to IT managers working in the area of RFID systems or involved in managing and overseeing RFID deployments. It is our hope that the work presented in this book will stimulate new discussions and generate original ideas that will further develop this important area.

We thank the authors for their outstanding and timely contributions. We would also like to thank the reviewers for their expertise and support in reviewing all submissions and for providing valuable feedback to the contributing authors. It has indeed been a great pleasure to work with such a dedicated group of world leading researchers who have contributed to make this book possible. Moreover, we are grateful to Springer for the opportunity to publish this book. Our special thanks go to Hermann Engesser for his continued support and Dorothea Glaunsinger of Springer for her support and professionalism during the whole publication process of this book.

February 2010   Damith C. Ranasinghe
                                                                              Quan Z. Sheng
                                                                              Sherali Zeadally

Contents

Contents

.

Contributors

Nova Ahmed College of Computing, Georgia Institute of Technology, Atlanta, GA 30332, USA, nova@cc.gatech.edu

Jasser Al-Kassab Institute for Technology Management (ITEM-HSG), Auto-ID Labs St. Gallen/Zurich, and SAP Research CEC St. Gallen, University of St. Gallen, St. Gallen, Switzerland, jasser.al-kassab@unisg.ch

Guillermo Arrebola AT4 wireless. S.A. C/ Severo Ochoa, n° 2, Parque Tecnológico de Andalucía, 29590 Campanillas (Málaga), Spain, garrebola@at4wireless.com

Janie Baños AT4 wireless. S.A. C/ Severo Ochoa, n° 2, Parque Tecnológico de Andalucía, 29590 Campanillas (Málaga), Spain, jbanos@at4wireless.com

Michael Beye Security Lab, Faculty of Electrical Engineering, Mathematics and Computer Science, Delft University of Technology (TU Delft), Mekelweg 4, 2628 CD, Delft, The Netherlands, M.R.T.Beye@tudelft.nl

Philipp Blome Supply Chain Innovation, MGI Metro Group Information Technology GmbH, Duesseldorf, Germany, phillip.blome@mgi.de

Dimitris Bochtis Department of Management Science and Technology, ELTRUN Research Lab, Athens University of Economics & Business, Athens, Greece, bochtis.d@gmail.com

José J. Cantero AT4 wireless. S.A. C/ Severo Ochoa, n° 2, Parque Tecnológico de Andalucía, 29590 Campanillas (Málaga), Spain, jjcantero@at4wireless.com

Sergio Cavalieri Department of Industrial Engineering, CELS, Research Centre on Logistics and After Sales Service, University of Bergamo, Viale Marconi, 5, 24044 Dalmine, Italy, sergio.cavalieri@unibg.it

Yoon S. Chang School of Air Transport, Transportation and Logistics, Ubiquitous Technology Application Research Center, Korea Aerospace University, Goyang, Republic of Korea, yoonchang@kau.ac.kr

Leonardo W. F. Chaves SAP Research, CEC Karlsruhe,
Vincenz-Priessnitz-Strasse 1, 76131 Karlsruhe, Germany,
leonardo.weiss.f.chaves@sap.com

Peter H. Cole Auto-ID Lab Adelaide, The University of Adelaide, Adelaide, SA
5005, Australia, cole@eleceng.adelaide.edu.au

Roberto De Virgilio Universitá Roma Tre, Rome, Italy, dvr@dia.uniroma3.it

Eugenio Di Sciascio Politecnico di Bari, Via Re David 200, Bari BA I-70125,
Italy, disciascio@poliba.it

Andriana Dimakopoulou Department of Management Science and Technology,
Athens University of Economics and Business, Athens, Greece,
andrianadima@aueb.gr

Stefano Dotti Department of Industrial Engineering, CELS, Research Centre on
Logistics and After Sales Service, University of Bergamo, Viale Marconi, 5, 24044
Dalmine, Italy, stefano.dotti@unibg.it

Stelios Eliakis Department of Management Science and Technology, ELTRUN
Research Lab, Athens University of Economics & Business, Athens, Greece,
eliakis@aueb.gr

Elgar Fleisch Institute for Technology Management (ITEM-HSG) and
Department of Management, Technology and Economics (D-MTEC), ETH Zürich,
Auto-ID Labs St. Gallen/Zurich, University of St. Gallen, St. Gallen, Switzerland,
elgar.fleisch@unisg.ch

Miguel A. Guijarro AT4 wireless. S.A. C/ Severo Ochoa, n° 2, Parque
Tecnológico de Andalucía, 29590 Campanillas (Málaga), Spain,
maguijarro@at4wireless.com

Oliver Günter Institute of Information Systems, Humboldt-Universität zu Berlin,
Berlin, Germany, guenther@wiwi.hu-berlin.de

H. Halfar Planning and Control of Production Systems, University of Bremen,
Hochschulring 20, D 28359 Bremen, Germany, hal@biba.uni-bremen.de

Zhonghao Hu Auto-ID Lab Adelaide, The University of Adelaide, Adelaide, SA
5005, Australia, nathanhu@eleceng.adelaide.edu.au

M.-A. Isenberg Planning and Control of Production Systems, University of
Bremen, Hochschulring 20, D 28359 Bremen, Germany, ise@biba.uni-bremen.de

Lenka Ivantysynova University Potsdam, Potsdam, Germany,
ivantysynova@cs.uni-potsdam.de

Angeliki Karagiannaki Department of Management Science and Technology,
Athens University of Economics and Business, Athens, Greece,
akaragianaki@aueb.gr

Elena Legnani Department of Industrial Engineering, CELS, Research CENTRE on Logistics and After-sales Service, University of Bergamo, Viale Marconi, 5, 24044 Dalmine, Italy, elena.legnani@unibg.it

Mikko Lehtonen ETH Zürich, Information Management, Scheuchzerstrasse 7, 8092 Zürich, Switzerland, mlehtonen@ethz.ch

Xue Li School of Information Technology and Electrical Engineering, The University of Queensland, St. Lucia, Brisbane, QLD 4072, Australia, xueli@itee.uq.edu.au

Tao Lin Amitive Inc., 1400 Fashion Island Blvd, San Mateo, CA 94404, USA, tao.lin@amitive.com

Chengfei Liu Centre for Complex Software Systems and Services, Faculty of Information and Communication Technologies, Swinburne University of Technology, Melbourne, VIC, Australia, cliu@groupwise.swin.edu.au

Peiya Liu Integrated Data Systems Department, Siemens Corporate Research, Princeton, NJ, USA, peiya.liu@siemens.com

Aikaterini Mitrokotsa Security Lab, Faculty of Electrical Engineering, Mathematics and Computer Science, Delft University of Technology (TU Delft), Mekelweg 4, 2628 CD, Delft, The Netherlands, A.Mitrokotsa@tudelft.nl

John P.T. Mo School of Aerospace, Mechanical and Manufacturing Engineering, RMIT University, P.O. Box 71, Bundoora, VIC 3083, Australia, john.mo@rmit.edu.au

Zoltán Nochta SAP Research, CEC Karlsruhe, Vincenz-Priessnitz-Strasse 1, 76131 Karlsruhe, Germany, zoltan.nochta@sap.com

Chang H. Oh EXIS Software Engineering, #1801, Building 2, DMC Iaan Sang-am, 1653 Sangam-dong, Mapo-gu, Seoul, Republic of Korea, mailbox@exis.co.kr

George Papadopoulos OTEplus S.A., Technical & Business Solutions, Athens, Greece, papadopoulos@aueb.gr

Antonis Paraskevopoulos OTEplus S.A., Technical & Business Solutions, Athens, Greece, aparask@aueb.gr

Pedro Peris-Lopez Security Lab, Faculty of Electrical Engineering, Mathematics and Computer Science, Delft University of Technology (TU Delft), Mekelweg 4, 2628 CD, Delft, The Netherlands, P.PerisLopez@tudelft.nl

Roberto Pinto Department of Industrial Engineering, CELS, Research Centre on Logistics and After Sales Service, University of Bergamo, Viale Marconi, 5, 24044 Dalmine, Italy, roberto.pinto@unibg.it

Antonio Plaza AT4 wireless. S.A. C/ Severo Ochoa, n° 2, Parque Tecnológico de Andalucía, 29590 Campanillas (Málaga), Spain, aplaza@at4wireless.com

Elias Polytarchos Department of Management Science and Technology, ELTRUN Research Lab, Athens University of Economics & Business, Athens, Greece, ipoli@intracom.gr

Katerina Pramatari Department of Management Science and Technology, Athens University of Economics and Business, Athens, Greece, k.pramatari@aueb.gr

Umakishore Ramachandran College of Computing, Georgia Institute of Technology, Atlanta, GA 30332, USA, rama@cc.gatech.edu

Damith C. Ranasinghe Auto-ID Lab Adelaide, The University of Adelaide, Adelaide, SA 5005, Australia, damith@eleceng.adelaide.edu.au

Yacine Rekik EMLYON Business School, 23 avenue Guy de Collongues, 69134 Ecully, France, rekik@em-lyon.com

Daniel Ronzani Copenhagen Business School, Centre for Applies ICT, 2000 Frederiksberg, Denmark, dan@zurich.ibm.com

Michele Ruta Politecnico di Bari, Via Re David 200, Bari BA I-70125, Italy, m.ruta@poliba.it

B. Scholz-Reiter Planning and Control of Production Systems, University of Bremen, Hochschulring 20, D 28359 Bremen, Germany, bsr@biba.uni-bremen.de

Floriano Scioscia Politecnico di Bari, Via Re David 200, Bari BA I-70125, Italy, f.scioscia@poliba.it

M. Teucke Planning and Control of Production Systems, University of Bremen, Hochschulring 20, D 28359 Bremen, Germany, tck@biba.uni-bremen.de

Frédéric Thiesse Institute for Technology Management (ITEM-HSG) and Auto-ID Labs St. Gallen/Zurich, University of St. Gallen, St. Gallen, Switzerland, frederic.thiesse@unisg.ch

Riccardo Torlone Universitá Roma Tre, Rome, Italy, torlone@dia.uniroma3.it

Leigh H. Turner Invertech Electronics Pty. Ltd, P.O. Box 3334 Rundle Mall, Adelaide, SA 5000, Australia, leigh.turner@ieee.org

Dieter Uckelmann Planning and Control of Production Systems, University of Bremen, Hochschulring 20, D 28359 Bremen, Germany, uck@biba.uni-bremen.de

Fusheng Wang Center for Comprehensive Informatics, Emory University, Druid Hills, GA, USA, fusheng.wang@emory.edu

Gerd Wolfram METRO AG, Duesseldorf, Germany, gerd.wolfram@metro.de

Xiaohui Zhao Information Systems Group, School of Industrial Engineering and Innovation Sciences, Eindhoven University of Technology, Eindhoven, The Netherlands, x.zhao@tue.nl

Holger Ziekow Institute of Information Systems, Humboldt-Universität zu Berlin, Berlin, Germany, ziekow@wiwi.hu-berlin.de

About the Editors

Dr. Damith Ranasinghe received a BE in Information Technology and Tele-communication Engineering from The University of Adelaide with first class honours, in 2002. Since graduation he has worked in Auto-ID Labs at MIT, USA, Cambridge University, UK and The University of Adelaide, where he completed his PhD degree in Electrical and Electronic Engineering in 2007. Dr. Ranasinghe is currently leading research at the Adelaide Auto-ID Lab where he is the Associate Director of the laboratory. His current research interests are focused in the areas of pervasive computing, distributed computing and information systems, in particular, the areas of RFID Technology and lightweight cryptography for low cost RFID systems.

Dr. Quan Z. Sheng is currently a senior lecturer in the School of Computer Science at the University of Adelaide. Sheng holds a PhD degree in computer science from the University of New South Wales (UNSW) and did his post-doc as a research scientist at CSIRO ICT Centre. His main research interests include Web services, service-oriented architecture, RFID, sensor networks, and pervasive computing. Dr. Sheng is the founding chair of the International Workshop on RFID Technology. Dr. Michael Sheng has edited four books and published more than 50 refereed technical papers in premier international journals and conferences including VLDB Journal, IEEE Internet Computing, Communications of the ACM, VLDB, WWW, ICDE, and ICSE. Dr Sheng is the recipient of the Microsoft Research Fellowship (2003–2004) and CSC Fellowship (1999–2000). He is a member of the IEEE and the ACM.

Dr. Sherali Zeadally is currently an Associate Professor in the Department of Computer Science and Information Technology at the University of the District of Columbia in Washington DC, USA where he is the Founder of the UDC's Network Systems laboraTory (NEST). Dr. Sherali Zeadally is the Co-Editor-in-Chief of the International Journal of Internet Protocol Technology, and he currently serves as Editor/Associate Editor/Editorial board member of 13 peer reviewed international journals (published by Springer, Elsevier Science, Academic Press, etc). To date, he has authored/co-authored over 120 refereed technical publications in books, journals, conference, and workshop proceedings. His recent RFID activities include RFID publications in IEEE Computer, IEEE Internet Computing, Journal of

Information System Frontiers, Co-Chairing of the "International Workshop on RFID Technology: Concepts, Applications, Challenges" held in 2008 and 2009, Co-Guest editor of Special issues in various International journals (IJIPT, Information Systems Frontiers) on RFID. He is a Fellow of the British Computer Society (FBCS), UK, and a Fellow of the Institution of Engineering and Technology (FIET), UK.

Part I
Future of RFID

The Next Generation of RFID Technology

Peter H. Cole, Leigh H. Turner, Zhonghao Hu, and Damith C. Ranasinghe

Abstract The next generation of RFID will be governed by developments which have occurred in the production of printed semiconductors and in the manufacturing techniques by which RFID tags can be produced using these new materials. The paper considers all of these matters as well as protocols that are appropriate for printed tags.

1 Introduction

This chapter will be concerned with the future of RFID. In considering drivers of the future we observe that the world is never static. We take the view that RFID involves:

1. manufacturing,
2. technology,
3. protocols and
4. applications.

We conclude that in the future, RFID will involve a change in one or more of these items. Often a new development in one area will demand new developments in some others.

We have observed the emergence of printed electronics as a new development which we believe will have a significant impact on the future, and in this chapter pursue its implications in the area of manufacturing, and in each of the other three interrelated areas.

The future of RFID will see the traditional silicon integrated circuit with its complex and costly manufacture and antenna interconnect challenges gradually relegated to high-end tags requiring only the utmost operational performance in terms

P.H. Cole (✉)
Auto-ID Lab Adelaide, The University of Adelaide, Adelaide, SA 5005, Australia
e-mail: cole@eleceng.adelaide.edu.au

D.C. Ranasinghe et al. (eds.), *Unique Radio Innovation for the 21st Century*,
DOI 10.1007/978-3-642-03462-6_1, © Springer-Verlag Berlin Heidelberg 2010

of achievable reading distance, memory features and data capacity, speed of tag anti-collision arbitration and overall sophistication. Those supply chain market segments and other demanding applications are seen as remaining the exclusive province of traditional integrated circuit based tags.

EPCglobal UHF and HF Gen 2 (see http://www.epcglobalinc.org/standards/) are one example of an application space or area of the supply chain that will likely remain the domain of traditional and now complex albeit high performance single-crystal IC based tags. Alternative printed semiconductor based tags are not expected to significantly encroach on and compete in that space, but rather seen to flourish in the low and middle end applications where tag or label cost is the principal determinant and barrier to adoption, and where a pragmatic and judicious trade-off between cost and tag functionality is tolerable.

Fully printed RFID can potentially be one tenth or less of the cost to manufacture compared to the conventional silicon chip based tags (Frequently asked questions 2010). The target very high volume markets for printed tags include consumer product brand protection and authentication, disposable mass transit fare card ticketing, consumer retail product promotions, embedded product intelligence, etc. Many of these applications will capitalise on the eventual ubiquity of low cost readers, NFC technology and NFC enabled mobile phones and their wireless 3G internet connectivity, i.e. printed RFID manufacturers are going after the enormous market applications space where there is a good fit between the functional capability of the printed semiconductor technology and the end-use requirements; and importantly, which cannot ever be met by traditional IC based tag manufacturing methods. There are no "free lunches" however; simple low-cost fully printable RFID tags and labels come at a significant cost, complexity and functionality trade-off.

The imminent arrival of an alternative commercial printed semiconductor industry is nevertheless seen to open hitherto unachievable low cost points for RF labels and is expected to trigger large scale mass market adoption. This anticipated ubiquitous adoption has not happened with tags of cost constrained conventional single-crystal IC design and construction because of the RFID industry's hitherto inability and failure to respond to and meet end user demand for truly low cost.

Printed electronics seeks to break the above nexus. The new and rapidly emerging field of printed electronics is an art of low cost semiconductor manufacturing which involves a combination of electronics and printing techniques. In recent years, there has been a significant level of interest in printed electronics since it is believed that it will realize substantially lower cost electronic systems than those available from conventional single crystal integrated circuit (IC) chip based circuit fabrication. Hence, printed electronics is often conceived as a feasible way to solve the high cost problem limiting widespread deployment of RFID tags through dropping the manufacturing cost per tag to the sub one cent level when processes and manufacturing plant is gradually up-scaled to high volume production. Manufacturers of printed RFID are projecting early selling prices of only a few cents in the billions of RFID labels and some foresee sub one cent pricing in much higher volumes.

Finally we can note that printed electronics is an enabling technology that also opens the future prospect for realizing integrated electronic article surveillance

(EAS) tags, i.e. labels with an added Electronic Product Code (EPC) or similar style product ID functionality. Enhanced performance 8.2 MHz EAS labels based on a high Q printed MOS capacitor element have also been demonstrated that, unlike conventional existing EAS labels, exhibit no Lazarus effect (tag function returns from the dead due to spontaneous healing of the capacitor dielectric layer) when cancelled with existing retail store EAS tag countertop infrastructure.

This chapter discusses in Sect. 2 the processes for low cost manufacturing in RFID. Section 3 makes comparison between organic and silicon ink printed technologies. In Sect. 4 printed semiconductor standards will be compared with those for single crystal semiconductors. Finally we present our conclusions on promising RFID technologies emerging in the 21st century.

2 Low Cost Tag Manufacturing

We commence this section by noting that costs for the kind of traditional IC based tags presently in volume production today have a significant assembly and ancillaries cost; e.g. in the chip attachment, and antenna manufacturing cost that collectively keep systemic tag costs high even if the IC chip component cost becomes vanishingly small. Additionally the cost of the tag substrate material is also a significant cost contributor, particularly for larger form factor tags. Future low cost manufacturing techniques and processes will need to address these presently limiting issues before significantly lower production costs of any circuit implementation can be commercially realized.

Currently, tag antenna manufacturing using material conserving additive processes such as printing electrically conductive inks or electroplating metals are estimated to make up less than 20% of the market today. Silver based inks are expensive, but newly developed metal nano-particle inks may show promise for lower cost. But that percentage is predicted to grow because printing antennas and selectively depositing conductors with plating processes offers significant cost savings over subtractive etching them from copper or aluminum clad laminate, which is the most common method of RFID antenna manufacture today. As a result of recent advances in materials science and nano-particle technology, ink design and printing processes, entire functional circuits and antenna can now be printed out of conductive and semiconductor polymer inks, which mean that entire RFID tags can now be printed.

Unlike the case with printed electronics RFID circuits, conventional CMOS IC based tag vendors can offer full-feature EPCglobal UHF Class I Generation 2 style tag functionality (particularly in the key areas of data security and high performance anti-collision protocols) with a small silicon die size that approaches the high speed mechanical handling limit for flip-chip naked die or strap or interposer style circuit attach. However, there are well understood and accepted asymptotic limits to the systemic manufacturing cost constraints of producing this kind of conventional non-printed RFID tags. Much of the mainstream tag manufacturing industry is today operating at or close to that limit; and the big manufacturers have practical strategies for scalable volume increases should market demand ever call for it. But that cost unfortunately does not downward scale with that volume up-scaling; that is the unique feature of printed electronics alone.

Regardless of the way in which the electronic circuit portion of the tag is implemented, manufacturers have hitherto failed to adequately address the holistic or systemic tag manufacturing costs in a demonstrable way. It is seen as a mistake to focus on one narrow aspect of the tag economics equation to the exclusion of these other factors.

We can introduce here the new nomenclature PIC, representing the acronym *Printed IC*. By way of representative example, the present PIC from printed electronics company Kovio Inc. is shown in Fig. 1.

This device has a long and thin aspect ratio form factor with electrical interface pads at each end with the structure designed to act as the crossover bridge for the HF antenna coil. This configuration eliminates the extra processing step and cost of printing a dielectric layer over the turns of the coil and subsequently laying down a conductive silver paste conductor to connect the antenna with the circuit pads.

Future RFID antenna manufacturing is likely to embrace large scale adoption of wide web continuous roll-to-roll additive processes based on the flash high speed electroplate deposition of copper onto polymer plastic or coated paper substrate materials on top of a direct digitally patterned seed layer. Such low cost antenna realization technologies have been in existence for several years but have not yet found widespread adoption against the backdrop of a hitherto constrained global demand for RFID labels, and where manufacturing volumes have not yet reached the sweet spot at which large economies kick-in with this process and provide compelling return on the plant capital investment.

Another recently developed very promising antenna mass production technology is based on wide web roll-to-roll high speed laser ablation/cutting of very low cost paper clad aluminum foil to fabricate HF and UHF antenna patterns. Although

Fig. 1 Printed IC with circa 1000 TFT devices (Photograph courtesy of Kovio Inc. © Kovio Inc.)

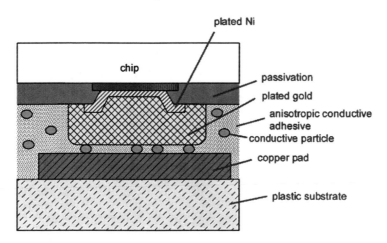

Fig. 2 Anisotropic Conductive Film (ACF) circuit attach

not an additive process, such metal clad foils are used in huge volumes in the consumer packaging industry and are very cheap raw material from which to fabricate antennas.

The Anisotropic Conductive Film (ACF) based thermosetting epoxy flip-chip IC bumped pad die or strap attach concept is shown in Fig. 2. This attachment method remains a major contributor of cost in today's mainstream tag manufacturing processes. Future low cost circuit attach techniques are likely to embrace electroplating bonding of the antenna metal directly to the IC or PIC pads. Such techniques have been shown to be commercially viable using substrate embossing and chip placement robotics. These kinds of new methods are likely to find commercial utility as future demand for very high volume low cost labels comes about.

As discussed above, there has been a significant level of interest by manufacturers in producing electronic devices through the use of a variety of printing processes. This has been to achieve substantially lower cost for realising mass produced electronic systems than can be achieved by conventional single crystal IC based fabrication. Such printing is expected to allow for the production of logic and analog circuit elements on flexible low-cost substrates such as paper, plastic and metal foils. Semiconductor circuit printing techniques potentially enable low manufacturing cost because of the greatly simplified process steps, low raw materials cost, and relatively small capital investment in the manufacturing facilities compared to a modern CMOS semiconductor fabrication plant. A comparison between conventional electronics and the printed electronics in terms of manufacturing process and costs is shown in Table 1.

As a result, printed electronics is considered to be the feasible way to solve the high tag cost problem in the RFID industry and to drop the cost per tag to one cent. However, currently and in the long term, the performance (possible operating frequencies) and compactness (number of transistors achievable) of printed electronics are much lower than those of conventional electronics. Therefore, the potential of

Table 1 Comparison between conventional electronics and printed electronics

	Conventional electronics	Printed electronics
Factory investment (Capital expenditure)	Very high for clean room and plant facilities (costing several 100s of million dollars)	No special needs
Processing	Photo-lithography etching based processing (complex processing steps)	Additive processing (simple processing steps, low raw material cost)
Substrate	Silicon wafer (many steps involved in preparation, high raw material cost)	Plastic, paper or metal foil (allows low-cost roll to roll or sheet feed processing

low cost of printed electronics can be realized only if it is achieved at an acceptable performance and compactness. These two features of performance and compactness are related to the combination of printing techniques and the printed electronic technologies applied.

In particular, the functionality and performance of an electronic circuit depends on the performance of each transistor composing the circuit and the number of these transistors, i.e. the circuit complexity. To enhance the performance of each transistor, the mobility of the charge carriers in the transistors should be maximized and the channel length between the drain and source of the transistors should be minimized. For organic printed electronics, the mobility is negatively related to the viscosity of the printed ink and for all printed electrics the achievable channel length is defined by the resolution of the adopted printing techniques (Sirringhaus et al. 2006; Knobloch et al. 2004; OE-A Roadmap for Organic and Printed Electronics 2009).

Furthermore, the resolution or minimum feature size also decides the circuit complexity. It is an application consideration rather than a technical issue because although you can have as many printed transistor integrated in the circuit as you want to increase the circuit functional-complexity, if the resolution is too low, the size of each transistor and also that of the complex printed circuit will be too big to implement in many applications.

Additionally, Subramanian (Sirringhaus 2009) concludes that the current throughput of organic printed electronics is still very poor which makes the cost of each printed transistor too high. Given the current large size and high cost per printed transistor, the printing technique is one of the key issues to fabricate a low cost RFID tag at an acceptably functional complexity and compact size. Therefore, it is worthwhile investigating further the printing techniques used in the fabrication of printed electronic systems.

Principally there are three kinds of printing techniques currently used in the fabrication of electronics:

Table 2 Printing techniques currently used for fabricating electronics

	Screen printing	Inkjet printing	Gravure printing
Ink Viscosity requirement	High (>1000 cP)[a]	Low (1–30 cP)[a]	Medium(40–200 cP)[a]
Throughput	Fast	Somewhat slower	Fast
Resolution	50 μm (commercial) 10 μm (laboratory)	< 20 μm	< 20 μm (commercial)

1. screen printing,
2. inkjet printing and
3. gravure printing.

Their features in terms of throughput, resolution and the requirement of the printing materials' viscosity are summarized in Table 2.

According to Table 2, the best of the achievable resolution in high speed screen printing techniques today is worse than 10 μm (Subramanian et al. 2008). Hence, the minimum channel length is about 10 μm which is several orders of magnitude less than that achieved by photo lithography based single crystal electronics (Sirringhaus et al. 2006). The low resolution of printed electronics leads to both substantially low performance (mobility and capacitance related) and compactness compared to the single crystal electronics just mentioned. Furthermore, in terms of material rheology and viscosity, for screen printing techniques in which the ink is pressed through a pattern screen on the substrate, an ink with high viscosity is required to prevent excessive spreading and bleedout (Subramanian et al. 2008). The required high viscosity inks are obtained by mixing polymer binders to the ink which degrades the mobility of printed transistors and the conductivity of conductors.

However, the manufacturing advantage of the screen printing technique is its high throughput. Conversely, slower inkjet printing meets the requirement of binder free low ink viscosity as shown in Table 2, but as a result of the low viscosity, the inkjet printing technique also suffers from drop to drop processing and drying phenomena and surface aberrations, line morphology variation, which factors impact upon the thin film deposition quality and the consequent tight control to attain high yield TFT devices slows throughput. However, the inkjet printing technique is widely used in both organic printed electronics and printed silicon ink electronics. Commercial printed silicon ink semiconductor developers are using highly customised inkjet print head delivery systems for applying their proprietary nano-particle formulation and solvent based ink technology.

Much cooperative research and joint product development effort is being undertaken by leading industrial MEMS based inkjet head technology vendors to address the demanding resolution and throughput requirements of the emerging printed semiconductor industry. Silicon ink developers like Kovio use high resolution inkjet printing of nano-crystalline compositions containing precursors to silicon, or germanium, metals and/or dielectrics to form TFT semiconductor channel, source drains and gates, conductor and/or oxide dielectric features to create their high carrier mobility CMOS printed ICs (PIC) with 10 μm design rules.

Reaching the goal of printing a few thousand TFT transistors per chip very cheaply requires an ability to deposit high-performance silicon inks quickly. Rather than focusing on the number of transistors per square centimeter, a key parameter metric in printed electronics is "how much material can be laid down in one place per second. Once you get the basic performance capability printed silicon ink technology has now achieved, it is all about how to do it faster. The primary goal going forward is to deposit more material very, very cheaply, instead of how small can you get the piece of silicon; the principal driver for the conventional single-crystal industry.

The gravure printing can simply be conducted between two rollers. One is etched or engraved with the pattern and loaded with ink, the other one is smooth. Once the substrate passes through these two rollers, the pattern is printed on it. This printing technique has drawn people's attention in printed electronics recently because of its high-throughput and relative low requirement on ink's viscosity (Sirringhaus et al. 2006; Subramanian et al. 2008). The disadvantage of this printing technique is its significant rate of defects which impacts upon circuit yield. It is reported that this printing technique has not been used in electronic fabrication yet (Sirringhaus et al. 2006; Subramanian et al. 2008).

As analyzed above, two conclusions can be drawn in this section:

1. The resolution of printed electronics is several orders lower than that of conventional electronics. This great divide makes the achievement of printed circuit with complex functionality within a tiny area impossible in the foreseeable future. The influence that the low functional-complexity of printed RFID tags has had upon current standards and protocols is discussed in Sect. 4.
2. The reason why researchers dedicate resources to research in printed electronics or in particular printed RFID tags is because its potential ability in lowering tag costs at a relatively acceptable functional complexity. However, as analyzed above, the realization of this expectation is highly constrained by the printing techniques, and research continues.

Besides the printing techniques, the semiconductor material technology of the printing electronics (organic or silicon printed electronics) applied is also critical to accomplish the expectation of adequate functional complexity at acceptable cost. Therefore, these two distinct branches or categories of printed electronics are investigated in Sect. 3.

3 Comparison Between Organic and Silicon Ink Printed Electronics

This section will also include some comparison with conventional integrated circuit electronics based on single crystal materials.

Organic and silicon printed ink electronics both adopt printing techniques in manufacturing but differ in the printed material (ink). The advantages and disadvantages of these two categories of printed electronic technologies, and their current development and applications, are discussed respectively in this section.

3.1 Organic Printed Electronics

Since the 1970s, certain organic materials attracted attention because of their conducting or semi-conducting properties. The researchers Alan J. Heeger, Alan G. MacDiarmid and Hideki Shirakawa then led the research and for their work won the Nobel Prize for Chemistry in 2000 (Knobloch et al. 2004).

More recently, interest in the semi-conductive properties of some organic materials, usually solution-processed molecular or conjugated polymer (Sirringhaus 2009) has continued and such materials have been deployed in various applications. Firstly, organic semiconductors can be made into diodes which can emit light. They can be processed to become displays, so called OLED (organic light emitting diodes) (Sirringhaus et al. 1998) displays. OLED displays are mostly on glass substrate and not printed yet (Das 2007). Conversely, they can also absorb light and transform the light into energy and therefore be used in organic photovoltaic cells, which is abbreviated as OPV. Secondly, various types of sensors can be made of organic materials (Crone et al. 2001; Liao et al. 2005; Someya et al. 2004). Last but not least, the organic semiconductor materials can be made into thin-film transistors (TFT) which have the ability to perform logic and analog circuit operations.

These printed thin film organic transistors (OTFT) can be integrated into RFID circuits (Chan et al. 2005; Steudel et al. 2006; Subramanian et al. 2005), and used as a basis of so-called organic printed tags. In the following paragraphs, the advantages and disadvantages of the organic printed electronics using in RFID are discussed and the current development of organic printed electronics in RFID industry are introduced.

As one type of the printed electronic technologies, organic printed electronics possesses the benefits of low manufacturing costs based on the printing techniques introduced in Sect. 2. Besides those benefits and compared with silicon printed ink electronics, organic printed electronics come with some other advantages, discussed as follows.

Firstly and apparently, the printed organic tags are environmental friendly. All the constituent semiconductor materials based on pentacene polymer plastics can be absorbed benignly and naturally in landfill after their disposal. This is a merit, but as discussed later, can also be a demerit.

Secondly, manufacturing allows low temperature processing (below 150°C) which makes the organic semiconductors compatible with flexible polymer plastic substrates, such as polyethylene terephthalate (PET) or polyethylene naphthalate (PEN), which materials cannot sustain high temperature but have low or moderate cost (Sirringhaus 2009).

However, there are some serious limitations and disadvantages which impede the application of organic printed tags.

First of all, the charge carrier mobility of the printed organic transistors is very low, compared with printed silicon transistors and not surprisingly even lower when compared with conventional single-crystal transistors. As stated before, mobility is the basic material parameter governing transistor performance. It is defined as the ratio of drift current density to internal electric field. The mobility parameter is critical for electronic circuits for following two important reasons.

1. It affects the performance of rectifiers (OE-A Roadmap for Organic and Printed Electronics 2009; Steudel et al. 2005) .
2. It governs the transistor switching speed and hence the maximum clock frequency that can be achieved in a tag circuit.

Most of the organic electronic devices are made using *p-type* accumulation mode. The mobility of such printed transistors is only of the order of 1 cm^2/Vs (Sirringhaus 2009). By contrast, the electron mobility for crystal silicon in conventional semiconductor electronics at room temperature (300 K) is 1350 cm^2/Vs and the hole mobility is around 480 cm^2/Vs, whereas the best mobility that printed silicon ink based transistors can presently achieve is believed to be about 200 cm^2/Vs for *n-type*.

The desire to achieve high mobility in organic printed transistors based on *n-type* accumulation mode is attracting researchers into that area. However, the low electron affinity of most organic semiconductors (3.5–4 eV) makes the formulation of an electron accumulation layer at the commonly used SiO_2-organic semiconductor interface difficult. Hence, the choice of gate dielectric becomes critical (Chua et al. 2005). The low mobility constrains the organic printed tags to working at LF (125 kHz) and HF (13.56 MHz) and with very low communication data rates on the tag return link. The highest achievable frequency for a rectifier based on an experimental organic printed tag is about 50 MHz (Steudel et al. 2006. To our knowledge, organic printed tags working at UHF have not been reported yet.

Furthermore, in the discussion in terms of the merits of printed organic electronics, one is that organic materials are environmentally friendly. All these materials can be absorbed naturally. However, because of that, the current organic printed devices based on both *p-type* and *n-type* are very sensitive to oxygen and moisture in the atmosphere, with *n-type* organic compounds being the worst (DeLeeuw et al. 1997). The working ability of exposed *p-type* based organic printed devices is restricted to a few months before degradation sets in. This working period drops to a few hours or days for unencapsulated *n-type* printed devices. Organic semiconductors are inherently ephemeral and suffer electrical parameter and performance stability issues. Much intensive research is underway to address and solve these device longevity limitations. On the other hand this acute transistor sensitivity to vapours and gaseous materials in the ambient environs can be constructively exploited by using exposed TFT channels for realizing low cost integrated organic sensor elements for future smart labels. These applications include food spoilage detection and novel biosensors.

In common with their traditional single-crystal IC counterparts, printed organic transistor based tags face difficult encapsulation and packaging challenges for very low-cost markets.

There are a few companies dedicating their R & D and commercialization resources to the applications of organic printed RFID tags. One of the leading companies in this area is PolyIC. In 2007, the company presented the first organic printed RFID tag working at the high frequency range of 13.56 MHz with a simple circuit and certainly low functionality (OE-A Roadmap for Organic and Printed Electronics 2009). It is only supposed to be used for brand protection and ticketing. Organic printed tags working in the LF band was obtained before the achievement of HF band tags. However, because the LF antenna element size is relative larger than that of HF antenna, the LF band tags are not applied as widely as the HF band tags. Philips also reported that a 64-bit tag composed of 1940 transistors is obtained based on organic printed electronics. The tag's data rate is 150 bits per second (Cantatore et al. 2007). As will be seen in the next section, this figure is much lower than that achievable with printed silicon ink CMOS technology, wherein a tag communication data rate of 106 kbps is believed to be achievable.

Organic semiconductor tags currently require a large power supply operating voltage VDD of 14 V and higher due to the high threshold voltage of the OTFT devices. This characteristic relegates high power consumption organic TFT based tags to inductive coupled LF and HF systems and where the tags consequently exhibit a requirement for a high strength energizing field from the reader system.

The key electrical parameters of the TFT devices such as VGS gate threshold voltage, VDS breakdown voltage, and leakage currents drift substantially with time and cause substantial problems in organic circuit design and in achieving predictable and repeatable operation; might be OK for a few hand-selected research laboratory prototypes but no good whatsoever in the commercial mass production of high-yield and reliable printed circuits. This manufacturing batch and product life cycle variation places significant demands on electrical parameter tolerant circuit design.

3.2 Silicon Ink Printed Electronics

Silicon printed ink electronics is a rapidly emerging environmentally friendly green semiconductor technology and is based on a much newer research and commercialization program than organic printed electronics. Processed silicon nano-particles are usually used in the printed ink formulation. There are very few publications in this research area and certainly very few commercial players, currently exemplified by Kovio Inc. of Milpitas California. However, we believe inkjet printed silicon ink based full complementary *n-type* and *p-type* (CMOS TFT) technology has some comparative advantage in terms of achieving low cost tag fabrication at an acceptable tag performance in the foreseeable future compared with organic printed tags.

Kovio has developed an enabling silicon ink in conjunction with other ancillary inks that are sequentially deposited on thin metal-foil substrates measuring 300–400 mm on a side. After the ink is printed on the substrate, it forms silicon islands that are annealed to drive out the solvents, leaving a poly-silicon crystal film. In addition to the primary "enabling ink" for the process, the company also developed oxide inks for the gate dielectric, inks for in situ *n-type* and *p-type* dopants and high conductivity metal inks for contacts to the silicon and forming really good quality interconnects having excellent electro-migration immunity properties.

Besides the benefits obtained with printed technologies in terms of low cost, there are a few more very attractive advantages associated with silicon printed tags.

First and foremost, the mobility in printed transistors based on silicon nano-particles is much higher (orders of magnitude) than that of printed organic transistors. For example, Kovio, one of the leading companies in the silicon printed electronics field of activity claim that their n-type products' mobility can reach as high as 200 cm^2/Vs compared with approximate 1 cm^2/Vs obtained by organic printed electronics. This high mobility is achieved by an inkjet printing process (Johnson 2007; Lammers 2007; O'Connor 2008). Resolution and achievable feature size from the company's inkjet printing process is 10 μm now with a near term roadmap to 4 μm. As stated before, electron mobility in silicon in conventional electronics at room temperature (300 K) is 1350 cm^2/Vs and the hole mobility is 480 cm^2/Vs. Clearly, today's silicon ink mobilities do not rival the single crystal mobilities, but improvement is expected, and most importantly the present day (January 2010) process dependent mobility of 80–200 cm^2/Vs is much better than that of the current organic printed electronics, which leads to a faster data rate and allows the tag rectifier working efficiently at 13.56 MHz and enabling synchronous protocols deriving internal tag circuit clocking from the energizing carrier. Rectification with printed silicon TFTs at UHF is feasible albeit at much lower efficiency with the current device feature sizes and associated parasitic capacitance. We can attribute the less than single crystal mobility to the fact that even after the laser polymerization of the ink jet deposited silicon inks to form the *n* and *p* material, the material is still polycrystalline.

The silicon ink technology approach produces transistors from polycrystalline silicon, which is laser-recrystallized amorphous silicon, typically with the gate on top of the channel ("top gate" technology). This technology allows the source and drain to be aligned with the gate and therefore gives lower overlap capacitances. It makes it possible to produce *p*MOS transistors just as well as *n*MOS transistors. The electron mobility in polycrystalline silicon is higher than in amorphous silicon and the much lower threshold voltage is stable during operation and facilitates low power supply voltage operation.

Secondly, the silicon ink material is not as sensitive as the organic materials to the atmosphere which enables the silicon ink printed tags to exhibit much better environmental and electrical parameter stability.

Thirdly, the substrate used in silicon ink printed electronics is metal foil, because of the heat generated by high temperature laser annealing and polymerization of the ink jet deposited or screen printed silicon inks. The price of the thin metal foil, typically stainless steel, presently used for the small area PIC substrate is substantially lower than that of plastic films used as the substrate in organic printed electronics. The cost of petroleum industry derived polymer films is highly related to the global commodity price of oil. Silicon ink devices and circuitry can be readily fabricated on either roll-to-roll or large sheet format printing equipment as illustrated in Fig. 3.

Kovio made the strategic decision to differentiate itself in the printed electronics space by putting more value on TFT device performance than on low-temperature processing. Their methods use higher temperature processing of the materials after printing; these can still work with a flexible substrate, but it must be a metal foil, rather than plastic.

Despite these benefits, silicon ink technology shares with organic transistor technology the disadvantages of limited transistor numbers, so there is still an impact and limitation on the complexity of protocols. Somewhere around 2000 transistors is believed to be the comfortable upper boundary zone for printed semiconductor tag circuits today; but that is arguably all that is needed for achieving useful tag functionality to satisfy many less demanding end-user requirements. It is a return to the original MIT Auto-ID Center concept and very well-founded roots for "minimalist architecture" ICs having simple, but adequate protocols that fit within a modest number of transistors; a fundamentally important precept for the commercially viable realization of truly low cost tags for the mass consumer markets. Somewhere along the EPC technology and product development evolutional pathway that important precept got lost and abandoned because the focus back then was on the traditional single-crystal IC and few folks anticipated the fully printed-CMOS semiconductor revolution that was on the near term horizon.

This situation will undoubtedly change with the industry having to come the full circle back to the simple license plate only RFID tag concept so wisely contemplated

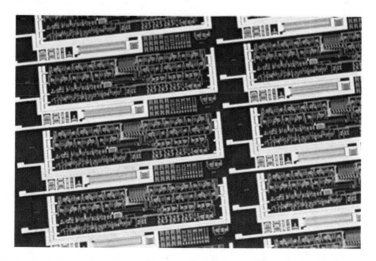

Fig. 3 PIC array in large sheet format (Photograph courtesy of Kovio Inc. © Kovio Inc.)

by the MIT Auto ID Center before the end user and technology vendor community bloated up the specification way beyond what is feasible for implementation with the then unforeseen low cost printed semiconductor technologies.

Silicon ink based tag memory today is based on either simple mask programmable ROM memory or OTP / WORM. Printed EEPROM memory cell structures, with their associated additional overhead of peripheral support circuitry, and with sufficient endurance and data retention time are down the commercialization roadmap for development and deployment in future generation printed tag products.

While printed silicon ink technology is well suited to implementing HF tags, it can also do UHF, but not at anywhere near the same long reading range traditionally associated with far field UHF tags of the conventional single-crystal kind. The reasons for this are threefold:

1. Rectifiers with the requisite high efficiency (RF input to DC output conversion) capable of operation at UHF are not yet possible or shown to be commercially practical with high yield printed semiconductor devices.
2. The VDD supply voltage needed to operate a printed silicon based TFT circuit is about two to three multiplies of today's single-crystal ICs due to significant differences in the printed transistor threshold voltages.
3. Higher circuit power consumption due to larger parasitic capacitances

These limitations mean printed silicon can do a UHF tag; but it will exhibit poor read range compared with established UHF products. The silicon ink technology could well do short-range proximity, near contact or few centimeter read UHF applications should a large commercial demand for low cost tags with that kind of operational characteristic emerge. On the other hand printed silicon could probably compete with near-field UHF tag systems should a significant demand for such tags take off some time in the future. This is because of the higher localized magnetic field strength / power density existing within the small volumetric confines of the true UHF near-field. This energy density is likely fortuitously conducive to overcoming to a useful extent the poor rectification efficiency and higher VDD and power requirements and hence the playing field somewhat leveled. However, this expectation would require empirical validation.

The relegation of printed silicon tag products to short range applications is viewed as not such a bad thing because having a small spatial volume of interrogation space also naturally limits the physical number of tags capable of being simultaneously in the field and thus greatly relaxes the anti-collision requirements to something that simple protocols such as Kovio and iPico (single-crystal counterpart with a simple *Tag Talks Only*, also called TTO, protocol) can deliver.

Whilst the above enumerated reasons also reduce the attainable read range of Kovio HF tags, vis a vis single-crystal HF tags, it is nowhere near as dramatic as with UHF. Hence printed TFT based tags are seen as much more competitive with applications for HF products.

Our conclusion after comparing organic and silicon ink printed electronics is that silicon ink printed tags are more likely than organic printed tags to replace conventional tags in many applications. This is because of their relative low price and high performance CMOS TFT circuitry that facilitate and supports fully synchronous HF protocols and fast communication data rates. Printed silicon ink technology can produce transistors that are fast enough for realizing high performance RFID tags.

We are not aware of any alternative high-performance based semiconductor enabling technology developments on the foreseeable horizon.

4 The Need for Printed Semi-conductors Standards as Opposed to Single Crystal Standards

Today's ISO/IEC 18000 EPC Gen2 specifications require circa 75,000–130,000 transistors to implement. That level of complexity is of course satisfactory for traditional semiconductor industry single-crystal IC chips, but is grossly incompatible and mutually exclusive for the now rapidly emerging new generation of very low cost fully printed RFID labels. An industry wide re-think is therefore necessary: and some new printed RFID centric and focused standards creation work is required if the piecemeal emergence of de facto standard(s) is to be averted. The now mature single-crystal standards were driven by the major stake holders of the global RFID technology vendors and semiconductor industry in response to perceived end user requirements. There is now a similar drive from end users in respect to standardizing low cost printed RFID.

The future challenge for the RFID industry is to develop in a considered and orderly manner some new standards based on minimalist protocols and tag architectures specifically targeted at printed TFT based semiconductor realizations. All of the technical standards in existence today have been designed around traditional single-crystal CMOS ICs where great complexity and large transistor counts is not a serious issue and can be tolerated without incurring significant penalty. However, the emergence of the printed CMOS semiconductor industry will bring a resurgence and urgent need to come the full circle back to minimalist principles based on simple, but efficacious system architectures. Such kinds of successful protocols have been exemplified by the iPico, EMarin, *Tag Talks First* (TTF) and TTO concepts and other published and unpublished proposals for simple *Reader Talk First* (RTF) terminating style protocols that are based on a limited command set. RTF protocols using a few very strategically chosen commands can be contemplated for use with printed tags, but doing so might arguably bring relatively few practical and operational advantages in many applications. The required tag State Machine digital logic can blow-out very quickly in transistor count whenever a reader command based protocol is adopted; even a modest one.

Although RFID has been strongly promoted for use in the supply chain (Cole and Engels 2006) we believe that emerging manufacturers of low cost RFID are planning to target the near term high volume markets of mass transit ticketing and rapidly emerging NFC applications, both of which have one tag in the interrogation field at a time. Low cost minimalist printed RFID HF label technology of the simple Tag Talks First (TTF), Tag Talks Only (TTO) "RF barcode" kind is attracting interest for deployment in brand authentication and protection, consumer product promotion, item level interactivity, asset management, event ticketing or wristbands, and retail loss prevention. Vendors of silicon ink based printed tags believe these highly cost sensitive markets will dwarf the size of the EPC market, so they are not currently very interested in taking on board the additional circuit complexity to address anti-collision issues and adaptive rounds. However, this chapter does mention those issues, which might become important in future applications, and might become practicable to implement as feasible transistor counts in printed technologies grow larger. However, protocols, anti-collision and security systems will still need a major paradigm change because of limitations on transistor count in printed tags. Until such time these protocol design and architectural aspects are simplified to only what is really necessary, the printed electronics industry is restricted from entering the supply chain EPC Item Level tagging market because of the yet to be addresses complexity issues. In the case of embracing fully printed electronics RFID labels the industry needs to experience an epiphany and associated shift towards minimalist thinking and low complexity performance oriented tag design.

4.1 A Simplified Adaptive Round Protocol

Many adaptive round protocols had been published for traditional single-crystal electronic implementation (Cole et al. 2002). But transistor counts are still outside the likely reach of printed electronic technologies. We believe that simplification of those protocols to eliminate such features as tag population selection and complex processes for detecting weakly replying tags, and exploiting deliberately limited tag operating dynamic range in the detection of collisions, could produce a simplified adaptive round protocol employing subcarrier modulation of four cycles in either the front half or back half of a bit period, and that this protocol might be achieved with about 2500 transistors. We believe that such a transistor count will become (or has become) commercially feasible in the near future.

4.2 An ISO Protocol

It is reported in (Harrop 2010) that Kovio have produced silicon ink based tags that adhere to the ubiquitous HF ISO/IEC 14443-A synchronous protocol, which has a return link data rate of 106 kbps. Bits are transmitted following the ISO/IEC 14443 Type A (Sect. 8.2) protocol specification, i.e. 106 kbps, Manchester encoding with

5% OOK (*on-off keying*) load modulation at 847 kHz. All the tag return-link parameters, subcarrier frequencies, modulation and data rates, etc are compliant with the published ISO specification. Notably, the Kovio tags use fully synchronous clocking which means they are easily read with traditional and existing 14443, 15693 HF reader equipment. The first generations of Kovio Printed IC (PIC) tags were designed to be a simplified sub-set of the ISO/IEC 14443A protocol; they did not implement any forward-link commands. The tags produced their 128 bit payload reply immediately on power up or power on reset and were TTO. The tags then continue to repeat transmissions after random sleep intervals for as long as they remain powered by the interrogation field and thereby facilitate a modest multiple tag read anti-collision capability. The associated absence of state machine logic and other peripheral circuitry to support commands allows a large reduction in transistor count down to approximately 1000. NFC reader hardware required minor firmware changes to accommodate the TTO communication protocol without the usual data flow control hand-shaking of the full protocol implementation. Significantly, the subcarrier frequencies and internal clocks are derived synchronously by direct countdown division of the 13.56 MHz interrogation carrier frequency. This clock synchronicity and very fast transmission data rate is something only high performance silicon-ink based TFT transistors and CMOS circuitry is currently capable of achieving.

Versions of printed silicon tags with this popular HF protocol have been demonstrated with NFC enabled mobile phone handsets and other standard legacy HF reader hardware. Leveraging on the installed reader and IT infrastructure and backward compatibility is foreseen as an important starting premise for any future printed electronics protocol standard. Perhaps all that might be required is a minor firmware upgrade change to read the simple PIC based labels. The reader infrastructure hardware or reader chip-sets including the latest embedded NFC readers remain the same.

Perhaps there are strong arguments in favour of adopting a printed electronics standards approach embracing and retaining core aspects, albeit simplified, of the ubiquitous 14443A HF protocol. The salient requirements being short proximity reading distance, modest anti-collision arbitration ability commensurate with a few cm distance and hence naturally self limiting number of tags simultaneously in the reader field.

4.3 The TOTAL Protocol

The ISO 18000-6(c) standard has a simple anti-collision protocol of about 50 tags/s (or maybe better) originating with IPICO (see http://www.ipico.com. The protocol being known as TOTAL – *Tag Only Talks After Listening*). This protocol, probably with the listen first aspect omitted, might be of interest to printed tag manufacturers when their interest moves from applications with single tags in the field to multiple tags in the field. Such a simple TTO protocol embodying just a few thousand transistors for its realization is deemed more than adequate for meeting many real-world applications. Tag architectural, protocol, and circuit implementation complexity is neither beneficial nor desired when contemplating a new printed electronics standard. This modified TTO approach is an exemplary example of a low transistor count protocol suitable for printed electronics, yet it still yields adequate fit-for-purpose performance for RFID applications.

Certain insightful and pragmatic sectors of the RFID industry today already understand this complexity / performance / cost trade-off and is a reason why the simple TTO protocols enjoy the popularity and support they do in retail item level and other applications. The TOTAL derivative of the simple TTO protocol has been integrated into ISO/IEC 18000-6 UHF standards revisions and update Committee Draft document being ratified by the SC31 technical subcommittee workgroup. The TTO or TOTAL protocol is based on an unslotted random hold-off and retransmission for tag collision arbitration, with a salient feature being no forward-link command transmissions emanating from the reader. Interoperability and non-interference with all other tags is achieved because a TOTAL tag, while powered, continually listens for the presence of reader modulation and only ever transmits its burst of randomised reply data in the absence of such modulated commands on an energizing carrier in its vicinity. Such quiescent stay-quiet operation inherently ensures the TTO tag never talks on top of an RTF tag such as EPCglobal UHF Gen2 (ISO/IEC 18000-6C) in mixed tag populations. One could argue such a talk after listening mode is not required for many restricted proximity distance HF applications.

5 Conclusions

RFID technologies based on microelectronic tags have made significant advances in terms of cost and performance. We have sought to acknowledge those developments and consider the next wave of technological advances expected to further drive the future growth of the industry through the reduction in cost of an RFID tag.

The road to realizing a really low cost tag has been identified: *fully printed tag including the antenna*. The realization of such a tag has been largely already solved by leading commercial players through low cost manufacturing techniques. Ink-jet printing is the outstanding example a simple low cost printing technique for realizing this vision. More significantly the technology has the capability for high volume production because it fully exercises the facility. However, what is needed is the requisite capital investment in plant to upscale and roll out "distributed" production facilities to match the upcoming demand expected in the following application areas.

- The EAS market because of the absence of the healing problem possessed by existing shortable capacitor labels.
- Potential application in the RFID area as security tags for anti-fraud and anti-counterfeit purposes.

- Disposable mass transit fare card ticketing.
- Most forms of near field communication (NFC) labels for the mobile phone based mass consumer space.

Based on various reasons outlined in the chapter the authors have sought to explore two key printing technologies suitable for printed tags: Organic Electronics and Silicon Ink Printed Electronics. There are a number of key issues that still needs to be solved prior to these technologies becoming a serious contender for single crystal silicon microelectronic tags. These issues are summarized in Table 3 below.

It is clear from Table 3 and our discussion that, for organic electronics technology to become a serious contender in the RFID application space there are significantly more hurdles to be overcome. Moreover, there does not appear to be clear solutions to the issues facing organic electronics in the near future. In contrast, the higher motilities supported by silicon ink printed technologies with support for both *n-type* and *p-type* transistors clearly offer a superior solution.

We see the clear need for a number of key developments outlined below to enable these emerging technologies to foray into the real world application space.

- In some application areas the present lack of an ISO/IEC air interface protocol standard that is specifically tailored and optimised to the requirements and transistor count limitation of printed TFT devices.
- The educational aspect of the need for end-users to appreciate the tremendous tag cost reduction opportunities if simpler less ambitious tag functionality were to be accepted and embraced.
- A requisite shift in mindset away from the entrenched architectural complexity of today's single-crystal products more towards simple "RF barcode" style labels of modest anti-collision capability.
- Solving a lightweight but efficacious and implementation compact data security mechanism.

Table 3 Key challenges and issues facing organic and printed silicon ink electronics

Silicon ink printed electronics	Organic electronics
Severely limited tag area for implementing security mechanisms in tags	
Need significant transistor numbers	
Assumption of only one tag in field made to keep protocols simple	
High temperature processing. Although this is not a serious impediment.	Only p-type can be fabricated while *n-type* is recognized as needed, but this does not appear to be in on the horizon. Low value for carrier mobility 1 cm^2/Vs now, with slow growth to 10 cm^2/Vs in 2016.

References

Cantatore E, Geuns TCT, Gelinck GH et al (2007) A 13.56-MHz RFID system based on organic transponders. IEEE J Solid-State Circuits 42:84–92

Chan YJ, Kung CP, Pei Z (2005) Printed RFID: technology and application. IEEE Int Workshop on Radio-Freq Integr Tech. doi:10.1109/RFIT.2005.1598894

Chua LL, Zaumseil J, Chang JF et al (2005) General observation of n-type field-effect behaviour in organic semiconductors. Nature. doi:10.1038/nature03376

Cole PH, Engels DW (2006) 21st century supply chain technology. White Pap Auto-ID Lab. http://www.autoidlabs.org/uploads/media/ AUTOIDLABS-WP-SWNET-015.pdf. Accessed 20 Feb 2010

Cole PH, Kauer C, Engels DW (2002) Standards for HF EPC tags. http://www.quintessenz.org/rfid-docs/www.autoidcenter.org/media/hf_cheap_tags.pdf. Accessed 21 Feb 2010

Crone B, Dodabalapur A, Gelperin A et al (2001) Electronic sening of vapors with organic transistors. Appl Phys Lett 78:2229–2231

Das R (2007) Organic & printed electronics forecasts, players & opportunities. Print Electron World. http://www.printedelectronicsworld.com/articles/organic_and_printed_electronics_forecasts_players_and_opportunities_00000640.asp?sessionid=1. Accessed 12 Feb 2010

DeLeeuw DM, Simenon MMJ, Brown AR et al (1997) Stability of n-type doped conducting polymers and consequences for polymeric microelectronic devices. Synth . Metals 87:53–59

Frequently asked questions (2010). RFID Journal. http://www.rfidjournal.com /faq/20. Accessed 10 Feb 2010

Harrop P (2010) Printed RFID in 2010. Print Electron World. http://www.printedelectronicsworld.com/articles/printed_rfid_in_2010_00001961.asp?sessionid=1. Accessed 21 Feb 2010

Johnson RC (2007) Silicon circuits made ink-jet printable. EE Times. http://www.eetimes.com/news/semi/rss/showArticle.jhtml?articleID=202805929&cid=RSSfeed_eetimes_semiRSS. Accessed 16 Feb 2010

Knobloch A, Manuelli A, Bernds A et al (2004) Fully printed integrated circuits from solution processable polymers. J Appl Phys. doi:10.1063/1.1767291

Lammers D (2007) Kovio inkjet prints fast silicon transistor. Semicond Int. http://www.semiconductor.net/article/206308-Kovio_Inkjet_Prints_Fast_Silicon_Transistor.php. Accessed 15 Feb 2010

Liao F, Chen C, Subramanian V (2005) Organic TFTs a gas sensor for electronic nose applications. Sens & Actuators B. doi:10.1016/j.snb.2004.12.026

O'Connor MC (2008) Kovio Unveils Printed-Silicon HF RFID, Chip Tag. RFID J. http://www.rfidjournal.com/article/articleview/4389/1/1/. Accessed 15 Feb 2010

OE-A Roadmap for Organic and Printed Electronics (2009) Organic electronics association white paper, 3rd edn. http://www.vdma.org/wps/portal/Home/en/Branchen/O/OEA/. Accessed 1 March 2010

Sirringhaus H (2009) Materials and applications for solution-processed organic field-effect transistors. 0018-9219. doi:10.1109/JPROC.2009.2021680

Sirringhaus H, Sele CW, Ramsdale CR et al (2006) Manufacturing of organic transistor circuits by solution-based printing. In: Klauk H (ed) Organic electronics. Wiley, New York

Sirringhaus H, Tessler N, Friend RH (1998) Integrated optoelectronic devices based on conjugated polymers. Science. doi:10.1126/science.280.5370.1741

Someya T, Sekitani T, Iba S et al (2004) A large-area, flexible pressure sensor matrix with organic field-effect transistors for artificial skin applications. Proc Nat Acad Sci. 101:9966–9970

Steudel S, Myny K, Arkhipov V et al (2005) 50 MHz rectifier based on an organic diode. Nat Mater 4:597–600

Steudel S, Vusser SD, Myny K et al (2006) Comparison of organic diode structures regarding high-frequency rectification behavior in radio-frequency identification tags. J Appl Phys. doi:10.1063/1.2202243

Subramanian V, Chang JB, Vornbrock AF et al (2008) Printed Eelectronics for low-cost electronic systems: technology status and application development. doi:10.1109/ESSCIRC.2008.4681785
Subramanian V, Frechet JMJ, Chang PC et al (2005) Progress towards development of all-printed RFID tags. Proc IEEE 93:1330–1338

Breakthrough Towards the Internet of Things

Leonardo W. F. Chaves and Zoltán Nochta

Abstract In this chapter we introduce the Internet of Things (IoT) from the perspective of companies. The Internet of Things mainly refers to the continuous tracking and observation of real-world objects over the Internet. The resulting information can be used to optimize many processes along the entire value chain. Important prerequisites for the IoT are that the objects of interest can be uniquely identified and that their environment can be monitored with sensors. Currently, technologies, such as different types of barcodes, active and passive Radio Frequency Identification (RFID) and wireless sensor networks play the most important role. However, these technologies either do not provide monitoring of their environment or they are too expensive for widespread adoption. Organic Electronics is a new technology that allows printing electronic circuits using organic inks. It will produce ultra-low cost smart labels equipped with sensors, and thus it will become an enabler of the IoT. We discuss how organic smart labels can be used to implement the Internet of Things. We show how this technology is expected to develop. Finally, we indicate technical problems that arise when processing large volumes of data that will result from the usage of organic smart labels in business applications.

1 The Internet of Things from Companies' Perspective

The Internet today can be described to a large extent as a ubiquitous infrastructure: it is always on and is always accessible from nearly any place of the world. After the initial era of connecting places and connecting people, the Internet of the future will also connect things (Cosnard et al. 2008).

The core idea behind the resulting Internet of Things (IoT) is to seamlessly gather and use information about the physical environment and about, potentially, any kind of object in the real world ("things") during their entire lifecycle. Physical objects,

L.W.F. Chaves (✉)
SAP Research, CEC Karlsruhe, Vincenz-Priessnitz-Strasse 1, 76131 Karlsruhe, Germany
e-mail: leonardo.weiss.f.chaves@sap.com

D.C. Ranasinghe et al. (eds.), *Unique Radio Innovation for the 21st Century*,
DOI 10.1007/978-3-642-03462-6_2, © Springer-Verlag Berlin Heidelberg 2010

including not only everyday products and goods but also different kinds of company assets, such as machines, tools, buildings, vehicles, containers, warehouse equipment, etc. will be augmented with sensing, computing, and networking capabilities and become active participants within business processes in future.

Making gathered information available, for example, about products' and goods' origin, movements, physical and chemical properties, usage context, etc. via the Internet will help enterprises improve existing business processes and also create new opportunities.

Companies will make use of the IoT in order to manage, i.e., to monitor and control their *internal* business processes, including the production, distribution, transportation, service and maintenance, and recycling of their products more effectively than today. Traditional enterprise processes and related software systems typically rely on manual data collection. Since manual data collection is, in many cases, error prone software systems often do not have the correct information to take the best possible decision in a given situation.

For example, a system that automatically orders spare parts for production machines and schedules the machines' condition-based maintenance requires accurate and timely information in order to ensure the continuous production and to optimize cost as well as the asset utilization of the company. In such business critical scenarios the usage of inaccurate data that do not adequately reflect the situation in the real world can lower both quality and performance of business processes and can lead people to make ineffective decisions in critical situations. Example negative implications are delayed order fulfillment, increasing costs, or out of stock situations.

The IoT will help companies capture the status of the entire enterprise and processes more accurately, in the ideal case exactly, where representations of the real world in software systems are an accurate, timely and complete reflection of the reality.

To achieve this goal, physical items of the real enterprise environment, such as the above mentioned machines and the corresponding spare parts, have to provide some "smart" functionality. For example, when arriving at or leaving the warehouse, machine spare parts can automatically reveal their identity to the respective warehouse gate without any human interaction. Based on this information, the parts warehouse inventory can always be up-to-date, helping avoid out of stock situations.

"Smart" physical items, or more precisely, miniature devices that are attached to or embedded into the items, should provide functionality that is useful but also affordable in the context of the envisaged novel business processes.

The functionalities that such smart items can offer for can be grouped into *five abstract categories* (Mühlhauser and Gurevych 2008). These categories are called:

1. Information Storage,
2. Information Collection,
3. Communication, Information Processing and
4. Performance of Actions.

A given smart item may offer any meaningful subset and combination of these functional elements depending on given requirements, available technologies and affordable costs. For example, a pallet that is equipped with an RFID tag offers information storage and communication functionality in order to automatically capture the pallet's unique Serialized Shipping Container Code at the relevant reading points.

1. *Information storage:* In companies operating with traditional information systems, data about business objects is usually stored in large centralized databases. Normally, there is no direct linkage between a physical object and the backend datasets associated with it. Smart items can help change this situation and establish a more direct linkage by storing and revealing different types of information either about themselves, or their environment. The information an item stores can be pre-determined and static (e.g., identifiers, production/expiry date, target location, weight, owner, etc.), or it can be dynamically updated during the life cycle of the item (e.g., tracking history, current location, critical temperatures the object has been exposed to, etc.). Depending on requirements, different types of memory components might be used to store the respective data on the item, such as read-only, write-once-read-many, or write-many-read-many memory modules. Information about objects can be stored in electronic devices, but also in printed labels, encoded as linear barcode or two-dimensional data matrices.

2. *Information collection:* A smart item may also be able to autonomously gather information either about itself, or its environment. Observation of different, dynamically changing parameters can be carried out by using various specialized sensors and respective technologies. One of the most important observable parameters of a potentially moving item is its location. Important location properties are objects' absolute position in a given coordinate system (2D or 3D) as well as their orientation. Objects' location can be determined by using explicit observation techniques and systems, such as the Global Positioning System (GPS). Location information may also be implicitly derived from the known position of the observing device, such as the known position of a stationary RFID reader. Knowing the identifier and the current location of a given object, a huge potential for optimizing business processes opens up. For example, based on accurate and timely location information of moving assets in a company, maintenance processes can be optimised. Objects' orientation is also measurable by multiple means. For example, when selecting read ranges carefully, a smart shelf in a store using RFID is able to determine whether a tagged product on the shelf is placed correctly or upside down. Besides the ability to determine objects' location, it may also be of interest to monitor their physical state. Measured by appropriate special sensors, temperature, speed, acceleration, motion, pressure, humidity, pressure, light intensity, mechanical stress and other parameters might be of interest. With today's sophisticated sensor technology it is also possible to determine and continuously monitor chemical properties of goods, mainly of fluids and gases. It is feasible to determine their composition and also the presence of chemicals residing in a "smart" sensor-equipped container, room, or a

chimney. This information can be useful for emissions management, but also to monitor chemical processes that take place in a barrel or container during transportation or storage to prevent the development of potentially dangerous compounds.

3. *Communication:* A fundamental capability of smart items is, of course, the ability to communicate. This capability is required whenever an item should interact either with other devices and items in its surroundings, or even with a business software system directly. In a majority of smart item systems known to us, items communicate with each other wirelessly, but wired solutions can also be found in practice. In wireless systems, including RFID, usually radio waves over various frequency bands are used as the communication medium. Other examples of media used to transmit information wirelessly are light waves, such as infrared light, and also sound waves. Information exchange between business software systems and smart items can be implemented by following the request-response scheme. Typically, the application is the requesting party. It expects responses from the items either in a synchronous, or an asynchronous mode. Another way of interaction is sending unidirectional messages from smart items to the backend system. These messages are also important building blocks of notification and alerting scenarios. In an example case, a smart room would only contact the backend system when the room temperature has reached or exceeded a certain pre-configured threshold.

4. *Information processing:* With the increasing number of smart items in a given environment especially the problem of how to handle the amounts of collected data may arise. In order to overcome such problems, smart items might (pre-) process the gathered information autonomously. Based on information processing capabilities provided by an integrated microprocessor or microcontroller, smart items may also adapt their state or behavior to the current context and environmental conditions. For instance, an item can automatically determine its expiry date in accordance with monitored storage conditions. Items may also be requested to aggregate the potentially large volumes of data they collect. The aim of data aggregation can be to deliver only the piece of information required by the relying business process. Information processing can also be carried out in a distributed and collaborative manner. Think, for example, of multiple items in a room, e.g., furniture, window, walls, each equipped with light, sound, and temperature sensors and a proper wireless communication interface, such as Zigbee. Based on the measured and collected sensor values, the items may be able to jointly discover whether there is an intruder in the room. In such application cases, single items only provide fragments of the data required to make the respective decision and draw the right conclusions, such as to alert the police.

5. *Performing actions:* Smart items may be able to actively control and change their own state or the state of the real world by performing physical actions if required. This capability becomes obvious when considering embedded systems, which are specifically designed to operate and control real-world objects, e.g., to effectively change the room temperature or to adjust the rotation speed of an engine. Proper actuators allow smart items to actively perform movements, for

example, in response to changing environmental conditions. Smart items can also interact with human beings: human readable information may be shown on a display and optical or acoustic warnings can be issued.

The influence and integration of smart items and the IoT on companies' business processes and underlying software systems can be illustrated and characterized by process integration patterns. In Fig. 1 three integration patterns, called *"Real-time Data Delivery"*, *"Process Control"* and *"Relocated Task Execution"* are shown (Mühlhauser and Gurevych 2008).

- *Real-time Data Delivery:* Enterprise processes require large volumes of information about the current or even past status of business relevant real world objects and their environment. Smart items may collect and deliver data in near- to-real-time to backend systems for further processing. Here, smart items play the most passive role from the business process execution point of view.
- *Process Control:* Since smart items are directly placed at the physical points of action, there is a potential to influence and indirectly or directly control the flow of the supported business processes that are implemented by backend systems. Depending on the current situation and context, as it is "seen", for instance, by distributed sensors, smart items can autonomously decide to start or stop the relevant process steps at the right time.
- *Relocated Task Execution:* The most complex usage pattern allows for well-defined parts of the business process, i.e., tasks or sub-processes, to be directly executed by smart items. The term "execution" basically means that data collection and transformation steps corresponding to the relocated process tasks are completely carried out by (collaborating) smart items.

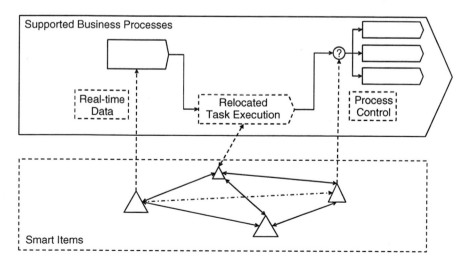

Fig. 1 Utilizing smart items in enterprise business processes (Reprinted with permission from the Publisher from Nochta, 2008)

Today, when items leave companies' internal process context, for instance, because a produced item has been sold, the item in question usually "disappears" from the issuing company's radar. In future, service providers will utilize information about real world items, information that is collected and managed by multiple independent business entities. Those services will rely on managed business applications which interested parties along the entire value chain can use for their respective purposes. Connecting today's isolated intra-company scenarios while using managed business data from multiple enterprises will not only be a major step towards the realization of the IoT, but has also significant business potential for participating enterprises. Here, we highlight one example application that follows this schema.

1.1 An Application Scenario

Product authentication services using the IoT can help brand owners protect their products against product counterfeiting and piracy as well as against illicit trading with originals (see Fig. 2). The service can use the data of participating brand owners

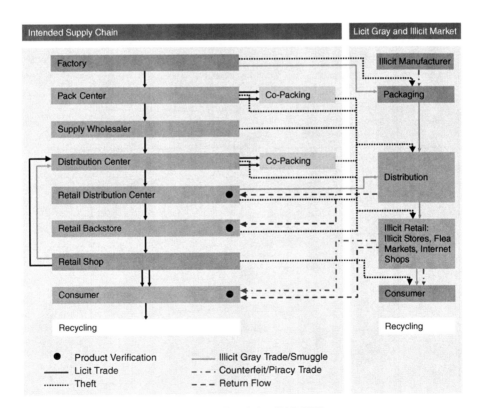

Fig. 2 Overview of illicit practices along value chains (SAP 2008).

Fig. 3 Printed security
marking as captured by a
cell-phone camera

in combination, whereas centrally managed application logic can be made available
via the Web for licit distributors, retailers, customs, and also for end-consumers that
are interested in buying only original products.

The service provides these legitimate parties along the product value chain with
a unique access point to distinguish between genuine products and counterfeits in
any situation by utilizing security techniques and product markings that cannot be
copied without being detected. Also, based on serialized tracking and tracing information provided by value chain participants the system will help identify anomalies
that indicate illegal parallel trading activities with originals, i.e., the injection of
stolen or diverted products into the licit value chain.

The following example scenario illustrates the service functionality from the
end-users' perspective. Before buying a counterfeit-prone product, a consumer can
photograph a security label which is printed on the product package with his or
her camera-equipped cell phone, see Fig. 3 below. The security label contains a
data-matrix (also known as a "2D barcode") that stores item level information about
the object (e.g., Serialized Global Trade Identifier, production lot number, expiry
date, etc.) and a so called Copy Detection Pattern (Picard 2004). After sending the
captured image to the system via the Internet, the product authentication service
will analyze the received information immediately and return a secure authentication result to the user together with the feedback that the product item in question
has been distributed in legal ways. An RFID enabled phone will be able to automate
the entire process by automatically recognising and communicating with the smart
security label.

2 Technologies to Implement the Internet of Things

The scenarios described in this chapter and the Internet of Things in general require
a tight coupling of information services with the real-world. That is, technologies that are used to identify and track objects and to sense their environment are

required. We will show that current smart item technologies used for this purpose today – including barcode, single crystal silicon based RFID, wireless sensor networks – face several inherent problems and are therefore unsuitable for realising the full potential from the Internet of Things today.

- *Barcode:* This is the simplest form of smart item. A one- or two-dimensional barcode can code an identification number for an item. The barcode can be printed onto a label which is later attached to an item, or it can be directly printed onto the item. It is very cheap, and it incurs negligible costs when it is directly printed onto an item. However, identification with barcodes faces many problems. Only items in line-of-sight can be identified, e.g., items inside boxes or pallets cannot be identified without significant manual work. Furthermore, only one item can be identified at once, further increasing the manual work when identifying items with barcodes.
- *Passive RFID:* When using Radio Frequency Identification (RFID), several items can be uniquely identified at once without line-of-sight contact. However passive RFID tags, i.e., the ones that operate without a battery, usually do not provide any sensor information. Even though passive RFID tags can provide immediate sensor information, they usually do not provide historical data. Furthermore, passive RFID tags are still too expensive for wide-spread market adoption. On average, current RFID tags that are fully converted to a label ready for attaching on to an object cost between 0.09 US$ and 0.15 US$ (ODIN Technologes 2010), depending on volume. Consider the supply chain of a retailer: the cost of an RFID tag is too high for large quantities of cheap products like milk or yoghurt. It appears that traditional RFID tags based on single crystal silicon chips will not reach the ultra-low costs required for applying the technology in many product segments. This is mainly because the production of RFID tags or labels requires many expensive steps that cannot be eliminated today: first the RFID chip is produced, and then it is typically attached to a strap. Also, the RFID antenna has to be produced, laminated, and finally connected to the RFID chip. Then the resulting "smart label" has to be attached to an item.
- *Active RFID:* Active RFID tags use batteries to power sensors and to aid the wireless communication. Such tags however are large and cannot be applied to every kind of object. And they are very expensive, with costs ranging from 10.00 US$ to 30.00 US$. Therefore active RFID tags are seldom attached to single items. Most of the time one active RFID tag is used to monitor one box, pallet or even container.
- *Wireless Sensor Networks:* These are networks consisting of several small computers equipped with sensors. They are similar to deployments of active RFID tags. However, in contrast to active RFID tags they can also communicate with each other. E.g., many wireless sensors deployed in a room can communicate with each other to calculate the average temperature in the room. This technology however faces the same problems as active RFID: sensors are large and expensive.

In the following section we present a new technology that combines the positive qualities of existing technologies, ultra-low costs, wireless identification and sensory information, and thus truly enable the Internet of Things.

3 Printed Organic Electronics

Printed electronics is a new technology to produce ultra-low cost smart labels, which can be a true enabler of the Internet of Things. Using conductive inks, electronic circuits can be printed (Leenen et al. 2009; Subramanian et al. 2008). This technology can produce ultra-low cost smart labels, which can be printed directly on the package of items in their manufacturing process (Das and Harrop 2007). Furthermore, the technology can help greatly extend the functionality of current smart labels by printing batteries, sensors and displays.

Organic (Subramanian et al. 2008) or inorganic (Leenen et al. 2009) materials can be printed by using standard industrial printers, e.g., printers that are also used to print newspapers and that can print several square meters per second (see Fig. 4). This results in ultra low-cost electronic components, and thus allows the use of electronic components like smart labels in scenarios where it was not possible before.

Most of traditional electronic components used today are based on single crystal silicon. These electronic components show high performance, are highly miniaturized and integrated, and their price per transistor is very low. However, they are complicated to manufacture, i.e., they require creating several masks and etching a single crystal silicon wafer (subtractive process). They also require costly clean room facilities, taking the costs of a microchip factory to several hundred millions of dollars.

With printed electronics several layers of different materials are printed onto each other to form electronic components (additive process). This reduces the overall number of steps in manufacturing, it reduces material costs and overall tooling costs, i.e., the production becomes very simple, fast and cheap. However, printed electronics show lower performance. And even though the price per transistor for printed electronics is higher than that of traditional electronics, the price per area is very low. Therefore, printed electronics are expected to replace traditional electronics in scenarios where electronics do not need to be very small or fast, and where the price of the electronic components has to be very low. This is the case with smart labels for products.

3.1 Organic Electronics and Inorganic Electronics

As already mentioned, printed electronics can use inorganic or organic materials. Inorganic materials, like zinc oxide (ZnO), show good environmental stability and performance (charge carrier mobility). However, inorganic particles are not soluble and therefore they are difficult to process. On the other hand, organic materials

Fig. 4 Organic electronics produced in a roll-to-roll printing process (PolyIC Press Picture. © PolyIC 2006)

like plastics and polymers in general have a good processability and their physical properties can be easily tailored chemically (Leenen et al. 2009). In this chapter, we focus on Organic Electronics.

With Organic Electronics, components like diodes, transistors, memories, batteries and sensors can be printed and easily integrated. This technology allows the production of new "organic" smart labels that are much cheaper than conventional smart labels. These organic smart labels can contain a multitude of sensors, like temperature, light, pressure and strain sensors. Furthermore, because of its ultra low-costs, several organic smart labels can be printed onto a single object (multi-tag). On the one hand, this increases the communication reliability of organic smart labels (refer to Bolotnyy and Robins, 2007) for similar experiments with RFID). On the other hand, it provides many sensor values for a single object. This is equivalent to each single object being a wireless sensor network, making data management more complex, but allowing fine grain monitoring of each object's condition. Table 1 compares organic smart labels to standard RFID labels based on single crystal silicon.

Table 1 Comparison of standard RFID with organic smart labels

	Passive RFID	Active RFID	Organic smart label
Sensors	Possible	Yes	Many
Battery	None	Yes	Yes
Price	0.20–1.00 US$	10.00–30.00 US$	TBD, expected to be more than 10 times lower than passive RFID (Finkenzeller 2003)
Range	< 10 m (UHF) < 1 m (HF)	In the order of 1000s of meters	not known yet
Frequency	LF, HF, UHF, Ghz	LF, HF, UHF, Ghz	HF (UHF by 2018)
Memory capacity	Up to 64 KB	Up to 1 MB	1 Bit (96 Bit by 2016)
Memory type	ROM, WORM, RW	ROM, WORM, RW	ROM (WORM by 2016, RW by 2018)

3.2 Roadmap for Printed Organic Smart Labels

The main concepts required for building organic smart labels have been researched, and demonstrators show the proof-of-concept. However, production parameters like yield have to be optimized for successful market introduction. Today, only simple organic smart labels exist, which only have 1–4 Bits of Read only Memory (ROM) and contain no sensors or batteries.

Figure 5 shows a roadmap for the general availability of printed organic smart labels. It is based on the data in (OE-A Roadmap for Organic and Printed Electronics 2009). As already mentioned, today only simple organic sensor exist. In 2012, the first fully integrated printed batteries will appear, allowing the organic smart label to be equipped with more complex sensors that can continuously monitor their environment, even in the absence of a reading device. Over the years, the memory capacity of the smart labels will increase until reaching the milestone of 96 Bits in 2016. Furthermore, memory of the type Write once Read many (WORM) will be introduced. This is a milestone in the development of organic smart labels, since they will be able to store an Electronic Product Code (EPC), which is an important standard for coding the identification number of an object. On the long term, organic smart labels are expected to implement the full EPC communication protocol. First

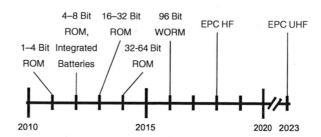

Fig. 5 OE-A roadmap for printed organic electronics

implementations will be for High Frequency (13.56 MHz) tags followed later by Ultra High Frequency (850–950 MHz).

Sensors are not shown in the figure to avoid clutter. Today, a large variety of sensors exist, like temperature, light, pressure and strain sensors while a plethora of other sensor types are being developed.

3.3 Challenges to Utilising Smart Organic Labels

The IoT consists of large-scale information systems, which encompass resource planning systems, database management systems, application servers and others. The IoT utilizes smart label technologies to couple real-world objects with business processes. The information systems within the IoT obtain data from an infrastructure network of devices, e.g., all tag reading devices from all stores of a retail chain. The main tasks are then to (1) process the acquired data, e.g., the identification information and additional sensor data, (2) perform actions accordingly, e.g., initiate an order process for replenishment. And (3) store the acquired data, often for several years, e.g., for the purpose of compliance in the food industry. However, the massive deployment of organic smart labels will result in data which cannot be efficiently processed by current systems. This is because the data volume will be orders of magnitude larger than the one that results from equivalent RFID installations, since each object will carry many smart labels. And therefore each object will contain many more sensors that may return conflicting data. Both aspects are discussed in the following.

1. *Large amounts of data:* With mass usage of organic smart labels the central challenge for information systems within the IoT is data processing and management of vast amounts of data. Identification information is always associated with metadata, such as location of an item or status within a business process. For instance, if Wal-Mart operates RFID at the item level, it is expected to generate 7 terabytes (TB) of data every day (Schuman 2004). When applying 10 s or 100 s of organic smart labels to each item, the data volume vastly increases. At peak load situations, e.g., when palettes of items arrive at a reader device, metadata changes dramatically as a flood of update operations propagate through the information systems, e.g., updating the items' locations metadata. Information systems must operate at high data rates to process the data fast enough. The massive data explosion will impose higher loads for middleware frameworks throughout the entire supply chain. Data from different sources will be combined to enable complex event processing along the supply chain. Simultaneously, the information systems are requested to provide real-time processing. Algorithms developed for RFID data compression (Hu et al. 2005) and processing (Wang et al. 2009) might provide a basis for coping with the huge amount of data resulting from organic smart labels.

2. *Data quality:* Data quality becomes a crucial aspect when multiple organic smart labels and their sensors are attached to an object. With 100 s or 1000 s of sensors per object, the data from a single smart label may move into the background. On the one hand, redundancy is provided. A failed smart label is not fatal and data from neighboring smart labels can be used for compensation. On the other hand, it is more complex to filter out inconsistent or conflicting readings. Erroneous sensors may trigger unnecessary or even costly processes and actions. The manifold relationships between information systems in the IoT make it hard to isolate the original cause. Related work to interpret (Cocci et al. 2008) and filter (Jeffery 2006) uncertain RFID data might be adapted to improve the data quality resulting from organic smart labels.

The integration of organic smart labels into the IoT presents challenges for additional research efforts. Information systems are requested to scale up with the vastly growing amount of data while simultaneously allowing real-time queries. The high load on middleware systems, event processing throughout the supply chain and the use of multiple organic smart labels per object require a flexible distribution of the data processing work load. It may be distributed among the organic smart labels, the reader device, middleware computer systems and database management systems. It also means that one needs to partly reconsider some established ways in order to accomplish these challenges.

4 Conclusions

In this chapter we introduced the Internet of Things (IoT) from companies' perspective. The Internet of Things allows tracking of real-world objects over the inter-net. It can be used to optimize many processes. However, today, current technologies used for this purpose today – including barcode, RFID tags based on single crystal silicon, wireless sensor networks – face several inherent problems as an enabling technology for Internet of Things. Organic Electronics is a new technology that allows printing electronic circuits using organic inks. It will produce ultra-low cost smart labels equipped with sensors, and thus it will truly enable the IoT. We show how the Internet of Things can benefit from such smart labels. Furthermore, we discuss how this technology is expected to develop. At the end, we point out technical problems that arise when processing huge amounts of data that will result from the usage of organic smart labels in business applications.

Acknowledgments The work presented in this chapter was partly funded by the German government (Bundesministerium für Bildung und Forschung) through the project Polytos.

References

Bolotnyy L, Robins G (2007) The case for multi-tag RFID systems. In: Proceedings of international conference on wireless algorithms, systems and applications. Chicago, IL, Aug. 1–3

Cocci R, Tran T, Diao Y, Shenoy P (2008) Efficient data interpretation and compression over RFID streams. In: Proceedings of the 2008 IEEE 24th ICDE, pp 1445–1447, Cancún, México

Cosnard M, Dickerson K, Jeffery K, Pogorel G, Prasad R, Sieber A, Weigel W (2008) ICT Shaping the world: a scientific view. Wiley-Blackwell, Chichester

Das R, Harrop P (2007) Organic & printed electronics – forecasts, players & opportunities 2007–2027. IDTechEx research report. http://media2.idtechex.com/pdfs/en/U3021T7639.pdf. Accessed 1 March 2010

Finkenzeller K (2003) RFID handbook: fundamentals and applications in contactless smart cards and identification, 2nd edn. Wiley, Chichester

Hu Y, Sundara S, Chorma T, Srinivasan J (2005) Supporting RFID-based item tracking applications in Oracle DBMS using a bitmap datatype. In: Proceedings of the 31st international conference on VLDB, **Trondheim, Norway,** pp 1140–1151,

Jeffery SR, Garofalakid M, Franklin MJ (2006) Adaptive cleaning for RFID data streams. In: Proceedings of the 32nd international conference on VLDB, Seoul, pp 163–174

Leenen MAM, Arning V, Thiem H, Steiger J, Anselman R (2009) Printable electronics: flexibility for the future. Phys Status Solidi (A) 206(4):588–597

Mühlhauser M, Gurevych I (2008) Ubiquitous computing technology for real time enterprises. Information Science Reference, IGI Global, Hershey, PA

Nochta Z (2008) Smart items in real time enterprises. In: Mühlhauser M, Gurevych, I (eds) Handbook of research on ubiquitous computing technology for real time enterprises. IGI Global, Hershey, PA.

ODIN Technologes (2010) RFID tag pricing guide^{TM} report. http://www.odintechnologies.com/rfid-tag-pricing-guide?cmp=unknown&panel. Accessed 10 March 2010

OE-A Roadmap for Organic and Printed Electronics (2009) Organic electronics association white paper, 3rd edn. http://www.vdma.org/wps/portal/Home/en/Branchen/O/OEA/. Accessed 1 March 2010

Picard J (2004) Digital authentication with copy-detection patterns. In: Rudolf van R (ed) Optical security and counterfeit deterrence techniques V, Proceedings of the SPIE 5310:176–183

SAP (2008) SAP research: SAP research report 2007/2008. http://www.sap.com/about/company/research/pdf/SAP_RR_2007-2008.pdf. Accessed 1 March 2010

Schuman E (2004) Will users get buried under RFID data? Ziff Davis internet, November 9. http://www.eweek.com/c/a/Enterprise-Applications/Will-Users-Get-Buried-Under-RFID-Data/1/. Accessed 1 March 2010

Subramanian V, Chang JB, Fuente V, Alejandro de L et al (2008) Printed electronics for low-cost electronic systems: technology status and application development. In: Proceedings of IEEE European solid-state device research conference, Edinburgh 15–19 September, 2008. doi: 10.1109/ESSDERC.2008.4681677

Staake T, Fleisch E (2008) Countering counterfeit trade. Springer, Germany

Wang F, Liu S, Liu, P (2009) Complex RFID event processing. VLDB J 18(4):913–931

Threats to Networked RFID Systems

Aikaterini Mitrokotsa, Michael Beye, and Pedro Peris-Lopez

Abstract RFID technology is an area currently undergoing active development. An issue, which has received a lot of attention, is the security risks that arise due to the inherent vulnerabilities of RFID technology. Most of this attention, however, has focused on related privacy issues. The goal of this chapter is to present a more global overview of RFID threats. This can not only help experts perform risk analyses of RFID systems but also increase awareness and understanding of RFID security issues for non-experts. We use clearly defined and widely accepted concepts from both the RFID area and classical risk analysis to structure this overview.

1 Introduction

RFID technology is a prominent area of research in ubiquitous computing. Its contactless nature and potential for data processing and storage gives it many advantages over existing machine-readable identification techniques (e.g. barcodes, optical recognition charters). Nevertheless, RFID systems have vulnerabilities making them susceptible to a broad range of attacks. In this chapter, we attempt to give a clear overview of existing threats against RFID technology. While privacy is an important issue and is extensively examined in the literature, security also needs considerable attention. A well-structured classification of RFID threats may help us to facilitate a thorough understanding of RFID security and thus, choose and develop effective countermeasures.

Overviews of security issues related to RFID systems have been presented before. More precisely, initial works simply listed common attacks in RFID systems (Juels 2006; Peris-Lopez et al. 2006). Other papers focused on privacy threats (Garfinkel et al. 2005; Avoine and Oeschlin 2005; Ayoade 2007), while yet others

A. Mitrokotsa (✉)
Security Lab, Faculty of Electrical Engineering, Mathematics and Computer Science, Delft University of Technology (TU Delft), Mekelweg 4, 2628 CD, Delft, The Netherlands
e-mail: A.Mitrokotsa@tudelft.nl

D.C. Ranasinghe et al. (eds.), *Unique Radio Innovation for the 21st Century*,
DOI 10.1007/978-3-642-03462-6_3, © Springer-Verlag Berlin Heidelberg 2010

proposed a more detailed taxonomy (Karygiannis et al. 2006; Mirowski et al. 2009; Mitrokotsa et al. 2009). Karygiannis et al. (2006) proposed an RFID risk model focusing on network, business process and business intelligence risks. Mirowski et al. (2009) focused on the RFID hardware layer and model attack sequences, while Mitrokotsa et al. (2009) discriminate RFID threats in four main layers: physical, network-transport, application and strategic layer. This chapter will build from the concept of layers but incorporate the cornerstones of information security and risk analysis (*confidentiality*, *integrity*, and *availability (CIA)*).

More precisely, we discriminate three main categories of RFID attacks, based on which part of the system they target: attacks that affect the *RFID Edge Hardware layer*, the *Communication layer*, and the *Back-end layer*. The *RFID Edge Hardware* consists of the RFID devices (tags and readers). The physical security of these devices is usually not very strong. Thus, they are vulnerable to tampering and other physical attacks; this is particularly true for tags, since their resources are often constrained due to cost and size limitations.

The *Communication layer* deals with the exchange of information. The main purpose of radio-based technology such as RFID is sending and receiving data. Thus, the radio link becomes a prominent point of attack – everyone can listen in, and signals are easily modified or jammed.

The last distinct part of the RFID system is the *Back-end layer* which is responsible for connecting RFID readers to databases and other supporting systems where RF transaction data are stored, analyzed and processed (Karygiannis et al. 2007). Since the *back-end layer* contains elements such as databases, webservers etc., many attacks on networking applications and systems can be launched. However, our analysis includes attacks that are specific to the RFID communication; many other sources (Kaufman et al. 2002) are available to give a better overview of attacks on networking systems.

In each of these layers, we subdivide three groups according to the security property that can be compromised. *Confidentiality* ensures that information or services cannot be accessed by unauthorised parties, *integrity* guarantees that information or services are not modified by unauthorised parties while *availability* ensures that information and/or services should be always available to all legitimate parties. Figure 1 depicts the proposed classification.

This chapter is organized as follows. Section 2 describes the threats related to the *RFID-edge hardware layer,* Sect. 3 describes the threats related to the *Communication layer* and Sect. 4 presents the threats related to the *Back*-end layer. We should note here that while the presented classification of threats covers all types of RFID systems, in some cases only a subset of RFID technologies is affected. This is indicated explicitly. More precisely, at the end of each of section we provide a table, which relates the threats of each layer to the associated damage caused by the threat, the cost to implement the attack/threat, the type of RFID tags most vulnerable to these threats as well as possible countermeasures and their associated costs. In each table we distinguish RFID tags into two main categories: *high* and *low* cost tags. *High-cost tags* include most *active* and *semi-active* tags, or more generally those tags, which are self-powered, have greater computational capabilities or radio range. On the other hand *low-cost tags* have more limited resources (gate equivalents

Fig. 1 Classification of
RFID threats

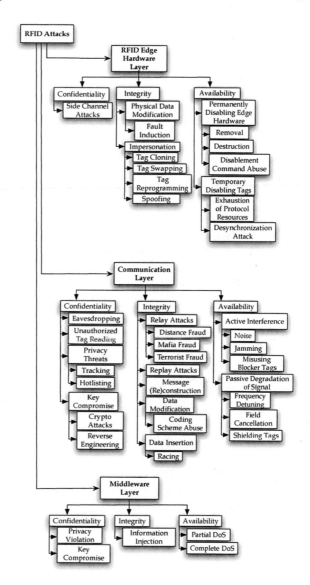

and power consumption) and computational capabilities (small memory). Finally
Sect. 5 concludes the chapter and presents some open issues.

2 RFID Edge Hardware Layer

In the *RFID Edge Hardware layer* we include attacks that may affect the RFID
devices (tags and readers) by exploiting their poor physical security and their inadequate resistance against physical manipulation. This layer includes *side-channel*

attacks, physical data modification, and impersonation as well as *temporary* or *permanent disablement of edge hardware*.

2.1 Confidentiality

To protect and preserve the *confidentiality* of information, it should be restricted only to authorised parties. The means of mounting an attack on the *confidentiality* of the system differ greatly, depending on which part of the system is targeted.

2.1.1 Side-Channel Attacks

When an adversary has physical access to the edge hardware, and sufficient time, he/she can attempt to extract information from it. This can be done by either deducing information from (side-channel) measurements, inducing errors in the operation of the hardware (fault attacks or "glitching"), thus influencing the output, or by directly reading out the memory (tampering). A brief overview is given by Berkes (2006).

Many of these attacks require specific equipment, and cannot be performed by a hobbyist, but rather by experts and well-equipped laboratories (Hancke and Kuhn 2008; Berkes 2006). The impact of a *side-channel attack* varies and depends on the application scenario. Thus, it may lead to retrieve the *kill* password of an RFID tag or even reveal the secret key used for authenticating a tag to a reader. In the former case an adversary could silence RFID tags permanently and thus steal valuable products. In the latter case where the secret key may be revealed, the adversary may gain access to restricted places, critical information or even conduct financial fraud.

Let us briefly discuss the various ways of mounting such attacks:

- *Side-Channel Analysis*: is a non-invasive type of attack that may be conducted when the attacker measures fluctuations in timing delays, power consumption or emitted signals and radiation – information that may reveal critical data about the input and the internal states of the RFID devices. For instance, fluctuations in the backscatter signal may yield information regarding changes in the internal resistance. Several types of *side-channel attacks* can be distinguished, depending on which measurements are made to disclose information. More precisely, *Simple Power Analysis (SPA)* (Oren and Shamir 2007) or *Differential Power Analysis (DPA)* (Hutter et al. 2007, 2009; Kocher et al. 1999; Plos 2008) attacks are based on the measured variations of the power consumption. *Timing* attacks exploit the fluctuation in the rate of computation of the target (possible delays), while in *Differential Electro-Magnetic Analysis (DEMA)* (Hutter et al. 2007; Agrawal et al. 2003) are measured the electromagnetic field variations caused while an RFID device performs cryptographic operations. In the latter case the adversary may reveal critical information such as secret cryptographic keys.
  Oren and Shamir (2007) have performed the first *SPA* attack, which reveals the *kill* password of *ultra-high frequency* (UHF) tags. Their attack was based on

observing the power trace reflected from the tag to the reader. Hutter et al. (2007) have performed the first *DPA* attack in RFID tags. More precisely, by performing power and electromagnetic analysis they were able to analyze hardware and software AES (Advanced Encryption Standard) implementations in *high frequency (HF)* tags. Even UHF tags have been proved to be vulnerable to DPA attacks by Plos (2008). Hutter et al. (2009) have also been able to perform a *DPA* attack in a public-key enabled RFID device.

- *Fault Attacks*: may be launched when the adversary tries to produce errors in the operation of the device targeting in illegally extracting information through regular output channels or side-channels. This is achieved by varying the input of the device, or by manipulating the environmental conditions (Hancke and Kuhn 2008) such as heat, cold, electromagnetic radiation (e.g. Ultraviolet (UV) light or X-rays), voltage levels, continuity of power supply or magnetism.

- *Physical Tampering*: When all else fails, one can try to physically take apart the device without destroying the data contained therein, in an attempt to (partially) recover it. Another type of physical tampering involves targeted manipulation or damage in specific hardware parts (Hancke and Kuhn 2008; Berkes 2006). An example would be changing individual cells in ROM (Read Only Memory) to change the behaviour or seed of a cipher. Other invasive attacks rely on retention and direct reading of memory cells. Volatile memory loses its data when power is interrupted. Some devices (like smartcards) have been protected with memory-erasing behaviour when attacks are detected. Some means of slowing this loss of memory can be to cool the hardware (Lowry 2004), to etch the state into the memory permanently (by means of radiation), or to tunnel into the hardware at great speeds, in order to cut off protective measures before they can react. Even without these exotic and often expensive methods, sometimes memory can be read after power-loss. If RAM (Random Access Memory) memory is powered at a constant state (0 or 1) for extended periods of time, an imprinting effect called "memory remanence" (Hancke and Kuhn 2008) will take place, which can allow (partial) recovery of data.

2.2 Integrity

Preserving the *integrity* of information can be defined as the assurance that all information and related methods that process it are accurate and complete (ISO 2005). *Integrity* can be easily subverted if *authentication* has been eschewed or once it is bypassed. Therefore, we will look at attacks that threaten proper authentication in the RFID systems, and their integrity. In the edge hardware this comes down to *physical data modification* and *impersonation* attempts.

2.2.1 Physical Data Modification

In RFID systems *physical data modification* can be achieved either by *fault induction* or by *memory writing*. *Fault induction* involves modifying data when it is being written or processed. *Memory writing* can be performed by using specialist equipment, such as laser cutting microscopes or small charged needle probes. This way, memory cells can be directly influenced and written to (the ROM attacks in Hancke and Kuhn (2008) are a good example). Although difficult and time-consuming, this bypasses any software protections. The impact of *physical data modification* depends on the application scenario and the criticality of the information that was modified; for sure such attacks cause inconsistency between the data stored on the tags and the objects to which the tags are attached. Consequently, such an attack may lead to severe medical mistakes (i.e. modifying the data stored on a tag attached to a drug) or even counterfeiting original products.

2.2.2 Impersonation

Impersonation of RFID tags can be achieved by *tag cloning*, *tag swapping*, *tag reprogramming* or *spoofing*. In all cases the adversary targets and imitates the identity of tags or readers. Through this attack an adversary may gain access to restricted areas, sensitive information and credentials.

- *Tag Cloning*: Replicating RFID tags has proven to be very easy, since it does not cost a lot of money nor requires a lot of expertise, while all the necessary equipment such as software and blank tags are freely available. A representative example has been demonstrated by a German researcher, who has proven that German epassports are susceptible to cloning attacks (Reid 2006). If the RFID tag does not use any security mechanism then cloning simply includes copying the tag's identifier (ID) and any other associated data to the rogue RFID tag. Nevertheless, if the tag does employ security features then a more sophisticated attack should be launched so that the cloned tag becomes indistinguishable from the legitimate one. The amount of effort needed to launch a cloning attack depends on the security mechanisms used. But cloning does not just mean copying a tag ID and data but creating an RFID tag that follows the original one even to the form factor. The human eye should not be able to discriminate a legitimate from a cloned tag.
- *Tag Swapping*: Another quite popular impersonation attack is *tag swapping*; quite simple in its tactic but a real threat in retail product tracking and automated sales processing. *Tag swapping* involves removing an RFID tag from a tagged object and subsequently attaching it to another one (just like "switching" price tags). An illustrative example of *tag swapping* is a thief in a retail shop who picks out a high-priced item and a cheap one and switches their tags so that he/she "buys" the expensive one and pays less at checkout. Thus, the integrity of the back-end system is violated since it cannot correlate correctly the tag's ID with the object.
- *Tag Reprogramming*: Some tags are reprogrammable, either directly (through the Radio Frequency (RF) interface, usually when a password is supplied), or through some (wired) interface, which is supposed to be used only in the construction process. Other tags cannot be reprogrammed, because they use ROM. In impersonation attempts, the cheapest way of recreating the form-factor of tags can

often be re-using existing tags, and simply reprogramming them if needed. One could steal and swap tags for this purpose, use discarded tags acquired through dumpster-diving, or buy tags of the same type if they are available on the open market (Juels 2005).

- *Spoofing*: Spoofing can be considered as a variation of tag cloning. However, their main difference is that spoofing does not involve the physical reproduction of an RFID tag. Furthermore, spoofing attacks are not limited to tags since the identity of RFID readers can also be spoofed. Successful deployment of this attack requires specialist equipment that allows the emulation of RFID tags or readers based on some data content. The adversary requires full access to legitimate communication channels as well as knowledge of the protocols and secrets used in the authentication process if there is any. The goal of the adversary is to imitate legitimate tags or readers elicit sensitive information and gain unauthorised access to services.

2.3 Availability

An asset is considered to be available when it can always be accessed and used by all authorised parties (ISO 2005). The availability of the *RFID edge hardware* can be compromised when it is permanently or temporarily disabled (through removal, destruction or sabotage).

2.3.1 Permanently Disabling Edge Hardware

RFID tags and readers may be rendered permanently inoperable vial removal or destruction. RFID tags may also become irreversibly disabled if specific disablement commands are abused. *Permanently disabling RFID hardware* leads to permanent untraceability of tagged objects and thus to great loss in a supply-chain organization or a retail shop, depending on the scale to which this attack is performed.

- *Removal*: Considering the poor physical security that RFID tags present, they can be easily removed from the associated items if they are not strongly attached to or embedded in them. Tag removal is a serious threat that can be easily deployed without the need of exceptional technical skills. It is a threat that leads to untraceable objects and suggests a significant security problem. Luckily, this kind of attack cannot be launched on a massive scale. RFID readers may also be removed if they are situated in unattended places. However, their size renders this attack hard to deploy.
- *Destruction*: Poor physical security leads not only to possible tag removal but also to easy tag destruction. Unattended RFID tags run the risk of being vandalised by malicious attackers through chemical exposure, application of increased pressure or tension loads or even by simply clipping their antennas off. Nevertheless, even

if RFID tags escape the violent intentions of vandals they might get destroyed by severe environmental conditions such as extreme temperatures or by abrasion produced by rough handling. Moreover, RFID tags may also be rendered deliberately inoperable by abusing privacy-enforcing devices such as the RFID Zapper (Collins 2006). This device's operation is based on the production of a strong electromagnetic field that burns out the tag's internal circuitry.

While it is easier to physically manipulate RFID tags due to their small size, RFID readers may undergo similar threats. Considering the fact that RFID readers often store critical security credentials (i.e. encryption keys) they may become subject to vandalization and physical destruction, especially if they are unsupervised. A compromised or stolen RFID reader may disrupt the RFID communication and violate the RFID system's availability.

- *Disablement Command Abuse*: Some tags have security features that permanently disable or lock them. The KILL command, a specification created by the Auto-ID center (2003) and EPC global is a command for permanently silencing a tag, thus rendering it unresponsive to requests. Several RFID standards use LOCK commands to prevent unauthorized writing to tags (Karygiannis et al. 2006). Usually a predefined password is used for authentication; the locking itself can be temporary or permanent (e.g. "permalock" in EPC tags). For practical reasons, multiple tags could share a password (i.e. all items in the same store). Otherwise, password management becomes problematic (large lists need to be shipped with the items). Although these features can be used for privacy-protection reasons, they can also be misused to render RFID tags permanently inoperable and sabotage RFID communications. In some cases, the password used is of low entropy (8 bits for 1 EPC Gen 1 tags, 32 bit for EPC Gen 2 tags), and could easily be brute-forced. Moreover, having one master-password may lead to successfully locking or killing a great number of tags and severely harming the system.

Temporarily Disabling Edge Hardware

RFID edge hardware may also be disabled temporarily by extreme environmental conditions (i.e. tags covered with water or ice), *exhaustion of protocol resources* or possible *desynchronization attacks*. Similar to *permanently disabling edge hardware*, even if the disablement is temporary, such an attack may disturb the whole RFID system and in the simplest case scenario can provide free goods to a shop-lifter.

- *Exhaustion of Protocol Resources*: Some protocols allow a tag to only be read a certain number of times, or allow a limited number of unsuccessful reads before rendering it inactive (Ohkubo et al. 2004). Some protocols use counters or timestamps with a maximum value. When this value is reached, the tags become unreadable. Other protocols protect a tag from tracking, but only for a limited number of reads. An example is hash-chain protocols, which use hash-chains (pseudonyms) of fixed length stored on a tag. When these tags run out of pseudonyms, they once more become vulnerable to tracking attacks. The OSK (Ohkubo-Suzuki-Kinoshita) protocol (Ohkubo et al. 2003) is an example of a hash-chain protocol.

Depleting the battery power of active tags and shortening their lifespan is another example of such attacks. Targeted attacks (but also repeated accidental reads by foreign systems on the same frequency) can cause tags to be "exhausted" and fail. In some cases this error is recoverable, in many it is not. In either case, it may have a financial impact due to interrupted operations and the costs associated with system recovery (Han et al. 2006).

- *Desynchronisation Attack*: Some protocols rely on a form of synchronization between tags and reader/server. This can be in the form of counters (number of reads), timestamps, or updated pseudonyms and keys. When the update does not take place on both sides, desynchronization can occur through adversarial reads or update prevention (e.g. protocol interruption). Unless the protocol is designed to handle this or recover from it, the server will no longer be able to read or recognize the tag. Even in protocols which allow a limited amount of desynchronization, attacks or repeated errors (reads by other systems) can lead to desynchronization (Radomirovic and van Deursen 2008).

2.4 Evaluation of Threats and Possible Countermeasures

Figure 2 relates the threats, associated with the *RFID-edge hardware layer*, to their potential *damage*, the *cost* of implementing each type of attack/threat and the class of tags that are more vulnerable against this type of attack. For each threat we also list possible countermeasures (*solutions*) that could be used to combat it as well as an estimation of the *cost* of deploying this countermeasure. We are actually presenting a qualitative estimate (*low*, *medium*, *high*) of the *cost* in both cases that takes into account the effort, time and financial cost required to launch or combat an attack correspondingly.

The attacks that are more expensive to launch are mainly *side channel attacks* and some advanced types of *impersonation attacks*. From the countermeasures against attacks in the *RFID-edge hardware layer*, the most expensive are those that require either special tags (i.e. more robust or tamper resistant tags) or increased physical security (i.e. guards, cameras, gates). More precisely, side channel attacks constitute one of the most difficult attacks to combat. The most reasonable method to counter these threats involves limiting the electromagnetic emissions of the RFID system. Nevertheless, this measure leads to narrowing the operational range of the system. Another high cost countermeasure is to increase the complexity of the RFID tag's internal circuit, thereby confounding attempts of reverse engineering; a task quite challenging considering the small size required for RFID tags. Use of tamper resistant tags (Swedberg 2006; SAG Security Assembly Group 2010) could also make harder the success of such an attack. However, as such attacks are particularly costly, they may not be very prevalent.

Attack	Potential Damage	Attack Cost*	Class of Tag	Solution - Cost*
Confidentiality — Side Channel Attacks	- Extract information (i.e. cryptographic keys).	H	Low Cost Tags	- Use of tamper resistant tags. **(H)** - Limit electromagnetic emissions. **(M)** - Increase complexity of the circuit. **(H)**
Integrity — Physical Data Modification	- Altering data stored on tag memory.	H	Low Cost Tags	- Memory protection. **(M)** - Secure cryptographic protocols. **(M)**
Impersonation	- Supplant legitimate tags. - Elicit sensitive information. - Gain unauthorized access to services.	M	Low Cost Tags	- Use of tamper resistant tags (i.e. Physical Unclonable Function (PUF)). **(H)** - Memory protection mechanisms. - Physical protection (against tag swapping). **(H)** - Use of encryption techniques. **(M)**
Availability — Permanently Disabling Edge Hardware	- Avoid identification. - Untraceability of tagged objects.	L	High/Low Cost Tags	- Rugged, flexible tags. **(M)** - Increased physical security. **(H)** - Efficient key management (regarding command abuse). **(M)**
Temporarily Disabling Edge Hardware	- Avoid identification. - Untraceability of tagged objects.	M	High/Low Cost Tags	- Have limited number of unsuccessful reads. **(L)** - Store both the old and the potential new key or pseudonym values. **(M)**

*Cost: **H** high, **M** medium, **L** low.

Fig. 2 RFID threats and countermeasures related to the RFID edge hardware layer

3 Communication Layer

This layer includes all the attacks that are based on the RFID communication and the transfer of data between the entities of the RFID network.

3.1 Confidentiality

Confidentiality in RFID communication is mainly violated by attacks such as *eavesdropping, unauthorised tag reading, privacy threats* as well as possible *key compromise*.

3.1.1 Eavesdropping

The open medium and the insecure nature of radio communication channel render eavesdropping one of the most critical encountered attacks against RFID systems. Eavesdropping in RFID communication is defined as surreptitiously listening and intercepting messages transferred between legitimate RFID entities. An adversary may eavesdrop on both channels reader-to tag (*forward channel*) and tag-to reader (*backward channel*). Nevertheless, the *forward channel* is more suscepti-ble to this threat since the readers' signal is much stronger. The success of the attack also depends on the location of the adversary, while the intercepted infor-mation may subsequently be used for the deployment of more sophisticated attacks. An eavesdropper may intercept messages and extract information that can be used for launching more sophisticated attacks. The same applies even if encryption and authentication techniques are used to protect the RFID communication (traffic analysis attacks).

Every type of RFID system has its own maximum range at which communication can take place. The *forward channel* (reader-to tag), especially in passive systems, has a longer range than the *backward channel* (tag-to reader). Also, since the passive tags use the *forward channel* to power themselves this creates an effective *third range* that of the maximum distance the received power is sufficient for the tag's operation. The *nominal read range* for a system is determined by the minimum of the aforementioned three ranges. Note that an eavesdropper does not have to power a tag, and can use a larger antenna than a regular tag or reader and operate at an increased *rogue reading range*. Finally, *detection range* can play a role. This is the maximum range at which a tag or reader can be detected, but no sense can be made of actual information being transmitted.

3.1.2 Unauthorised Tag Reading

Unfortunately, RFID tags lack an on/off switch that would allow or prevent reading. Even worse, not all types of RFID tag are able to use secure authentication protocols that prevent unauthorised reading. Hence, in many cases RFID tags can be read without authorization and without any indication that they were read.

Both *eavesdropping* and *unauthorized tag readings* are widely-deployed attacks with considerable negative effects for the victim. These attacks, when performed for competitive espionage purposes may reveal secrets and sensitive information such as marketing strategies or availability of stocks.

3.1.3 Privacy Threats

There are several proposals that attempt to formalize privacy in RFID protocols. Initially, privacy is formalized by the ability to distinguish two known tags (Avoine 2005). However, the model rules out the availability of side-channel information (e.g. knowing the success/failure of a protocol instance on the reader). Juels and Weis extended this model using the side-channel information and allowing the two tags to be chosen by the adversary (Juels and Weis 2007). In (Vaudenay 2007), a hierarchy of privacy models is presented and the restrictions of RFID systems regarding tag corruption and availability of side channels are studied. Specifically, the proposed model captures the notion of a powerful adversary who can monitor all communications, trace tags within a limited period of time, corrupt tags, and get side-channel information on the reader output. Adversaries who do not have access to this side-channel information are called *narrow* adversaries. Depending on the amount of corruption, adversaries are called *strong*, *destructive*, *forward*, or *weak adversaries*.

We describe in detail two of the most significant privacy threats: *tracking* and *hotlisting*.

- *Tracking*: The response of RFID tags to authorized or unauthorized readers is performed silently without giving any sign of activity. This characteristic can be exploited in order to surreptitiously collect personal information that may be used to create profiles and track users. Collected information may vary from purchasing preferences, critical personal information such as medical data to location of individuals. For instance RFID tags produce traces that may subsequently be used to track the position of individuals. Even if these data are "anonymized" they can still give indication about the location of users and create movement profiles. Adversaries may even exploit the unused memory storage of multiple tags in order to create illegal communication channels and transfer information covertly (Karygiannis et al. 2006). It is hard to detect this unauthorized transfer of information. An example could be the use of RFID tags, whose normal use is the identification of people, to reveal personal information related to social activities.
- *Hotlisting*: Information related to the location of a user or the association of an individual with an object could be used by an adversary to enable other more direct attacks. More precisely, an adversary may target and rob people that collect valuable items (e.g. jewellery), scan the contents of house before breaking into it, pick pocket purses with tagged banknotes or scan cargos of valuable or sensitive items. Considering that passports are also tagged they might be used by terrorists to detect people of specific nationalities and trigger "RFID bombs" against them (European Commission 1995).

3.1.4 Key Compromise

An attractive target for adversaries is always information related to encryption techniques and key material. Knowledge of these kinds of information would allow them to easily impersonate tags and readers or to access other information through elevation of privilege. For instance, it may enable them to read sensitive information stored in e-passports and identify nationalities.

- *Crypto Attacks*: Sensitive data stored on RFID tags are usually protected by employing encryption techniques. However, a determined adversary could mount crypto-attacks in order to break the employed cryptographic algorithms and disclose or manipulate data. Targets of attack include password authentication schemes, ciphers, pseudo-random number generators, hash-functions. Examples of classic attacks are *brute force* (password/cipher), *chosen-ciphertext* or *known-plaintext* attacks (ciphers), *first pre-image* or *collision* attacks (hashes). A representative example of RFID crypto attacks was the demonstration that the Dutch passport can be broken via *brute force* attack (Riscure 2006). Furthermore, researchers from the Radboud University of Nijmegen (Garcia et al. 2008) have performed an attack against the crypto-1 algorithm of the MIFARE card based on an exploit of the proprietary algorithm. The same type of card is used in the Dutch public transport protocol.

- *Reverse Engineering*: Reverse engineering is a term used to describe attacks that attempt to model the inner workings of a device or piece of software, usually in order to mimic its behaviour, or to be able to attack it more efficiently. Uncovering the detailed workings of a proprietary cipher, hash or protocol, can be a first step towards finding weaknesses. If "security through obscurity" has been relied upon, or the algorithm has not been rigorously tested, this can have a severe impact on security. The recent publications concerning the Mifare Classic tag and its proprietary Crypto-1 algorithm are a good example (de Koning Gans et al. 2008).

Reverse engineering can also be an invaluable tool in *side-channel analysis*, probing or "glitching". A full understanding of the inner workings of the device is often required in order for such attacks to be successful. Even impersonation attempts can benefit from reverse engineering; knowing how the original device behaves is key in replicating it.

3.2 Integrity

The *integrity* of the RFID communication channel can be compromised through *relay, replay* attacks, *message reconstruction* or *modification/insertion of data*.

3.2.1 Relay Attacks

In a relay attack an adversary acts as a man-in-the-middle. An adversarial device is placed surreptitiously between a legitimate RFID tag and reader to intercept (and possibly modify) the communications between tag and reader. Tag and reader are fooled into thinking that they are communicating directly with each other. A large distance between a tag and a reader can be bridged by using two devices: one for the communication with the reader (the "ghost") and one for the communication with the RFID tag (the "leech"). Note that these devices may operate at larger ranges than the *nominal read* or power up ranges, especially the "ghost" as it does not actually rely on power from the reader.

Recently, a German MSc. Student (Tanenbaum 2008) proved the vulnerability of the Dutch public transport by performing a relay attack on the Dutch transit ticket. The student just implemented the "ghost and leech" model as described by Kfir and Wool (2005) and created great concerns for the $2 billion Dutch public transport system. Another example of a high-impact attack would be to charge a payment to a victim's RFID-enabled credit card (Heydt-Benjami et al. 2008) without his knowledge.

Depending on which party is committing the fraud, different names are given to relay attacks.

- *Distance Fraud*: In this attack the adversary uses a rogue tag and tries to convince a legitimate reader that she is nearer than she really is.
- *Mafia Fraud*: This attack involves three main parties: a legitimate tag T, a legitimate reader R and the attacker A. The attacker has at her disposal a fraudulent tag T' and reader R' and attempts to convince the legitimate reader R that she is communicating with the legitimate tag T while in reality the reader R communicates with the attacker A. Nevertheless, no disclosure of the private keys shared between the legitimate and reader is made.
- *Terrorist fraud*: "Terrorist fraud" involves a fully cooperating tag (owner), who does not share secret key material with the relaying party. This means that it will compute responses to challenges, but will not give the attackers the means to perform the computations on their own (Tu and Piramuthu 2007; Desmedt 1998).

3.2.2 Replay Attacks

Replay attacks are impersonation attacks that involve the re-sending of valid replies at a later time. These replies can be obtained through eavesdropping or adversarial sessions. They are related to relay attacks, but they take place "offline" in the sense that there is a clear delay between time of obtaining and time of re-sending the message. The simplest application scenario of a replay attack is the replay of an intercepted message transmitted from a legitimate tag to a legitimate reader in an RFID based access control system or in an RFID identification system (which does not use an advanced challenge-response authentication protocol). In these scenarios the adversary is able to gain access in a specific building or supplant the presence of a particular item just by replaying the intercepted message. Even, if the messages are encrypted a successful attack may be launched easily. The inclusion of a source of freshness (random number) in the messages is a necessary condition but by itself does not guarantee the protection against replay attacks.

3.2.3 Message (Re)construction

Protocols that include a random session variable (a nonce) are usually resistant to replay attacks. In some cases, this "freshness" can be eliminated though, by combining or analyzing several messages. This allows for the (re)construction of new valid messages which can be used in future impersonation attempts. Thus, this attack may enable an adversary to perform a more sophisticated impersonation attack (e.g. gain unauthorized access to a restricted place) in case that a simple replay attack is not sufficient.

3.2.4 Data Modification

Since RFID tags are usually equipped with writable memory, adversaries can exploit this to transform or erase data. The feasibility of this happening highly depends on the employed READ/WRITE protection as well as the used RFID standard. The impact of the attack depends on the application as well as the degree of modification. For instance, the modification of tags used in medical applications (e.g. carrying a medicine's recommended dosage or a patient's history) may have dreadful implications. Sophisticated adversaries may modify critical information without transforming the tag's ID or any security related info such as encryption keys or credentials. Thus, the reader is not able to indicate alterations. One of the main approaches used to modify data in RFID communication is by *abusing the coding scheme*.

- *Coding Scheme Abuse*: One way to modify data in transit is to replace or flip bits in transmissions. Some coding schemes are more vulnerable to this type of attack that others. For instance the NFC (Near Field Communication) tags (Haselsteiner and Breitfuß 2006) use "modified Miller coding" with 100% modulation ratio at 106 KBaud, while above 106 kBaud they switch to "Manchester" with a 10% modulation ratio. With 100% modulation ratio, meaning that a "0" and a "1" are encoded as "no signal" and "full signal" respectively. An adversary can change a "0" into a "1" bit by sending out a signal of his own, but she cannot change a "1" into a "0" because she has no effective way of reducing the signal strength from the legitimate reader. However, with a 10% modulation ratio a "0" is encoded as a weak signal (82%), while a "1" is a stronger signal (100%). Now an adversary can flip bits as she desires, by adding to both signals. Signals of 80% will seem like baseline noise, 100% will seem like a "0" and 125% will seem like a "1".

3.2.5 Data Insertion

Attacks that do not modify parts of a transmission, but add new data (or even whole messages), are termed data insertion rather than data modification attacks. This can take place at the edge hardware or in the back-end through various means, but the easiest and most common place to execute such attacks is in the communication protocols themselves. Possible application scenarios of such an attack vary. For instance, this attack may enable an adversary to insert information and thus alter fields such as the prices of goods in a department store or a warehouse. One of the most challenging ways to perform data insertion is the approach called "*racing*".

- *Racing*: A specific way of inserting data is for an adversary to quickly send a response before the legitimate reader. This need for split-second timing gives rise to the term "racing". By doing so, one can let the reader perform part of the protocol (i.e. unlocking a tag) and hijack the session (for instance write different information to the tag, or close the session without updating pseudonyms).

3.3 Availability

Denial of Service (DoS) attacks are one of the most challenging threats against RFID communication layer, since they can be easily deployed while they are hard to defend against. This type of attacks can be discriminated to attacks that passively degrade the RF signal, and attacks that actively jam or disrupt communications.

3.3.1 Active Interference

Active interference may be the result of a *noisy* environment, of intentional *jamming* or the abuse of privacy protection approaches such as the *blocker tags* (Juels et al. 2003) or the *RFID guardian* (Rieback et al. 2005).

- *Noise*: Since RFID systems usually operate in an inherently noisy and unstable environment. Thus the RFID communication is vulnerable to possible interference and collisions caused by various sources of radio interference such as power switching supplies and electronic generators.
- *Jamming*: In RFID communication in some cases, where authentication mechanisms are not employed, RFID tags listen indiscriminately to every radio signal within their range. This can be exploited by adversaries to disrupt communication between legitimate tags and readers by deliberating causing electromagnetic interference via a radio signal in the same rage as the reader.
- *Malicious Blocker Tags*: The normal operation of RFID tags may be interrupted by deliberately blocking access to them. Intentionally blocked access and subsequent denial of service for RFID tags may be caused by abusing *blocker tags* (Juels et al. 2003) or the *RFID guardian* (Rieback et al. 2005). Both approaches were proposed to protect RFID communications against privacy threats. However, both of them could also be employed by adversaries to perform a deliberate denial of service.

3.3.2 Passive Degradation of Signal

The presence of metal compounds, water and other materials can also negatively impact radio communications. The *passive degradation of signal* can be caused by *basic degradation* or *more complex propagation effects* or even by *shielding tags*.

- *Basic Degradation Effects*: The presence of water, metal but also the human body and some types of plastics can interfere with radio transmissions. The negative effects vary from absorption (water and conductive liquids) and reflection/refraction (metal objects and surfaces) to dielectric effects/frequency detuning (dielectric materials like plastics or living tissue) (Sweeney 2005). Generally speaking, the higher the frequency, the greater the impact of metals and liquids on performance.

- *Complex Propagation Effects*: These effects take place when high frequency (UHF/Microwave) radio waves bounce off surfaces and collide in certain places, cancelling each other out if their phases happen to be opposite at that particular point in space. This creates standing waves and multi-pathing/field cancellation. These dead zones have very bad reception, but by slightly moving the source of transmission or the receiving antenna, these problems van be solved (the dead zones will shift or change). Such effects are very unpredictable and thus a naturally occurring problem rather than an attack since the effect is hard to produce in any controlled manner (Oertel et al. 2004).
- *Shielding Tags*: Faraday cages such as aluminium foil-lined bags can shield tags from electromagnetic waves, thus disrupting the communication between tags and readers. This can be exploited by a prospective thief to avoid the checkout reader and steal any product undisturbed.

Both passive and active interference could lead to interruption of RFID communications or even to a complete crash of the identification systems deployed in companies, organizations and merchant stores. The goal of the adversary would be either to sabotage the victim or to perform malicious actions (e.g. steal goods etc.) undisturbed and undetected.

3.4 Evaluation of Threats and Possible Countermeasures

Similarly with Figs. 2 and. 3 summarizes the possible threats related to the RFID *communication layer* and refers to their main impacts (*damages*) as well as the possible defense mechanisms (*solutions*) that could be used in each case. Both *low* and *high cost* RFID tags are vulnerable to almost all threats included in this layer.

The cost for implementing an attack varies from *low* (i.e. *eavesdropping* or *passive/active interference*) to *high* for more sophisticated threats (i.e. relay attacks). From the countermeasures against attacks in the *Communication layer*, in most cases the use of efficient encryption and authentication protocols can significantly safeguard an RFID system with a moderate (medium) additional cost.

However, in other cases such as relay attacks more sophisticated defense mechanisms are required. A good way to guard against relay attacks is to use the distance between tag and reader as a measure of security. If the tag is very close to the reader, one can assume that no adversary is able to get between them and relay the messages without being detected.

Several techniques can be employed to measure this distance: measuring signals strength, the time-delay or round-trip-time, or the orientation of the tag to the reader (triangulation, angle-of-arrival). Ideally, one should make protocols as tight as possible on their timing requirements, and tags should respond as fast as possible, to allow very little room for cheating. Distance bounding protocols (Hancke and Kuhn 2008; Clulow et al. 2006; Singlee and Preneel 2005; Hancke 2005; Kim et al. 2008) are based on this principle and were introduced to combat relay attacks.

	Attack	Potential Damage	Attack Cost*	Class of Tags	Solution - Cost*
Confidentiality	**Eavesdropping**	- Case A: Intercept messages. - Case B: Extract information.	Case A:**L** Case B:**H**	High/Low Cost Tags	- Store critical data on the back-end. **(M)** - Shielding. **(M)** - Use of encryption techniques. **(M)**
Confidentiality	**Unauthorized Tag Reading**	- Extract information.	**L**	Low Cost Tags	- Use of authentication protocols.**(M)**
Confidentiality	**Privacy Threats**	- Traceability. - Collection of personal information.	**M**	High/Low Cost Tags	- Killing tags. **(L)** - Blocking access. **(M)** - Relabeling, use of pseudonyms. **(M)** - Use of encryption techniques.**(M)**
Confidentiality	**Key Compromise**	- Impersonate. - Access to sensitive information. - Break the whole system.	**H**	High/Low Cost Tags	- Strong & published, well-known cryptographic algorithms. **(M)** - Long keys. **(L to H)**
Integrity	**Relay Attacks**	- Manipulate communications. - Deception regarding its location (distance).	**M**	High/Low Cost Tags	- Distance bounding protocols (use of round-trip-time). **(M)** - Measure signal strength and triangulation. **(H)**
Integrity	**Replay Attacks**	- Impersonation. - Desynchronization.	**L**	High/Low Cost Tags	- Use of key updating schemes.**(M)** - Use of timestamps. **(L)** - Use of challenge-response protocols (with nonces, clock synchronization, counters). **(M)**
Integrity	**Message (Re)constrution**	- Impersonation. - Desynchronization.	**M**	Low Cost Tags	- Use of strong cryptographic techniques. **(M)**
Integrity	**Data Modification/ Insertion**	- Alter data on the tag or the back-end data.	**M**	High/Low Cost Tags	- Use read-only tags. **(M)** - Data Modification: use of efficient and secure coding schemes. **(M)** - Data Insertion: dependence between the challenge and the response in the authentication process. **(M)**
Availability	**Active/Passive Interference**	- Interruption of Communication.	**L**	High/Low Cost Tags	Active: - Open problem. - Use of opaque walls. **(H)** - Establish regulations. **(L)** Passive: - Select appropriate frequencies and RFID reader's location. **(L)**

***Cost: H** high, **M** medium, **L** low.

Fig. 3 RFID threats and countermeasures related to the communication layer

4 Back-End Layer

The *Back-end layer* is also vulnerable to a broad range of attacks but consists of many different (non specific to the RFID communication) components. The *Back-end layer* generally consists of three components: the "server", the "middleware" and the "application software" (O'Brien 2008). The "server" is responsible for collecting RF transaction data from the readers, while the "middleware" sanitizes and converts the data as needed for further processing by the rest of the back-end, which is usually performed by the "application software" such as databases or business specific software. Attacks against the *back-end layer* may have severe implications.

The most likely points of entry to launch an attack are the information flowing in from the edge hardware and any existing networked connections (EPC Object Name Service or internet access, linking or sharing of databases with other parties etc). We will address only the threats that originate from the side of the edge hardware.

4.1 Confidentiality

Most of the information in a typical RFID system is kept at the *back-end* (e.g. key information, read history, additional information related to tags, items or owners). This renders the *back-end* an attractive target for disclosure attacks.

4.1.1 Privacy Violation

Transaction histories or person-related information could be (mis)used by the owner of the system or attackers for tracking, "hotlisting", preferential or blackmail attacks.

4.1.2 Key Compromise

Key material is often stored on the server a good design choice in terms of security. However, even here secret keys are not entirely safe; if an adversary could gain control over the server or access its storage, a leak can occur. Aside from more generic attacks such as network attacks, social engineering, insider attacks etc, the RFID tags and readers themselves can be a potential point of entry to mount an attack. Obtaining key information directly from the *back-end* could enable an attacker to track, access or impersonate tags and readers at will. Forward security is essential to allow systems to recover from such high-impact attacks (Avoine et al. 2009). The impact of the key comprise through an attack to the back-end layer is similar to that when the key compromise is performed through the communication layer. This attack enables an adversary to gain access in critical information that can be used for his own benefit (e.g. perform financial fraud).

4.2 Integrity

The integrity of the information kept by the *back-end* is most directly threatened by *information injection* attacks. Again, these could occur through networked attacks or through the information flowing in from the readers and tags.

4.2.1 Information Injection

Code insertion attacks on the middleware can originate from the tag/reader side. One way of mounting such an attack is through exploits. Considering the fact that middleware applications often use multiple scripting languages such as Javascript,

PHP, XML, SQL etc. there is plenty of opportunity for security holes to exist. An example SQL injection attack is described in Rieback et al. (2006).

Another avenue for information injection are *buffer overflow attacks*. These are a major threat and are among the hardest security problems in software to guard against. Buffer overflow exploits store data or code beyond the bounds of a fixed-length buffer. When the system attempts to normally process this content, data will flow beyond the buffer and onto the stack, causing it to be executed. Considering the limited memory storage of RFID tags, this may not be trivial, but there are still commands that allow an RFID tag to send the same data block repetitively in order to overflow a buffer in the back-end. Other options include the use of other devices with more resources such as smart cards or devices that are able to emulate (multiple) RFID tags (e.g. RFID guardian). More sophisticated attacks involve using RFID tags to propagate hostile code that subsequently could infect other entities of the RFID network (readers and connecting networks). In these cases we speak of worms (replicating code requiring external data) and viruses (self-sustained replicating code). Again, an adversary could use either sophisticated tags or additional hardware in order to store and spread a virus or other RFID malware. The impact of such attacks can be extremely serious. For instance, an injected virus or malware could crash the whole backend system in a hospital identification system or alter the private information of patients.

Although these types of attacks are not widespread, and rather unlikely to be performed in real-life scenarios through simple passive RFID tags, laboratory experiments (Rieback et al. 2006) have proved that they are feasible. We should not forget after all that instead of using simple passive RFID tags to perform code injection other more powerful devices (instead of passive RFID tags) or access to the back-end database though other means could be used to successfully inject a virus or another type of malware.

4.3 Availability

Successful Denial of Service (DoS) attacks against the back-end of an RFID system can have a system-wide impact (unless adequate backup measures are in place). A primary point of attack is again the information flow from the edge-hardware.

4.3.1 Partial DoS

Flooding and spamming attacks can cause temporary failure in the back-end of an RFID system, or provide such a workload that processing of legitimate requests is slowed or delayed. One could think of using *blocker tags* (Juels et al. 2003) to spam readers with answers upon attempted read (these requests are all passed on to the server, unless care was taken to detect and prevent this sort of attack by the reader). Also, open systems which allow untrusted readers to interface with the server could be spammed with (many) malicious readers, generating a large amount of traffic at one specific time (Burmester et al. 2006).

4.3.2 Complete DoS

If the chain of programs running in the *back-end* is hit by a successful DoS attack (the database is taken down, the server is crashed by bad input, etc.), the whole system will suffer from DoS, unless backups are in place and working. As mentioned before, we do not go into the plethora of attacks that could lead to DoS in databases, Operating Systems or networks, but again, the RFID edge hardware can be an extra point of entry that should be properly secured (Rieback et al. 2006).

The effects of a *partial/complete DoS* attack in a real application scenario vary from temporary interruption of services to the complete disruption of an RFID system. For instance, such an attack may enable an adversary to crash the RFID system of a rival company and thus resulting loss of revenue.

4.4 Evaluation of Threats and Possible Countermeasures

Figure 4 depicts the association between the threats related to the *Back-end layer*, their potential impact (*damage*), the *cost* required to perform these attacks as well as possible defense methods (*solutions*) and their associated *cost*. In all cases the cost required to perform an attack is medium, as these are common attacks in network systems in general, while their impact can be very large in an RFID application scenario. The countermeasures against such threats mainly include standard security mechanisms such as access control systems, intrusion detection systems and firewalls. The most expensive attack to combat is *information injection*, which requires detailed data and code checking; a very hard and time consuming task.

	Attack	Potential Damage	Attack Cost*	Solution - Cost*
Confidentiality	**Privacy Violation/Key Compromise**	- Tracking, "hotlisting". - Access to private information.	**M**	- Access Control Mechanisms. **(L to M)** - Firewalls, Intrusion Detection Systems. **(L to H)**
Integrity	**Information Injection**	- Manipulation/ erase of data.	**M**	- Data and code checking. **(H)**
Availability	**Denial of Service Attacks**	- Interruption of Services. - Crash of the whole RFID system.	**M**	- Access Control Mechanisms. **(L to M)** - Firewalls, Intrusion Detection Systems. **(L to H)** - Efficient search protocols. **(M)**

*Cost: H high, **M** medium, **L** low.

Fig. 4 RFID threats and countermeasures related to the back-end layer

5 Open Issues and Discussion

In this chapter we presented a detailed overview of the most prominent RFID threats by dividing them in three main layers and then considering which of the three security principles (*confidentiality*, *integrity* and *availability* (*CIA*)) is being compromised in each case. This is the first time, to the best of our knowledge, that the concept of the main security principles has been used as a criterion to classify RFID threats. We believe that this point of view provides a structured description and global perspective of the problem. Additionally, we relate the treats at each layer to their impact, their delivery cost, the type of RFID tags most vulnerable to these threat, as well as possible countermeasures and their associated cost.

Many defense mechanisms have already been proposed to safeguard RFID systems against possible attacks. Some of these attacks are easy to combat (i.e. *unauthorized tag reading* and *tracking*) by using efficiently designed protocols and cryptographic primitives as well as implementing appropriate software. Other threats are harder or more costly to defend against (i.e. hardware-related threats, like tampering attacks or signal degradation), while others are still open problems and subject to research (i.e. attacks that compromise the *availability*).

It is obvious that there is a need for effective defense mechanisms to guarantee the reliability and security of RFID systems. In this chapter we provide a list of the main solutions used to combat these threats.

Acknowledgments We would like to thank Christos Dimitrakakis for additional proofreading. This work was partially supported by the Netherlands Organization for Scientific Research (NWO) under the RUBICON "Intrusion Detection in Ubiquitous Computing Technologies" grant and the ICT talent grant supported by the Delft Institute for Research on ICT (DIRECT) under the grant "Intrusion Detection and Response in Wireless Communications" awarded to Aikaterini Mitrokotsa.

References

Agrawal D, Archambeault B, Rao JR et al (2003) The EM Side-Channel(s). In: CHES '02: Revised Papers from the 4th international workshop on cryptographic hardware and embedded systems, London, UK. Springer, Heidelberg, pp 29–45

Auto-ID Center (2003) Draft protocol specification for a 900 MHz Class 0 Radio Frequency (RF) Identification Tag. http://www.epcglobalinc.org/standards/specs/900_MHz_Class_0_RFIDTag_Specification.pdf. Accessed 15 Feb 2010

Avoine G (2005) Cryprography in radio frequency identification and fair exchange protocols. PhD thesis, No. 3407, Ecole Polytechnique Fédérale de Lausanne, Switzerland, December 2005

Avoine G, Lauradoux C, Martin T (2009) When compromised readers meet RFID – extended version. In: Workshop on RFID security – RFIDSec'09, Leuven, Belgium

Avoine G. Oechslin P (2005) RFID traceability: a multilayer problem. In: Patrick A, Yung M (eds) Financial cryptography and data security, 9th International conference, FS 2005, LNCS 3570.. Springer, Heidelberg, pp 125–140

Ayoade J (2007) Privacy and RFID systems, roadmap for solving security and privacy concerns in RFID systems. Comput Law Secur Rep 23:555–561

Berkes J (2006) Hardware attacks on cryptographic devices. Technical Report ECE 628, University of Waterloo

Burmester M, van Le T, de Madeiros B (2006) Provably secure ubiquitous systems: universally composable RFID authentication protocols. In: 2nd IEEE/CreateNet international conference on security & privacy in communication networks (SECURECOMM 2006), Baltimore, MD, USA, IEEE Computer Society, pp 1–9

Clulow J, Hancke GP, Kuhn MG et al (2006) So near and yet so far: distance-bounding attacks in wireless networks. In: Proceedings of the European workshop on security and privacy in Ad Hoc and sensor networks (ESAS'06), Hamburg, Germany, pp 83–97

Collins J (2006) RFID-Zapper shoots to kill. RFID Journal. http://www.rfidjournal.com/article/print/2098. Accessed 15 Feb 2010

de Koning Gans G, Hoepman JH, Garcia FD (2008) A practical attack on the MIFARE classic. In: Grimaud G, Standaert FX (eds) Smart card research and advanced applications, 8th IFIP WG 8.8/11.2 International conference, CARDIS 2008, London, UK, September 8–11, 2008, Proceedings, Series LNCS, Subseries Security and Cryptology, vol. 5189, Springer, Heidelberg

Desmedt Y(1988) Major security problems with the "unforgeable" (Feige-)Fiat- Shamir proofs for identity and how to overcome them. In 6th worldwide congress on computer and communications security and protection (Securicomm'88), Paris, 15–17 March, pp 147–159

European Commission (1995) Directive 95/46/EC of the European parliament and of the council of 24 October 1995 on the protection of individuals with regard to the processing of personal data and on the free Movement of such data. Official Journal of European Communities L.281:31

Garfinkel S, Juels A, Pappu R (2005) RFID privacy: an overview of problems and proposed solutions. IEEE Secur Priv 3(3):34–43

Garcia FD, de Koning Gans G, Muijers R et al (2008) Dismantling MIFARE Classic. In: Jajodia S, Lopez J (eds) ESORICS 2008, LNCS 5283. Springer, Heidelberg, pp 97–114

Han DG, Tagaki T, Kim HW et al (2006) New security problem in RFID systems "tag killing". In: Computational science and its application – ICCSA 2006, workshop on applied cryptography and information security (ACIS 2006), LNCS 3982. Springer, Heidelberg, pp 375–384

Hancke GP (2005) A practical relay attack on ISO 14443 proximity cards. Technical Report, University of Cambridge, Computer Laboratory

Hancke GP, Kuhn MG (2008) Attacks on time-of-flight distance bounding attacks. In: 1st ACM conference on wireless network security, ACM, New York, NY, pp 375–384

Haselsteiner E, Breitfuß K (2006) Security in near field communication (NFC) – strengths and weaknesses. In: Workshop on RFID security, Graz, Austria, 12–14 July 2006, pp 1–9

Heydt-Benjami TS, Bailey DV, Fu K et al. (2008) Vulnerabilities in first generation RFID-enabled credit cards. Financial cryptography and data security, 11th international conference, FC 2007, and 1st international workshop on usable security, USEC 2007, Scarborough, Trinidad and Tobago, February 12–16, 2007, LNCS 4886. Springer, Heidelberg, pp 2–14

Hutter M, Mangard S, Felhofer M (2007) Power and EM attacks on passive 13.56 MHz RFID devices. In: Paillier P, Verbauwhede (eds) CHES 2007, LNCS 4727. Springer, Heidelberg, pp 320–333

Hutter M, Medwed M, Hein D et al (2009) Attacking ECDSA-enabled RFID devices. In: Abdalla M et al (eds) ACNS 2009, LNCS 5536. Springer, Heidelberg, pp 519–534

ISO (International Organization for Standardization) (2005) ISO/IEC 27001: 2005 information technology – security techniques – Specification for an information security management system. http://www.iso.org/iso/catalogue_detail?csnumber=42103. Accessed 15 Feb 2010

Juels A (2005) Strengthening EPC tags against cloning. In: Jacobson M, Poovendran R (eds) ACM workshop on wireless security (WiSe'05), LNCS 3982. Springer, Heidelberg, pp 67–76

Juels A (2006) RFID security and privacy: a research survey. In: IEEE J Sel Areas Commun 24(2):381–394

Juels A, Rivest R, Szydlo M (2003) The Blocker Tag: selective blocking of RFID tags for consumer privacy. In: Proceedings of the 10th ACM conference on computer and communication security. ACM New York, NY, USA, pp 103–111

Juels A, Weis S (2007) Defining strong privacy for RFID. In: Proceedings of the 5th annual IEEE international conference on pervasive computing and communications Workshop (PercomW'07), March 19–23, White Plains, NY, pp 342–347

Karygiannis T, Phillips T, Tsibertzopoulos A (2006) RFID Security: a taxonomy of risk. In: Proceedings of the 1st international conference on communications and networking in China (China'Com 2006), October 2006. IEEE Press, pp 1–8

Karygiannis T, Eydt B, Barber G et al (2007) Guidelines for securing Radio Frequency Identification (RFID) systems. Special Publication 800–98, National Institute of Standards and Technology, Technology Administration U.S. Department of Commerce, csrc.nist.gov/publications/nistpubs/800-98/SP800-98_RFID-2007.pdf. Accessed 15 Feb 2010

Kaufman C, Perlaman R, Speciner M (2002) Network security: private communication in a public world, 2nd Edn. Prentice Hall, Upper Saddle River, NJ

Kfir Z, Wool A (2005) Picking virtual pockets using relay attacks on contactless smartcard. In: Proceedings of the 1st international conference on security and privacy (SECURECOMM'05) . IEEE Computer Society Press, September 5–9, Athens, Greece, pp 47–48

Kim CH, Avoine G, Koeunem F et al. (2008) The Swiss-Knife RFID distance bounding protocol. In: Lee PJ, Cheon JH (eds) International conference on information security and cryptology – ICISC, LNCS 5461. Springer, Heidelberg, pp 98–115

Kocher PC, Jaffe J, Jun B (1999) Differential power analysis. In: Wiener M (ed) Advances in Cryptology – CRYPTO '99, vol. 1666. Springer,, Heidelberg, pp 388–397

Lowry J (2004) Adversary modeling to develop forensic observables. In: 4th annual digital forensics research workshop 2004, Baltimore, MD, pp 204–213

Mirowski L, Hartnett J, Williams R (2009) An RFID attacker behavior taxonomy. IEEE Pervasive Computing, doi:doi.ieeecomputersociety.org/10.1109/MPRV.2009.68

Mitrokotsa A, Rieback MR, Tanenbaum AS (2009) Classifying RFID attacks and defenses. Special issue on advances in RFID technology, Inf Syst Front, Springer. doi: 10.1007/s10796-009-9210-z, July 2009

O'Brien DF (2008) RFID: an introduction to security issues and concerns. In: Wiles J, Rogers R (eds) Techno security's guide to managing risks for IT managers, auditors, and investigators, Syngress Press, Burlington, MA

Oertel B, Wölk M, Hilty L et al (2004) Security aspects and prospective applications of RFID systems. Federal Office for Information Security. www.rfidconsultation.eu/docs/ficheiros/RIKCHA_englisch_Layout.pdf. Accessed 15 Feb 2010

Ohkubo M, Suzuki K, Kinoshita K (2004) Hash-chain based forward-secure privacy protection scheme for low-cost RFID. In: Proceedings of the symposium on cryptography and information security (SCIS 2004), vol. 1, Sendai, Japan, January 2004, pp 719–724

Ohkubo M, Suzuki K, Kinoshita S (2003) Cryptographic approach to "privacy-friendly" tags. In: RFID privacy workshop, MIT, MA

Oren Y, Shamir A (2007) Remote password extraction from RFID tags. In: IEEE Transactions on Computers. 56(9): 1292–1296. doi:10.1109/TC.2007.1050

Peris-Lopez P, Hernandez-Castro JC, Estevez-Tapiador JM et al. (2006) RFID systems: a survey on security threats and proposed solutions. In: Cuenca P, Orozco-Barbosa (eds) PWC 2006, LNCS 4217. Springer, Heidelberg, pp 159–170

Plos T (2008) Susceptibility of UHF RFID tags to electromagnetic analysis. In: Malkin, T.G. (ed.) CT-RSA 2008. LNCS, vol. 4964. Springer, Heidelberg, pp 288–300

Radomirovic S, van Deursen T (2008) Vulnerabilities in RFID protocols due to algebraic properties. In: 3rd Benelux workshop on information and system security, Eindhoven, The Netherlands

Reid D (2006) ePassport "at risk" from cloning. http://news.bbc.co.uk/2/hi/programmes/click_online/6182207.stm. Accessed 15 Feb 2010

Reid JT, Tang T, Gonzalez Nieto JM (2007) Detecting relay attacks with timing-based protocols. In: 2nd ASIAN ACM symposium on information, computer and communications security, Singapore, 2007. ACM New York, NY, USA, pp 204–213

Rieback M, Crispo B, Tanenbaum A (2005) RFID Guardian: A battery-powered mobile device for RFID privacy management. In: Mu Y, Susilo W, Seberry J (eds) Information security privacy, 13th Australian conference, (ACISP 2008), Wollonong, Australia, July 7–9, 2008, Proceedings, LNCS 5107. Springer, Heidelberg, pp 184–194

Rieback M, Crispo B, Tanenbaum A (2006) Is your cat infected with a computer virus? In: Proceedings of the 4th IEEE international conference on pervasive computing and communications (PerComm 2006), IEEE Computer Society, Washington, DC,

Riscure (2006) Privacy issue in electronic passport. http://www.riscure.com/contact/privacy-issue-in-electronic-passport.html. Accessed 15 Fe 2010

SAG Security Assembly Group (2010) SAG RFID tamper proof label. http://www.sag.com.tw/index.php?_Page=product&mode=show&cid=7&pid=32&SetLang=en-us. Accessed 15 Feb 2010

Singlee D, Preneel B (2005) Location verification using secure distance bounding protocols. In: Proceedings of the IEEE international mobile ad hoc and sensor systems conference, Washington, DC, 7–7 November, pp. 840–847

Swedberg C (2006) Broadcom introduces secure RFID chip. RFID Journal, 29 June 2006. http://rfidjournal.com/article/view/2464/1/1. Accessed 15 Feb 2010

Sweeney PJ (2005) RFID for dummies. Wiley, Indianapolis, IN

Tanenbaum AS (2008) Dutch public transit card broken: RFID replay attack allows free travel in the Netherlands, http://www.cs.vu.nl/˜ast/ov-chip-card/. Accessed 20 Nov 2009

Tu YJ, Piramuthu S (2007) RFID distance bounding protocols. In: 1st international EURASIP workshop in RFID technology, Vienna, Austria, 24–25 September

Vaudenay S (2007) On privacy models for RFID. In: Proceedings of ASIACRYPT'07, vol. 4833, LNCS. Springer, Heidelberg, pp 68–87

Part II
Data Management

Temporal and Location Based RFID Event Data Management and Processing

Fusheng Wang and Peiya Liu

Abstract Advance of sensor and RFID technology provides significant new power for humans to sense, understand and manage the world. RFID provides fast data collection with precise identification of objects with unique IDs without line of sight, thus it can be used for identifying, locating, tracking and monitoring physical objects. Despite these benefits, RFID poses many challenges for data processing and management. RFID data are temporal and history oriented, multi-dimensional, and carrying implicit semantics. Moreover, RFID applications are heterogeneous. RFID data management or data warehouse systems need to support generic and expressive data modeling for tracking and monitoring physical objects, and provide automated data interpretation and processing. We develop a powerful temporal and location oriented data model for modeling and querying RFID data, and a declarative event and rule based framework for automated complex RFID event processing. The approach is general and can be easily adapted for different RFID-enabled applications, thus significantly reduces the cost of RFID data integration.

1 Introduction

Radio frequency identification (RFID) is an Automatic Identification and Data Capture (AIDC) technology that uses RF waves to transfer data between a reader and a tagged object. The objective is to identify, track and monitor objects around the world. RFID is automatic and fast, and does not require line of sight or contact between readers and tagged objects. Tag ID is encoded through EPC standards (TDS 2008) to provide global identification.

The fundamental operations in RFID applications are observations. The data product generated from observation is a record with the reader's EPC, the tag's

F. Wang (✉)
Center for Comprehensive Informatics, Emory University, Druid Hills, GA, USA
e-mail:fusheng.wang@emory.edu

D.C. Ranasinghe et al. (eds.), *Unique Radio Innovation for the 21st Century*,
DOI 10.1007/978-3-642-03462-6_4, © Springer-Verlag Berlin Heidelberg 2010

EPC, and the observation timestamp. EPC is a family of coding schemes created as an eventual successor to the barcode. An EPC includes a header with a version number, domain manager with manufacturer information, object class (SKU) and serial number, which uniquely identify an object. While the observation itself looks simple, there are several fundamental characteristics for RFID data.

- *Temporal and History Oriented.* RFID observations are always timestamped. Observations either generate new events, or generate state changes. Besides, locations and aggregations change along the time. RFID data management systems have to be temporal oriented, and supports expressive temporal queries for tracking and monitoring on collected histories.
- *Multi-Dimensional.* Besides the temporal dimension, RFID data also have the spatial dimension: it can be either symbolic locations or continuous physical locations (e.g., physical locations from active RFID or GPS enabled RFID.) The third dimension is the containment dimension, for example, the relationships among truck, pallet, case and item. The EPC itself has its own hierarchy: manufacturer, SKU, and serial number. RFID data semantics from multiple dimensions have to be considered, and RFID data management system or data warehousing system needs to consider all the dimensions to optimize the data model, storage and queries.
- *Implicit Semantics.* An observation always indicates change of location, change of container, or occurrence of certain event/operation. A sequence of readings may form certain pattern to define a complex event. It becomes an essential question on how to automatically interpret the semantics from RFID readings or complex event. RFID data processing needs to be able to process complex RFID events, and automate the semantics interpretation.
- *Heterogeneous.* RFID applications can have fixed readers, movable readers, or 802.11 wireless access point "readers" (Cisco Location 2009). Tags can be read-only, read-write – where additional data can be written to a tag by a reader, or sensor-write – where a tag comes with onboard sensors that write sensor data to the tag. The diversity of RFID applications leads to complex semantics of RFID data. This poses another challenge for generic RFID data modeling.
- *High Integration Cost.* RFID data need to be integrated into existing business applications, and such integration cost is a big hurdle for the adoption of RFID. RFID middleware platforms need to provide high adaptability and interoperation for different applications.

In this chapter, we will discuss our work on an expressive temporal and location based data model of RFID data, and a declarative event and rule based framework for automated RFID data transformation. The approaches are generic and can be easily customized for different RFID applications.

2 Temporal and Location Based Modeling of RFID Data

2.1 An Example of RFID-Enabled Supply Chain Application

Figure 1 shows a simplified example of an RFID-enabled supply chain application, with the following workflow:

1. Carriers containing items move along the production line for processing, where each item is associated with a read-only tag and each carrier has a read-write tag. An RFID reader is mounted at each station and writes the information about the performed steps onto the tags of the carriers;
2. Readers at the end of the production line check the completeness of the production steps performed on the items. If complete, items are scanned and packed into cases;
3. Cases are packed into pallets in a warehouse;
4. Pallets are loaded onto a truck, which contains a sensor-write tag and a reader with GPS on board;
5. When the truck is en route, its reader together with the GPS sensor periodically sends the exact location information plus environment information by radio (e.g., cellphone-based networks) to a central computer;
6. Pallets are unloaded from the truck and unpacked into cases, which are scanned and moved into a retail store;
7. Items are scanned by RFID readers and checked out at registers of the retail store.

In this sample RFID application, we observe that while objects carrying RFID tags move around, data are exchanged between the tags and readers. Such raw data, however, provide no explicit meaning and must be transformed into a semantically

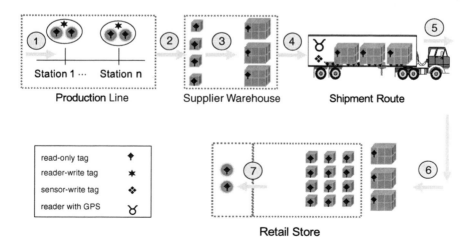

Fig. 1 An example of RFID-enabled supply chain application

explicit data model before they can be used for identifying, locating, tracking and monitoring objects. We propose a general data model for RFID applications, which is based on extension of ER data model (Chen 1976). We will first discuss the fundamental entities and relationships in RFID applications.

2.2 Fundamental Entities in RFID Applications

While there can be many entities in RFID applications, only a few of them are directly related to RFID. We consider such entities as fundamental entities in RFID applications.

- *Physical Objects*. These are EPC-tagged physical objects, such as items, carriers, cases, pallets and trucks in Fig. 1. Objects are uniquely identified by their surrogates – EPC tags.
- *Readers*. These refer to readers that use radio-frequency signals to communicate with EPC tags and read data from tags, or write data to read-write tags. A reader can also have multiple antennas. Each RFID reader (or its antenna) is also uniquely identified by its EPC code.
- *Sensors*. These are onboard sensors of RFID tags, such as temperature sensors, motion sensors, or location sensors. A sensor measures a target and then writes the measurement to its master RFID tag.
- *Locations*. A location represents where an object is (or was), and can be either a physical location – such as a point detected by GPS, or a symbolic location. The granularity of locations can be defined according to application needs. Moreover, one location may contain another. For example, in Fig. 1, the locations include Station 1 to Station n, production line end checkpoint, supplier warehouse, route from the warehouse to a retail store, retail store, and customers. In a smart box application (Lampe and Flörkemeier 2004), locations are simply "in-box" and "out-of-box."
- *Transactions*. These refer to business transactions in which EPC-tagged objects are involved. For example, a checkout involves a credit card transaction with many EPC readings. (Transactions are not considered in many RFID applications, and here we consider them for completeness and RFID-enabled business applications).

2.3 Relationships Between Entities

While the entities in RFID applications are static in general, relationships among these entities can be either static or dynamic. Static entity relationships are similar to traditional relationships in Entity-Relationship Model (ER model). For example, the relationship between an object and its on-board sensors – OBJECTSENSOR – is a static relationship. Due to the temporal nature of RFID data, most relationships in RFID applications, however, are dynamic and history-oriented. Entities interact with each other and generate movement, workflow, operations, and business logic. The interactions can be either event or state changes.

Events are generated when entities interact, which include:

- *Observations.* These are generated when readers interact with objects.
- *Sensor Measurements.* These are generated when a sensor on a tag senses a target (e.g., temperature).
- *Property Values.* These are generated when a writer (i.e., a reader) writes the value of an object property into the object's tag. For example, a reader may write the processing steps performed on an object to the tag attached to the object. The property here is a processing step.
- *Transacted items.* These are generated when an object participates in a transaction.

Besides event occurrences, there are also state changes when entities interact. State change history, i.e., the information about during which period an object is in a certain state, is essential to tracking and monitoring applications and has to be well captured in RFID data models. In RFID applications, state changes include:

- Change of object locations. For instance, the truck and its loaded pallets leave a warehouse.
- Change of object aggregation relationships. For example, cases are packed onto pallets, as shown in Fig. 1.
- Change of reader locations. For instance, a reader is deployed at a new location, or a moveable reader enters/leaves a location.

2.4 A Generalized Data Model for RFID

Based on the discussions above, we summarize a generalized data model for capturing additional information and relationships, including temporal and physical location data, in RFID-enabled applications by using combinations of various types of RFID tags, readers and sensors. This model is based on ER model and inherits the modeling of static entities and static relationships. The new features extended are discussed as follows.

- *Temporal Relationships.* There are two types of temporal relationships among RFID entities: relationships that generate events, and relationships that generate state histories. Here we use dash-dot lines and dash lines to represent these two relationships respectively, as shown in Fig. 2. For an event-based relationship, we use an attribute *timestamp* to represent the occurring timestamp of the event. For a state-based temporal relationship, we use attributes *tstart* and *tend* to represent the lifespan of the state.

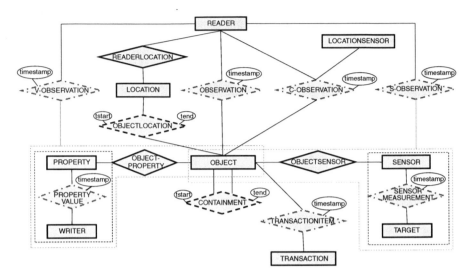

Fig. 2 The RFID data model

- *Nested Relations.* Nested relationship is a new characteristic in read-write RFID applications. For example, for an application with sensor-write tags, an onboard sensor records the temperature measurement history in the tag. Thus a reader observation contains both the EPC of the tag and the measurement history. That is, the observation relation includes a nested relation: the sensor measurement history. While a nested relation can be directly supported in some commercial databases, here we expand such relations into flat tables for generality.

The model is illustrated in Fig. 2, and can be implemented as following relational tables: static entity tables, static relationship tables, and dynamic relationship tables.

2.4.1 Static Entity Tables

- READER(reader_epc, reader_name, description)
  The READER table records the EPC, name and description of a reader, as shown in example Table 1.
- OBJECT(object_epc, object_name, description)
  The OBJECT table includes the EPC, name, and description of an EPC-tagged object.

Table 1 Sample READER table

Reader_epc	reader_name	Description
1.100.1	R1	Reader 1
1.100.2	R2	Reader 2
1.100.3	R3	Reader 3

Table 2 Sample LOCATION table

Location_id	location_name	Owner
LOC1	Production line E	Manufacturer D
LOC2	Warehouse A	Supplier A
LOC3	Route to retail C	Carrier B

- LOCATION(location_id, name, owner)

The location table defines symbolic locations used for tracking, including id, name, and owner of a location. Table 2 shows a sample location table.

- TRANSACTION(transaction_id, transaction_type)
  Since transaction data is business specific, here we simplify a transaction as a record with transaction id and a transaction type.

2.4.2 Static Relationship Tables

- READERLOCATION(reader_epc, location_id)
  This table keeps the location of a fixed reader including the EPC of a reader, and the location id.
- LOCATIONCONTAINMENT(location_id, parent_location_id)
  This table records the containment relationship among locations. For example, a warehouse may contain a loading zone and a departure exit.
- LOCATIONSENSOR(sensor_id, description)
  This table describes location sensors for identifying locations, such as GPS.
- OBJECTPROPERTY(object_epc, property_id, name)
  This table describes the properties defined for an object, including the object's EPC, property's id and name, which is a weak entity dependent on the OBJECT entity.
- OBJECTSENSOR(object_epc, sensor_id, name, type, measurement_ unit, description)
  This table describes the information about a sensor, including the EPC of the object containing the sensor, sensor ID, name, type, description and measurement unit.

2.4.3 Temporal Relationship Tables

- OBSERVATION(reader_epc, object_epc, timestamp)
  This table records the raw reading data generated from readers, including the EPC of the reader or antenna, the EPC of the tag, and the reading timestamp.
- CONTAINMENT(epc, parent_epc, tstart, tend)
  This table records in what period [tstart, tend) an object (identified by its EPC) is contained in its parent object (identified by the EPC of its parent). Table 3 shows a sample content of CONTAINMENT table.

Table 3 Sample CONTAINMENT table

object_epc	parent_epc	Tstart	tend
1.1.1.1	1.1.2.1	2008-11-01 10:33:00	2008-11-07 11:00:00
1.1.1.2	1.1.2.1	2008-11-01 10:33:00	2008-11-07 11:00:00
1.1.2.1	1.1.3.1	2008-11-01 10:35:00	2008-11-07 10:59:00

Table 4 Sample OBJECTLOCATION table

object_epc	location_id	tstart	Tend
1.2.3.4	LOC0	2004-10-10 12:33:00	2004-10-30 17:33:00
1.2.3.4	LOC1	2004-10-30 17:33:00	2004-11-01 10:35:00
1.2.3.4	LOC2	2004-11-01 10:35:00	2004-11-07 11:00:00
1.2.3.4	LOC3	2004-11-07 11:00:00	2004-11-08 15:30:00
1.2.3.4	LOC4	2004-11-08 15:30:00	UC

- OBJECTLOCATION(object_epc, location_id, tstart, tend)
  This table preserves the location history of each object, including an object's EPC, location id, and the period [tstart, tend] during which the object stays in that location. For example, Table 4 shows a sample OBJECTLOCATION table, where *UC* denotes 'now'.
- TRANSACTIONITEM(transaction_id, epc, timestamp)
  This table records information about a transaction including the transaction id, EPC of the object in the transaction, and the timestamp when the transaction occurs.
- C-OBSERVATION(reader_epc, object_epc, timestamp, sensor_id, x, y, z)
  This physical location observation relation table is generated by combining the interaction among the READER, the OBJECT, and the LOCATIONSENSOR. The observation captures the physical location information through a sensor at the timestamp when an object is observed by a reader.
- V-OBSERVATION(vo_id, reader_epc, object_epc, timestamp)
  This property value observation table represents reader observations, where vo_id is the unique ID for each observation. vo_id forms the foreign key linking V-OBSERVATION and PROPERTYVALUE table discussed below.
- PROPERTYVALUE(vo_id, writer_id, timestamp, property_id, value)
  This table represents the history of property values with parent key vo_id, writer_id, the writing timestamp, the attribute's property_id and value. Note that the WRITER entity is omitted here, since it is a reader. Its ID is represented as writer_id in the PROPERTYVALUE relation.
  V-OBERVATION and PROPERTYVALUE tables are used to represent the following nested relations in an efficient way.
  N-V-OBSERVATION(reader_epc, object_epc, timestamp,
  N-PROPERTYVALUE(property_id, writer_id, value, timestamp)).

•

This nested table represents the observation of a tag, including the reader EPC, the object EPC, and observation timestamp, together with the nested history of property values written by a writer. The nested relations may be supported in RDBMS.

- S-OBSERVATION(so_id, reader_epc, object_epc, timestamp)
  This sensor measurement observation table represents reader observations, where so_id is the unique ID for each observation. This so_id forms the foreign key linking S-OBSERVATION and SENSORMEASUREMENT table as follows:
- SENSORMEASUREMENT(so_id, sensor_id, target, value, timestamp)
  This measurement table preserves the history of sensor measurements including the parent key so_id, the sensor id, the target (e.g., temperature), value, and the sensing timestamp.
  Here the TARGET entity is omitted and the target information is included in the SENSORMEASUREMENT relation.

S-OBSERVATION and SENSORMEASUREMENT tables are use to represent the following nested relations in an efficient way.

N-S-OBSERVATION(reader_epc, object_epc, timestamp, N-SENSORMEASUREMENT(sensor_id, target, value, timestamp))

This nested relation table represents the observation of a tag, including the reader EPC, the object EPC, and the observation timestamp, together with the nested history of sensor measurement written by sensors.\

Figure 3 shows a summary of tables for the supply chain application. The "X" symbol means that at current location, the corresponding tables are populated or updated.

Location / Table	Deployment	Production Line	Warehouse	Retail Store (back-store)	Retail Store (front-store)
Static Entity Table:					
READER	X				
OBJECT		X	X		
LOCATION	X				
TRANSACTION					X
Static Relationship Table:					
READERLOCATION	X				
OBJECTSENSOR	X				
OBJECTPROPERTY	X				
Dynamic Event-based Relationship Table:					
OBSERVATION		X	X	X	X
S-OBSERVATION				X	
SENSORMEASUREMENT				X	
V-OBSERVATION		X			
PROPERTYVALUE		X			
TRANSACTIONITEM					X
Dynamic State-based Relationship Table:					
OBJECTLOCATION		X	X	X	X
CONTAINMENT		X	X	X	

Fig. 3 Summary of tables for the example supply chain application

3 Expressively Querying RFID Data

A significant benefit of our RFID model is the power to support complex RFID queries. Most RFID queries are temporal queries with temporal constraints such as history, temporal snapshot, or temporal slicing, mixed with location constraints. There are more complex ones such as temporal joins and temporal aggregates. Queries can be summarized as two major categories: *RFID Object Tracking* to track RFID objects including missing objects; and *RFID Object Monitoring* is to monitor the states of RFID objects and the RFID system. Next we demonstrate the expressive power of the data model through several query examples.

3.1 RFID Object Tracking

Q1. Find the location history of an object with EPC value 'EPC'.

```
SELECT * FROM OBJECTLOCATION WHERE epc='EPC'
```
Q2. Get the latest temperature of the truck with EPC '123'.

```
SELECT m.value, m.timestamp
FROM SENSORMEASUREMENT m, S-OBSERVATION o
WHERE  m.target='temperature'
AND o.object_epc='123'
AND m.so_id = o.so_id AND m.timestamp=(
     SELECT max(m2.timestamp)
     FROM SENSORMEASUREMENT m2
     WHERE m2.so_id = m.so_id)
```

3.2 Missing RFID Object Detection

Q3. Find when and where object 'MEPC' was lost.

```
SELECT location_id, tstart, tend
FROM  OBJECTLOCATION
WHERE epc='MEPC' and tstart =(
    SELECT MAX(o.tstart)
    FROM  OBJECTLOCATION o WHERE
o.epc='MEPC')
```

3.3 RFID Object Identification

Since every RFID object is uniquely identified by its EPC, it is easy to identify an object:

Q4. A customer returns a product with EPC 'XEPC'. Check if this product was sold from this store (location 'L003').

```
SELECT *
FROM OBJECTLOCATION
WHERE epc='XEPC' AND location_id='L003'
```

3.4 RFID Object Snapshot Query

By specifying a snapshot timestamp, it is easy to monitor the snapshot information of any RFID objects, including snapshot locations, containment, observations, or transactions.

Q5. Find the direct container of object 'EPC' at time T.

```
SELECT parent_epc
FROM    CONTAINMENT
WHERE   epc='EPC' AND
        tstart <= 'T' AND tend >= 'T'
```

3.5 RFID Object Temporal Join Query

Temporal join query will retrieve information by joining multiple relations on certain temporal constraint.

Q6. This case (with epc 'TEPC') of meat is tainted. What other cases have ever been put in the same pallet with it?

```
SELECT c2.epc
FROM CONTAINMENT c1, CONTAINMENT c2
WHERE c1.parent_epc = c2.parent_epc
AND c1.epc = 'TEPC' AND overlaps(
c1.parent_epc.tstart,c1.parent_epc.tend,
overlapinteval(c1.tstart,c1.tend,c2.tstart,c2.tend))
```

where *overlaps ()* is a user-defined scalar function to check if two intervals overlap, and *overlapInterval()* returns overlapped internal. User-defined scalar temporal functions can be defined to simplify temporal queries.

While this query can be well expressed in a single query with our data model, it takes eight separate steps with an event based model proposed in (Harrison 2003), which is cumbersome and inefficient.

3.6 RFID Object Containment Queries

RFID containment queries are queries that retrieve the containment relationships between RFID objects. These queries are normally interleaved with other temporal RFID queries. Two special cases are recursive containment queries: *RFID Object Sibling Search*: find all the sibling objects of a container object, and *RFID Object Ancestor Search*: find all the ancestor container objects of an object. The following shows an example of sibling search (ancestor search can be done similarly by switching parent and child attributes).

Q7. RFID Sibling Object Search. Find all objects contained in object 'PEPC'.

```
WITH RECURSIVE all_sub(parentepc, epc) AS
   (SELECT parentepc, epc
   FROM CONTAINMENT
   WHERE parentepc = 'PEPC'
   UNION
   SELECT a.parentepc, c.epc
   FROM all_sub a, CONTAINMENT c
   WHERE a.epc = c.parentepc
   )
SELECT*
```

Based on the data model, we can also specify constraints and business intelligence queries, such as automatic shipping notice, low inventory alert, and trend analysis.

4 Modeling RFID Applications with Complex RFID Events

One essential goal for RFID applications is to map objects and their behaviors in the physical world into the virtual counterparts and their virtual behaviors in the applications by semantically interpreting and transforming RFID data. Application logic can often be devised and engineered as complex RFID events, and once such complex events are detected, the semantics can be automatically interpreted. Figure 4 shows the workflow of RFID event data processing, where primitive RFID events are collected and filtered through RFID edge servers. Once complex events are detected, RFID rules either transform RFID events into RFID data represented in RFID data models or send messages to applications.

The following shows two examples with complex RFID events.

- *Example 1: Data Aggregation.* In Fig. 5a, on a packing conveyer, a sequence of tagged items move through Reader A and are observed by the reader as a sequence of observations, and then a tagged case is read by Reader B as another observation. After that, all items of this sequence are packed into the case. The

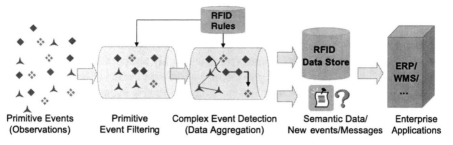

Fig. 4 Overview of complex RFID event processing

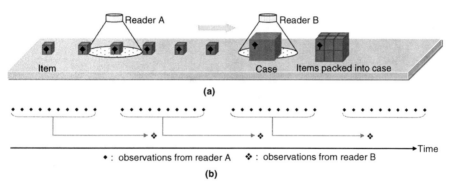

Fig. 5 Sample aggregation in RFID. (**a**) packing of items into its container case; (**b**) complex events used for aggregation

event pattern will be a sequence of event A (by reader A) followed by an event B (by reader B) (Fig. 5b).

- *Example 2. Real-time Monitoring.* A company uses RFID tags to identify asset items and employees in the building, and only authorized users (superusers) can move the asset items out of the building. When an unauthorized employee or a criminal takes a laptop (with an embedded RFID tag) out of the building, within 5 s, the system will send an alert to the security personnel for response. The event pattern is an asset event with a negation event of security, where the temporal constraint is 5 s.

We formulate a declarative rule based approach to provide powerful support of automatic RFID data transformation. Immediate benefits are (i) automated processing, (ii) configurable rules for different applications, and (iii) powerful support of data filtering, data aggregation, data transformation, and real-time monitoring.

4.1 Overview of RFID Events

Next, we formalize the semantics and specification for RFID events. In particular, we will discuss temporal RFID events, which are highly temporally constrained. We will first formalize RFID events by extending traditional events (Chakravarthy and Mishra 1994), and then study the characteristics of the new events.

An *event* is defined to be an occurrence of interest in time, which could be either a *primitive* event or a *complex* event. A primitive event occurs at a point in time, while a complex event is a pattern of primitive events and happens over a period of time.

In the following discussion, we use E to represent an event type, and e to represent an event instance. By default, a lower case representation means an event instance, e.g., e is an instance of event type E.

We first define several functions in time dimension used in our event expressions (Fig. 6).

- $t\_begin(e)$ returns the starting time of event e;
- $t\_end(e)$ returns the ending time of event e;
- $interval(e)$ returns the interval of an event and is equal to $t\_end(e) - t\_begin(e)$;
- $dist(e_1, e_2)$ returns time distance between two events e_1 and e_2 and is equal to $t\_end(e_2) - t\_end(e_1)$;
- $interval(e_1, e_2)$ returns the interval between two events e_1 and e_2 and is equal to $max\{t\_end(e_2), t\_end(e_1)\} - min\{t\_begin(e_2), t\_begin(e_1)\}$.

The following is a list of functions in space dimension:

- $s\_begin(e)$ returns the starting GPS location of event e;
- $s\_end(e)$ returns the ending GPS location of event e;
- $s\_interval(e)$ returns the GPS distance interval of an event and is equal to $|s\_end(e) - s\_begin(e)|$;
- $s\_dist(e_1, e_2)$ returns the GPS distance between two events e_1 and e_2 and is equal to $|s\_end(e_2) - s\_end(e_1)|$;
- $s\_interval(e_1, e_2)$ returns the interval between two events e_1 and e_2 and is equal to $max\{s\_end(e_2), s\_end(e_1)\} - min\{s\_begin(e_2), s\_begin(e_1)\}$.

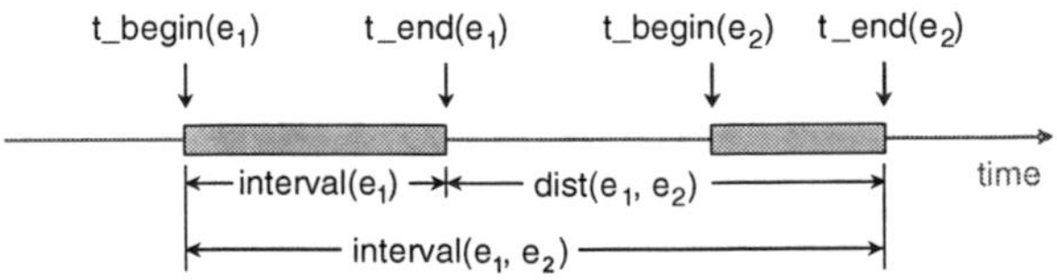

Fig. 6 An illustration of functions used in event expressions

4.2 Primitive RFID Events

Primitive events in RFID applications are events generated during the interaction between readers and tagged objects. There are two types of primitive events, events without physical location information, or events with physical location information. Thus a primitive event is represented in the format of observation(r, o, t) or c-observation(r, o, t, s, g), where r represents the reader EPC, o represents the object EPC, s represents sensor ID, g represents GPS location, and t represents the observation timestamp. For example, observation('r₁', o, t) represents events generated from a reader with EPC 'r₁'. Primitive events are instantaneous, that is, given any primitive event e, t_begin(e) is equal to t_end(e). Primitive events are also atomic: a primitive event either happens completely or does not happen at all.

4.2.1 Definition of Primitive Event Types

While primitive events are all from observations, they can be of different types according to the reader EPC or tag EPC. We first present two user-defined functions on primitive event attributes used to define primitive event types.

group(r). Multiple readers can be deployed into the same group to perform the same functionality;

type(o). The type can be extracted from its EPC value with a user-defined extraction function or specified by a user with a mapping function. For example, type ('8E5YUK691I0J60KDN')='laptop' while type ('UH7JEFU63MAW6I610')='pallet'.

With above functions, we can define primitive event types. For example, the primitive event type E is defined as:

E = observation(r,o,t),group(r)='g1',type(o)='case'

That is, observations of 'case' by readers in group 'g₁' are of type E.

If group () and type () functions are not explicitly specified, the default primitive event type is a group with the reader itself, i.e.,

E = observation('r', o, t)
E = observation('r', o, t), group(r)='r'

4.3 Complex RFID Events

A complex event is usually defined by applying event constructors to its constituent events, which are either primitive events or other complex events. There are two types of RFID event constructors: non-temporal and temporal, and the latter contains order, temporal constraints or both. While complex events defined with non-temporal event constructors can be detected without considering the orders among constituent events, complex events defined with temporal event constructors cannot be detected without checking the orders or other temporal constraints among their constituent events or both.

4.3.1 Basic Non-Temporal, Non-Spatial Complex Event Constructors

- *OR ($\vee$)*: Disjunction of two events E_1 and E_2 ($E_1 \vee E_2$) occurs when either E_1 or E_2 occurs;
- *AND ($\wedge$)*: Conjunction of two events E_1 and E_2 ($E_1 \wedge E_2$) occurs when both E_1 and E_2 occur disregarding their occurrence orders;
- *NOT($\neg$ or !)*:Negation of an event E ($\neg$E or !E) occurs if no instance of E occurs.

Negated events themselves are non-spontaneous and they are usually combined with other events or with some temporal constraints or both. Here we only consider the above three basic non-temporal complex event constructors, which can express general non-temporal events. For example, ANY and ALL events can all be expressed with above operators:

- ALL(E_1, E_2, ..., E_n) occurs if all E_1, E_2, ..., E_n occur irrespective of their orders. This is equivalent to $E = E_1 \wedge E_2 \wedge \ldots \wedge E_n$.
- ANY(m, E_1, E_2, ..., E_n) occurs if any m events out of the n distinct events occur disregarding their orders. This is equivalent to $E = ((E_1 \wedge E_2 \wedge \ldots \wedge E_m) \vee \ldots \vee (E_{n-m+1} \wedge E_{n-m+2} \wedge \ldots \wedge E_n))$.

4.3.2 Temporal and Spatial Complex Event Constructors

- *SEQ(;)*: Sequence of two events E_1 and E_2 (E_1;E_2) occurs when E_2 occurs given that E_1 has already occurred. Here we assume that E_1 ends before E_2 starts;
- *TSEQ(:)*: Distance-constrained sequence of two events E_1 and E_2 (TSEQ(E_1;E_2, τ_1, τ_u)) occurs when E_2 occurs given that E_1 has already occurred and that the temporal distance between the occurrences of E_1 and E_2 is bounded by [τ_1, τ_u]. That is, $\tau_1 \leq \text{dist}(E_1, E_2) \leq \tau_u$;
- *SEQ$^+$(;$^+$)*: The aperiodic sequence operator SEQ$^+$(E) allows one to express one or more occurrences of event type E;
- *TSEQ$^+$(;$^+$)*: The distance-constrained aperiodic sequence operator TSEQ$^+$(E, τ_1, τ_u) allows one to express one or more occurrences of an event E such that the temporal distance between any two adjacent occurrences of E are bounded by [τ_1, τ_u];
- *WITHIN*: An interval-constrained event WITHIN(E, τ) occurs if event e of event type E occurs and interval(e) $\leq \tau$;
- *S-WITHIN*: An interval-constrained event S-WITHIN(E, τ) occurs if event e of event type E occurs and s_interval(e) $\leq \tau$.

4.3.3 Temporal and Spatial Constraints

These temporal event constructors are essential for RFID applications. Most temporal event constructors use temporal constraints to specify temporal complex events. These include *distance constraint*: minimal distance (τ_1) between two events in a temporal sequence TSEQ and maximal distance (τ_u) between two events in a temporal sequence TSEQ; and *interval constraint*: maximal interval size (τ) of a complex event as in the WITHIN or S-WITHIN constructor.

Temporal constraints have been considered in past work in different spaces (Chakravarthy and Mishra 1994; Motakis and Zaniolo 1997; Hinze 2003; Wu et al. 2006). They are either not relevant to RFID events such as timestamp based constraints, or not sufficient for RFID events such as interval constraints. Hinze considers composite events with maximal temporal distances. Our approach is much more general and covers both lower bound and upper bound temporal constraints, and applies to aperiodical sequence events in addition to binary sequence events.

4.3.4 Examples of Complex Events

In Example 1, the complex event is:
$\text{TSEQ}(\text{TSEQ}^+(E_1, \tau_{11}, \tau_{u1}); E_2, \tau_{12}, \tau_{u2})$,
where event types $E_1 = \text{observation}(r_A, o_1, t_1)$,
$\text{group}(r_A) = \text{'}r_A\text{'}$ and $E_2 = \text{observation}(r_B, o_2, t_2)$,
$\text{group}(r_B) = \text{'}r_B\text{'}$.

In Example 2, the complex event is:
$\text{WITHIN}(E_1 \wedge \neg E_2, 5\text{sec})$,
where $E_1 = \text{observation}(\text{'}r_2\text{'}, o_1, t_1)$, $\text{type}(o_1) = \text{'laptop'}$ and $E_2 = \text{observation}(\text{'}r_2\text{'}, o_2, t_2)$,
$\text{type}(o_2) = \text{'superuser'}$.

5 Rules Based RFID Event Data Processing

5.1 RFID Rules

Based on event specification described above, we now define RFID rules. We first introduce the syntax of RFID rules as follows:

```
CREATE RULE rule_id, rule_name
ON event
IF condition (optional)
DO action₁; action₂; ...; actionₙ
```

Here rule_id and rule_name stand for the unique id and name for a rule; event is the event part of the rule, condition is a boolean combination of user-defined boolean functions and SQL queries; and action_1; action_2; ...; action_n is an ordered list of actions, where each action is either a SQL statement or a user-defined procedure, e.g., to send out alarms.

An alias of an event can be defined for reuse in the following form:

```
DEFINE event_name = event specification
```

Next, we show that with declarative RFID rules, we can provide powerful support for RFID data processing including data filtering, data transformation, data aggregation and real-time monitoring.

5.2 Data Transformation and Aggregation

One significant benefit of RFID rules is that data transformation and aggregation is simplified in a declarative way. With a set of data transformation and aggregation rules, RFID observations are automatically interpreted and mapped into their data models and stored in RFID data store, as demonstrated in the following examples.

- *Rule 1. Location Transformation.* Any observation by a reader r will change the location of the observed object o and update the object's old location to current location. This can be done by first updating the history of last location to end at current timestamp t, and then inserting a new location loc for this object with its starting timestamp as t and ending timestamp as "UC."

```
CREATE RULE r1, location_change_rule
ON observation(r, o, t)
DO
UPDATE OBJECTLOCATION SET tend = t
WHERE object_epc = o and tend = "UC";
INSERT INTO
OBJECTLOCATION VALUES(o, "loc", t, "UC");
```

- *Rule 2. Containment Relationship Aggregation.* If a distance-constrained aperiodic sequence of readings from reader "A" is observed followed by a distinct reading from a reader "B" (Fig. 5), it implies that objects observed by 'reader 'A" are being packed in the object observed by reader "B." Then the rule will insert new containment relationships into the OBJECTCONTAINMENT table.

```
DEFINE E1 = observation("A", o1, t1)
DEFINE E2 = observation("B", o2, t2)
CREATE RULE r2, containment_rule
ON TSEQ(TSEQ⁺(E1, 0.1sec, 1sec); E2, 10sec, 20sec)
DO
BULK INSERT INTO CONTAINMENT
VALUES (o2, o1, t2, "UC")
```

The keyword "BULK" will enforce a bulk insertion of all contained objects into the container.

5.3 Real-time Monitoring

RFID rules can also provide effective support of real-time monitoring, as shown in the following asset monitoring examples.

- *Rule 3. Real Time Monitoring.* As shown in Example 2, if reader "r_4" mounted at a building exit detects a tagged laptop but does not detect any tagged superuser (who is authorized to move asset items out of the building) within certain time threshold, e.g., 5 s, then it implies that the laptop is being taken out illegally, and an alert is sent to a security personnel.

  DEFINE E4 = observation("r4", o4, t4), type(o4) = "laptop"
  DEFINE E5 = observation("r4", o5, t5), type(o5) = "superuser"
  CREATE RULE r3, asset_monitoring_rule
  ON WITHIN(E4 ∧ ¬E5, 5sec)
  DO *send_alarm*

- *Rule 4. Real-time Position Monitoring.* Assume reader "r_5" monitors money transport shipping truck to a bank, if the shipping truck moves away from the destination (the bank) beyond a security distance, e.g., 5 miles, then an alert is required to send to a security personnel.

  DEFINE E6 = c-observation("r5", o6, t6, sensor1, gps1),type(o6) = "car"
  DEFINE E7 = c-observation("r5",o7, t7, sensor2, gps2),type(o7)
    ="bank"
  CREATE RULE r4, position_monitoring_rule
  ON S-WITHIN(E6 ∧ E7, 5miles)
  DO *send_alarm*

5.4 Event Detection for RFID Rules

RFID rules differ from traditional ECA rules in several ways. RFID events have temporal and spatial constraints distance constraints between events or interval constraints on single events, which need to be checked in event detection phase instead of in condition checking phase. RFID events support negation events in broader scope, which can be detected through temporal constraints. RFID rule actions generally will not generate new primitive events thus it is more scalable.

We develop a graph-based event processing engine by aggregating complex events from multiple rules into an event graph. In this event processing engine, temporal and location constraints are modeled as first-class constructs and are checked in the event detection phase. When new events come in, they propagate from both bottom-up for spontaneous events and top-down for non-spontaneous events – events that cannot detect their occurrences by themselves unless they either get expired or are explicitly queried.

Non-spontaneous events such as negate events will not be detected by their child events. In this case, we inject *pseudo events* into the incoming event queue at these events' expiration time. A pseudo event is a special artificial event used for querying the occurrences of non-spontaneous events during a specific period, and is scheduled to happen at an event node's expiration time. We can determine which event node needs to generate pseudo event for a target event in a top-down fashion.

6 Related Work

RFID technology has emerged for years and poses new challenges for data management (Chawathe et al. 2004; Bornhoevd et al. 2004). However, little research has been done on how to effectively modeling RFID data. In (Harrison 2003), some data characteristics of RFID data are summarized with some reference relations to represent the data. In this model, RFID data are modeled as events, thus, the state history and the temporal semantics of business processes are implicit. This data model is not effective on supporting complex queries such as RFID object tracking and monitoring. A query often needs to be divided into numerous steps, which is indirect, inefficient, and not natural for users. EPC Global (EPCGlobal 2009), the current EPC standard group, defines the networks for RFID data and product data, but RFID data modeling is not a task in its working group. A data warehousing model is proposed for the object transition and a method to process a path selection query in (Gonzalez et al. 2006), where multiple joins are needed for aggregated measures of the path. A path encoding and storage scheme is proposed in (Lee and Chung 2008) to encode the flow information for products in supply chains where queries have to be translated. In (Wang and Liu 2005), a data model is discussed for RFID applications, where two types of history data is proposed: event based and state based. However, this model is limited to the scenario of fixed readers with read-only tags, and will not support other use cases discussed here.

Recently, major IT vendors are providing sophisticated RFID platforms, including SAP Auto-ID Infrastructure (SAP Auto-ID 2009), Oracle Sensor Edge Server (Oracle Sensor Edge 2009), IBM WebSphere RFID Premises Server (IBM Premises 2009), Sybase RFID Solutions (Sybase RFID 2009), Microsoft RFID Middleware (Microsoft Biztalk RFID 2009), UCLA's WinRFID Middleware (WinRFID 2009), and others. These platforms provide a general interface to collect RFID data from readers and then forward the data to applications. These systems, however, support only primitive and predefined RFID events or their simple combinations. Thus it is up to users' applications to define the data model and detect complex events.

Event processing has been studied extensively in the past (Gatziu and Dirtrich 1994; Gehani et al. 1992; Chakravarthy and Mishra 1994; Widom and Ceri 1996), in the context of active databases. These systems normally use Event-Condition-Action (ECA) rules for event processing. RFID events differ from traditional events in several ways, including the highly temporal nature and existence of non-spontaneous events. Thus it is difficult for traditional event detection systems to support RFID event detection. Bai et al. (Bai et al. 2007) propose to extend a stream language to support RFID events. How the event processing can be implemented is not discussed.

7 Conclusions

We develop a comprehensive temporal and location oriented RFID data model by integrating the semantics of RFID applications. One significant benefit of the data model is that tracking and tracing is made very effective through expressive queries. We then propose a declarative event and rule based framework to automate the processing of RFID data, with a temporal sensitive RFID complex event detection engine. The framework can automate RFID data filtering, transformation, aggregation, and monitoring. The approaches are generic and can be easily adapted into diverse RFID applications, thus substantially reduce the cost of managing and integrating RFID data.

References

Bai YJ, Wang F, Liu P, Liu S (2007) RFID data processing with a data stream query language. In: ICDE, Istanbul, 15–20 April, pp 1184–1193

Bornhoevd C, Lin C et al (2004) Integrating Automatic Data Acquisition with Business Processes – Experiences with SAP's Auto-ID Infrastructure. In: VLDB, Toronto, Canada, August 31 – September 3, pp 1182–1188

Cisco Location Solution Overview (Cisco Location 2009) http://www.cisco.com/en/US/solutions/collateral/ns340/ns394/ns348/ns753/net_brochure0900aecd8064fe9d_ps6386_Products_Brochure.html. Accessed 30 Nov 2009

Chakravarthy S, Mishra D (1994) Snoop: an expressive event specification language for active databases. Data Knowl Eng 14(1):1–26

Chawathe SS, Krishnamurthy V, Ramachandrany S, Sarma S (2004) Managing RFID data. In: VLDB, Toronto, Canada, August 31 – September 3, pp 1189–1195

EPC Tag Data Standards (TDS) (2008) Version 1.4. EPCGlobal technique report. http://www.epcglobalinc.org/standards/tds/tds_1_4-standard-20080611.pdf. Accessed 30 Nov 2009

EPCGlobal (2009) The EPCglobal network and the global data synchronization network (GDSN). http://www.epcglobalinc.org/about/media_centre/EPCCglobal_and_GDSN_v4_0_Final.pdf. Accessed 30 Nov 2009

Gatziu S, Dirtrich KR (1994) Detecting composite events in active databases using petri nets. In: Workshop on research issues in data engineering: active database systems. Houston, TX, USA, 14–15 February, pp 2–9

Gehani NH Jagadish HV, Shmueli O (1992) Composite event specification in active databases: model and implementation. In: VLDB

Gonzalez H, Han J, Li X, Klabjan D (2006) Warehousing and analyzing massive RFID data sets. In: ICDE, Atlanta, Georgia, USA, 3–7 April, pp 83–93

Harrison M (2003) EPC information service – data model and queries. Auto-ID Center Technical Report

Hinze A (2003) Efficient filtering of composite events. In: BNCOD, Coventry, UK, 15–17 July, pp 207–225

IBM Websphere Premise Server (2009) http://www.ibm.com/software/integration/premises_server. Accessed 30 Nov 2009

Lampe M, Flörkemeier C (2004) The smart box application model. International conference on pervasive computing, Linz, Vienna, Austria, 18–23 April

Lee C, Chung C (2008) Efficient storage scheme and query processing for supply chain management using RFID. In: SIGMOD

Motakis I, Zaniolo C(1997) Formal semantics for composite temporal events in active database rules. J Syst Integrat 7:291–325

Oracle Sensor Edge Server (2009) http://www.oracle.com/technology/products/sensor_edge_server. Accessed 30 Nov 2009

Sybase RFID Anywhere (2009) http://www.sybase.com/products/mobilesolutions/rfid_anywhere. Accessed 30 Nov 2009

UCLA WinRFID (2009) http://winmec.ucla.edu/rfid/. Accessed 30 Nov 2009

Wang F, Liu P (2005) Temporal management of RFID data. In: VLDB, Trondheim, Norway, 30 August–2 September, pp 1128–1139

Wu E, Diao Y, Rizvi S (2006) High-performance complex event processing over streams. In: SIGMOD, Chicago, IL, 27–29 June, USA, pp 407–418

Widom J, Ceri S (1996) Active Database Systems: Triggers and Rules for Advanced Database Processing. Morgan Kaufmann, San Francisco, CA

Event Management of RFID Data Streams: Fast Moving Consumer Goods Supply Chains

John P. T. Mo and Xue Li

Abstract Radio Frequency Identification (RFID) is a wireless communication technology that uses radio-frequency waves to transfer information between tagged objects and readers without line of sight. This creates tremendous opportunities for linking real world objects into a world of "Internet of things". Application of RFID to Fast Moving Consumer Goods sector will introduce billions of RFID tags in the world. Almost everything is tagged for tracking and identification purposes. This phenomenon will impose a new challenge not only to the network capacity but also to the scalability of processing of RFID events and data. This chapter uses two national demonstrator projects in Australia as case studies to introduce an event management framework to process high volume RFID data streams in real time and automatically transform physical RFID observations into business-level events. The model handles various temporal event patterns, both simple and complex, with temporal constraints. The model can be implemented in a data management architecture that allows global RFID item tracking and enables fast, large-scale RFID deployment.

1 Introduction

Radio Frequency Identification (RFID) is a wireless communication technology that uses radio waves to transfer information between tagged objects and readers without line of sight. RFID has been around for more than half a century (Landt 2005). Recent advancement in the technology creates tremendous opportunities for linking real world objects into a world of "Internet of things" (Borriello 2005; Gershenfeld et al. 2004). In coming years, there will be billions of RFID tags in the world tagging. Almost everything is tagged for tracking and identification purposes.

J.P.T. Mo (✉)
School of Aerospace, Mechanical and Manufacturing Engineering, RMIT University, P.O. Box 71, Bundoora, VIC 3083, Australia
e-mail: john.mo@rmit.edu.au

D.C. Ranasinghe et al. (eds.), *Unique Radio Innovation for the 21st Century*,
DOI 10.1007/978-3-642-03462-6_5, © Springer-Verlag Berlin Heidelberg 2010

The importance of RFID technology development can be seen from recent international conferences and workshops on RFID technology including IEEE International Conference on RFID[1] and International Workshop on RFID Technology.[2] RFID technology already has significant impact on business and society at large. It is gaining momentum and there is still a lot of potential for its applications.

Currently, the Fast Moving Consumer Goods (FMCG) sector is still using Electronic Data Interchange (EDI) to communicate. EDI works by sending purchase order information electronically to members of the supply chain, such as supplier, importer and retailer. A purchase order can then be downloaded into a handheld device where the warehouse operator creates a number of Serial Shipping Container Code (SSCC) labels. The SSCC label and the product's barcode are then scanned and sent to the retailer as an Advance Shipping Notice (ASN) so that they know what is coming in. The freights are then delivered into the retailer's warehouse, where the SSCC is re-scanned, the packaging information is verified, and payment is made to the supplier. This EDI based process is very labour intensive and time consuming. Integration with most enterprise management systems is also difficult (Mo et al. 2009a).

Many promising RFID applications in the FMCG sector use passive devices, due to their low cost. However, passive RFID technology has several limitations: signal interference, poor speed of tag reading, and limited memory size. To investigate how passive RFID technology can be applied to the FMCG supply chain, two national demonstrator projects were conducted in Australia to investigate the effectiveness of using RFID to improve business efficiency. The first project namely, National Electronic Product Code (EPC) Network Demonstrator Project (NDP) tracked pallets and cartons through the supply chain (GS1 Australia 2006). EPC is an identification standard administered by EPCglobal[3] using a specific type of passive RFID. The second project is the NDP Extension Project which explored the issues related to the application of EPC data to support paperless deliveries (GS1 Australia 2007). In Europe, the "Building Radio frequency IDentification solutions for the Global Environment" (BRIDGE) project was developed to resolve the implementation barriers of the EPCglobal Network for FMCG sector in Europe (Lehtonen et al. 2007). Subsequently, the Commission to the European Parliament (2007) actively pursued actions and legislations to govern the use and data management of RFID applications. The EPCglobal Network standard provides a promising architecture for tracking and tracing objects over the Internet in these projects.

Application of RFID will introduce billions of RFID tags for supply-chain management, environmental monitoring and control, pharmaceutical pedigree tracing, baggage handling, and global tracking of moving objects (Chawathe et al. 2004). This phenomenon will impose a new challenge not only to the network capacity

---

[1] http://www.ieee-rfid.org/2007/

[2] http://www.iceis.org/workshops/iwrt/iwrt2007-cfp.html

[3] http://www.epcglobalinc.org/

but also to the scalability of event processing of RFID applications. The situation becomes more complicated when RFID tags are applied in large-scale applications. Issues such as missed and unreliable RFID readings, redundant RFID data, in-flood of RFID data, spatial and temporal event management are typical data management functions that are required as back end support of large scale RFID systems applications.

This chapter uses the two national projects as case studies and introduces a data stream system model that handles various event patterns, both simple and complex, with temporal constraints. The model addresses these challenges and can be implemented in a RFID data management architecture that allows global RFID item tracking, thereby enabling fast, large-scale RFID deployment.

2 FMCG Case Studies

The two national projects were based on the EPCglobal Network standard, which provides a promising architecture for tracking and tracing objects over the Internet. However, there were no large scale commercial projects that could provide guidance to public on how the standard could be implemented. Consequently, practitioners only have their own limited deployment experiences to learn. To gain first-hand, practical experience with using the EPCglobal Network standard, the national projects were developed with a total of sixteen industry and four non-industry partners, working between the cities of Melbourne and Sydney.

The first project, the National EPC Network Demonstrator Project (NDP), aimed to identify the business benefits of sharing information securely using the Electronic Product Code (EPC) Network, provided authentication to interacting parties, and enhancing the ability to track and trace movement of goods within the entire supply chain involving transactions among multiple enterprises (Mo 2008). The NDP pioneered the use of EPCglobal-defined protocol's full stack to enable inter-organizational transactions and supply-chain management. When a given tag was detected, instead of having each company storing this information and communicating to the next partner, the EPCglobal model adopted in this project defined one authoritative registry of numbers that could be queried for links to access detail information from a global server.

Items of interest were assigned a registered number: products were identified by a Serial Global Trade Item Number, or SGTIN; pallets by a Global Returnable Asset Identifier, or GRAI; and unit loads identified by an SSCC. A key process that bound these items into a business transaction is a "pick face" process (Fig. 1). The EPC on the products were read by the RFID reader while it was packed to a shipment in the distribution centre. This data was shared with all other partners of the FMCG supply chain through the querying portal.

The second NDP was the National EPC Network Demonstration on Business Information Integration (NDP Extension). This project aimed to develop a paperless system to reduce human errors (for example, in data entry) and the subsequent corrective costs. Achieving this goal required participating companies to integrate

Fig.1 FMCG "pick face"
process ensuring relationships
between EPC items

the EPC Network with their own business systems. The NDP Extension particularly focused on asset management - in this case, pallets (Fig. 2). Since more than 10 million pallets were in circulation throughout Australia, the potential of loss was enormous, and efforts to manage assets have been proved to be extremely costly.

The NDP Extension consortium performed a series of paperless test runs on pallet dispatch and delivery transactions among six sites between Melbourne and Sydney. An overall efficiency gain of 18.1% was achieved. Most of the productivity gains were cost reduction due to the elimination of data entry, verification and recon-ciliation processes were significant, especially for the pallet supplier. Furthermore, improvements in inventory accuracy as well as in quality area were significant, for example, accuracy and transparency of information and real time processing had great impact on the other logistics operations such as planning and forecasting.

Fig. 2 Pallets were stacked
with all tags showing at one
side

Experience in the national demonstrator projects shows that transactions in the FMCG supply chain are complex events and can occur in different forms. Data in RFID network is constantly evolving due to RFID tag mobility, change of item ownership, change of workflows, change of nature of transactions and event patterns. Hence, both the accuracy of RFID data as well as the context in which the data is captured should be well managed, viz, the complete set of information associated with the RFID event. This means an RFID data management system is required to allow companies in FMCG supply chain to easily track and trace RFID data among these different representation levels.

3 Principles of Event Management

The primary focus of the RFID enabled FMCG supply chain system is event management. An event is an occurrence of interest happening in a particular time and location in the real world and is recorded in RFID data management systems. RFID events can be divided into different types according to applications. We adopt the data structure of event records from the EPC Information Services (EPCIS) standard, that is, each core event type has the fields that represent four key dimensions of any EPCIS event:

1. The object(s) or other entities that are the subject of the event;
2. The date and time;
3. The location at which the event occurred; and
4. The business context.

These four dimensions usually represent information about RFID tagged items on "what, when, where, and why", respectively. A primitive event is an RFID reader observation. In particular, each primitive event has a timestamp, representing the time point that an RFID reader reads an RFID tag. A complex event is a pattern of primitive events and happens over a period of time. A complex event usually is composed of a few primitive events, and has a start time and an end time.

RFID events have their own characteristics that cannot be supported by traditional event systems (Derakhshan et al. 2007; Rizvi et al. 2005). RFID Application Level Event (ALE) standard specified by EPCglobal has a common interface to process raw RFID events, also emphasizes the importance of RFID data stream processing. RFID data is time-dependent, dynamically changing, in large volumes, and are carrying implicit semantics. To support these characteristics, a general RFID data processing framework is required to process high volume RFID data streams in real time, and automatically transform the physical RFID observations into the virtual counterparts linked to business applications (Wang et al. 2006; Wu et al. 2006). Among various RFID applications, event detection plays an important role. Several event processing systems that execute complex event queries over real-time streams of RFID readings have been proposed in recent years (Bai et al. 2007; Wang and Liu 2005). The complex event queries in these systems can filter and correlate events to match with specific patterns, and transform the relevant events into high-level business events that can be used by external applications. However, most RFID applications have time restrictions on target events. An event is valid only if it happens within or after a time limit. For example, check-in baggage at an airport should arrive at a certain place for the aircraft loading 30 min before the scheduled flight taking off. Unfortunately, this kind of time-restriction issues are largely ignored by most of the current event processing systems and left as an individual application designing problem. A full treatment on time restriction issues could be found in Li et al. (2009).

To handle various temporal event patterns, both primitive and complex, with temporal constraints, it is important to develop a novel RFID event management system. RFID event processing is a key component to all RFID applications. Complex event patterns of RFID streaming data represent business activities that may be of business importance. This chapter presents a novel theory and methodology that can be applied to streaming data management of complex events for RFID applications.

4 Requirements of RFID Data Management

Based on first-hand experiences of the two NDPs, we identify several key requirements for an effective RFID data management system. It is necessary to note that a typical log in an EPCIS record is a random event stream as shown in Fig. 3.

The data stream in Fig. 3 exists in all EPCIS at the company. EPCglobal's core services are used to share data since supply chains such as FMCG are operating globally. Therefore the use of and hence the use of EPCIS facilitate the sharing and exchange of event data that occur in all locations (Armenio et al. 2007). The key to success is to structure the RFID data appropriately to support these services.

4.1 RFID Data Quality

In RFID data management, the whole process is data-centric. RFID data pre-processing and cleaning are difficult because RFID devices may generate errors. RFID data has some unique features compared with other traditional data:

```
urn:epc:id:sgtin:49871766.20144.3, Fri Feb 10 12:52:23 EST 2006, Metcash Pick Face Smartshelf 1,
urn:epc:id:sgtin:49024300.24741.19, Fri Feb 10 12:43:00 EST 2006, Metcash Pick Face Smartshelf 2,
urn:epc:id:sgtin:49024300.14979.10, Fri Feb 10 12:52:43 EST 2006, Metcash Pick Face Smartshelf 2,
urn:epc:id:sgtin:49871766.20144.3, Fri Feb 10 12:52:30 EST 2006, Metcash Pick Face Smartshelf 2,
urn:epc:id:sgtin:49024300.14975.17, Fri Feb 10 12:52:30 EST 2006, Metcash Pick Face Smartshelf 3,
urn:epc:id:sgtin:49024300.14979.10, Fri Feb 10 12:42:19 EST 2006, Metcash Pick Face Smartshelf 3,
urn:epc:id:sgtin:49024300.14979.10, Fri Feb 10 12:42:18 EST 2006, Metcash Pick Face Smartshelf 4,
```

Fig. 3 Metcash smart shelf records

1. *Simplicity* – they are simple tuples of (EPC, Time, location). An RFID tuple represents the temporal and spatial information about an item. A tuple is unique in the given application.
2. *Implied semantics* – The same EPC appeared in different RFID tuples may imply different semantics. For example the same EPC read in a supermarket at different times or locations may imply the events of different context such as shop-lifting, check-out, or on-the-shelf.
3. *Duplication or omission* – an RFID tagged item may be repeatedly read multiple times by the same reader or in inverse, could not be read properly by a reader.
4. *Noise and inconsistency* – One major problem for RFID data quality is about tracing of the paths for specific RFID tagged items. The RFID data sequence may be out-of-order, or have inconsistent readings (e.g., the same EPC was read/found at different locations at the same time). Given a set of items, tracing their locations in a certain time period is a major challenge.

In RFID applications, many RFID tagged items are read together, as they are transported together in batches, in large containers, trucks, ships, and pallets. The challenge here is how to process and manage this data effectively and efficiently. Understanding the relationship between a batch and the constituent items is a key challenge in order to enable tracking at both levels of granularity (Gonzalez et al. 2006).

In the NDP, an order reconciliation capability has been implemented on the global EPCIS that used successfully read EPCs to link the cartons or pallets within a particular shipment.

Figure 4 shows reconciliation of a Gillette order. The result shows that the read rate was not 100% when the goods passed through the Metcash dock door. However, since the order was packed with 100% certainty and registered as a containment, it was possible for the global EPCIS to associate the missing EPCs with the shipment order through some successful reads. There were 13 items in the shipment and they were registered in a pick face process. At the dock door read point, the pallet tag (item 13) and one of the product tags (item 5) were not read. The other tags were read successfully and reconciled as a shipment by the order reconciliation capability.

Various scenarios could exist for possibilities of reading SGTIN tags compared to the GRAI tag containing them. The options are shown in Table 1.

This shows that some data processing rules should exist to handle cases such as reconciliation when the data quality is uncertain.

4.2 Unordered Event Stream(UnES)

When processing complex events, the sequence and the length of sequence are two important characteristics of a complex event. However, in general the order of arriving events may not match with the order of the occurrence of the events in the real world. We use an example in NDP (Fig. 5) to illustrate. Several events have been

Fig. 4 Data reconciliation screen on the visualisation tool

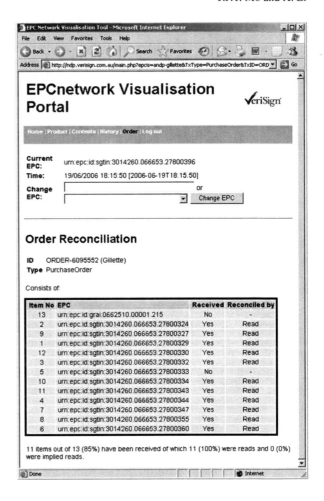

recorded on the item urn:epc:id:sgtin:0041333.141501.27900731. Following a few records at the pick face and wrapping station, the item was recorded 3 times at the dock door. These events occurred in a short time period (within 4 min). One would then question what has actually happened. Did the item go in or out of the dock door? Did the event recorded at 11:44:29 mean that the shipment left the site at that point? Between this last event and the next event on 19/06/2006, what happened to this EPC?

The above question indicates a problem, named as Unordered Event Stream (UnES). It can be caused by network routing delay or by an arbitrary selection of records when a single tag is read by multiple readers simultaneously. An analogy can be found in the Internet communications when a datagram arrives at the system application level being out of sequence. The UnES problem is about managing the event records that arrive the system out of their real-time order. Existing works

Table 1 Cases for order reconciliation

Case	Container (GRAI)	Item (SGTIN)	Likely
1	N	N	If no container tag and no SGTIN are read it is not possible for the system to infer anything. No data will be recorded about this event for the project.
2	Y	Y	If the container and all SGTINs are read the read rate will be 100% for that container.
3	Y	N	Based on containment information of at least one EPC read, it will be possible to come up with a read rate for that container by looking at what has been contained, or what has contained the EPC.
4	N	Y	Similar to case 3.

Fig. 5 Track and trace for urn:epc:id:sgtin: 0041333.141501. 27900731

address the UnES problem in two ways: one is to simply ignore the problem by assuming that the event record enters into the system with the same order of their occurrences in the real world, such as SASE (Wu et al. 2006). The other is to sort event data according to their timestamps before extracting complex event patterns from them, such as Cayuga (Brenna et al. 2007), using the technique proposed by Srivastava and Widom (2004).

Both approaches, however, have disadvantages. For the first approach, since RFID readers are widely distributed and each one may have a delay, it is obvious that event records cannot always enter into the system in the order of their occurrences. Moreover, RFID tags can be read by different readers simultaneously, but the readers may send these primitive events to the system at different times. Also, even after the event record enters into the system, they may still be subject to queuing of different processes. Clearly, ignoring the UnES problem is impractical to most RFID applications. For the second approach, it requires the time point that an event record enters into the system must be no later than a bound of time delay, cope with all applications by the same bound. If there are many applications simultaneously handled by the system, this single bound cannot be sufficient to deal with the combined effect of all incoming data.

4.3 Tracing Large Number of Complex Events

Generally an RFID reading can be considered as a primitive event. Wide deployment of reader devices will generate a huge number of events. In order to detect complex events that are appropriate to end-user applications at a semantic level, we need to filter and correlate such huge number of primitive events. These complex events are implemented as groups of continuous queries that run against incoming primitive events. In the process of complex events, all the new incoming primitive events that might lead to the completion of complex events need to be recorded in a structure for the construction of sequences. The problem here is on how to effectively retrieve all intermediate states of complex events fast enough when primitive events are coming into system faster than the speed of that the intermediate states of complex events are brought back to system memory.

Figure 6 shows an example from the NDP that item 601, `urn:epc:id: sgtin:49024300.24737.601` under the class object 24737 and manager ID 49024300 was first recorded in the EPCIS on 8/10/2005.

After a number of transactions, information about item 601 could be retrieved in different associations (Fig. 7):

(a) By association within containment of a pallet
(b) As an individual item detected at the pick face and dock door
(c) As an individual item detected within a SSCC
(d) As a group contained within a pallet

There is a clear need for a complex event processing function in RFID event management systems. Franklin et al. (2005) proposed the HiFi system architecture. HiFi aggregates events along a tree-structured network on various temporal and geographic scales. Mansouri-Samani and Sloman (1997) also considered temporal restrictions. However, those temporal restrictions cannot be used to support the special RFID event patterns such as sequence with time constraints on the intervals of events or on the intervals of negative events (i.e., the events that cannot occur). Furthermore, Brenna et al. (2007) proposed the Cayuga system, to process complex events in publish/subscribe systems. Cayuga offers an expressive language similar to the one required in RFID data management systems. Cayuga can deal with a

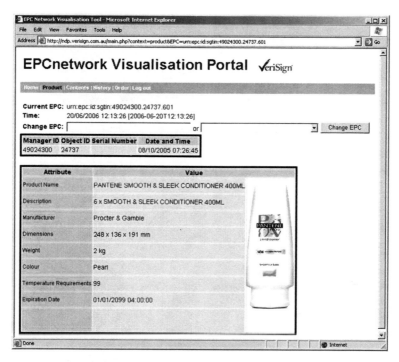

Fig. 6 EPC urn:epc:id:sgtin:49024300.24737.601 was first record in product information screen on global EPCIS

wide range of complex event patterns, including sequence of Kleene closure patterns (Gyllstrom et al. 2008). However, Cayuga cannot support the constraints on time intervals between two successive events.

4.4 Pattern Recognition on Streaming Data

Due to system requirements, the filtering rules in the EPCIS can vary in different implementations. One of the rules was a filtering rule whereby the EPCIS could be set to ignore the same tag's information within a defined time window. The length of this "ignore" time window was critical to capture as many tags as possible while minimising the possibility of missing some tags. If it was set too long, the EPCIS would ignore too many tag reads and severely affected the read rate. On the other hand, if it was set too short, multiple events were recorded. To ensure maximum chance of capturing tags, the project team decided to reduce the "ignore" time.

During the NDPs, experiments with various time values were conducted. In the case that no filtering was applied, a maximum of 18 read events on a tag (among around 100 tags) were recorded within 1 s. This was seen in the EPC data stream where the same tag was read multiple times in the same second. In other words, two

(a) Within a pallet containment

(b) Detected at pick face and dock door

(c) Contained in a SSCC

(d) Displayed in a group within a pallet

Fig. 7 Shipment information of item 601 in different views

tag-read events occurred and there was no time (zero second) between these two events. This problem did not pose significant issue to the data logging process but had adverse effect when the data was processed.

Hence, the queries in the RFID data management system should be designed to find complex event patterns. The answer to a query will be the event patterns for the given query conditions. The execution of a query is to search through RFID data streams over the network. The event patterns can be represented in a variety of formats such as associations, correlations, clusters, sequences, or dependencies between primitive events. Pattern recognition on RFID data streams can be characterized by following three properties:

1. Data must be processed faster than it arrives. Therefore, the pattern recognition algorithms should scan the data only once. The construction of complex event patterns therefore must be maintained incrementally in order to avoid re-scanning old data.
2. Sufficient amount of data must be seen during the process for the statistical significance of patterns.
3. Algorithms must be able to deal with changes – the discovered patterns may change over time. In other words, they are only true at the time when they are identified. So the system must be able to constantly monitor the data streams.

5 RFID Data Stream Modelling

For any piece of RFID data, it is necessary to see its physical values (for example, location and time), as well as its logical ones (such as its event pattern, ownership, and business transaction). RFID data has different representation levels. At the physical level, data represents floor plans and geographical locations. At the network level, it represents connections between mobile readers and networks. At the business level, data represents business events. Subsequently, the business partners have to specially develop every component to match their different business practices and system requirements.

Considering the requirements discussed in previous section, an RFID data stream model is developed.

5.1 Model for Large-Scale RFID Deployment

Enabling fast, large-scale RFID deployment requires a novel model — the system model developed for the global RFID item tracking. In particular, our experience within the NDPs demonstrated that the data in an RFID network is constantly evolving due to RFID tag mobility, change of item ownership, and change of workflows and event patterns. The constant changes inherent in RFID systems call for a novel RFID data model that lets us track and trace RFID data among these different representation levels. A Data Lineage Model (DLM) is developed to support tracking and tracing RFID data efficiently at these three layers, as Fig. 8 shows (Mo et al. 2009). These three layers in the model are logically independent from each other.

The event layer is responsible for modelling business transactions. In an RFID enabled FMCG supply-chain management system, we can define business events by

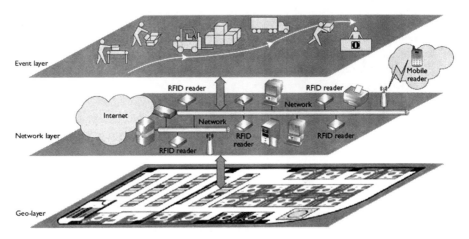

Fig. 8 A data lineage model for global item tracking (Mo et al. (2009) RFID infrastructure design: A case study of two Australian RFID projects, IEEE Internet Computing. © 2009 IEEE)

pre- or post-conditions and triggers that describe under which (spatial or temporal) conditions such events will occur and the actions the system should take. For example, we might model an ownership change as both a pre- and post-condition for different owner IDs when a distributor has an order on the product from the manufacturer and the retailer has a payment record in the distributor's accounts-receivable. The event could be a reading at a warehouse dock door, and the action would be registering the shipment and checking the product quantity on hand.

At the network layer, the system maps all business events from the event layer to network configurations to make them traceable. For each business event, the system registers the associated network devices. This layer also protects the data lineage from illegal data injection and readings.

Finally, the geo-layer ties all network devices to their geo-properties (for example, in a floor plan with 3D locations) and their business ownerships. The system can further map business events from the network devices' configurations to their corresponding physical locations. Given that we can use a mobile network to track moving objects, the system should dynamically map network devices' logical connections to their new locations. In this case, the mapping between the network layer and the geo-layer would represent a tight coupling for tracing business transactions at the event layer.

A logic independent hierarchy guarantees that changes occurring in one layer (such as a change to a business process) will be independent from other layers within the RFID system. Hence, changes in the lower layers (for example, a change in the network configuration) won't affect the business logic. Note that although the RFID Application Level Event standard that EPCglobal proposed emphasizes the importance of RFID data stream processing, it relies heavily on system developers to implement application specific processes into the infrastructure.

5.2 RFID Data Cleaning and Pre-processing

Low-level RFID data has problems such as duplication, noise, and uncertainties. A data filtering process is considered, not only for cleaning the data but also for capturing the statistical features of the data.

5.3 Complex Event Processing and Pattern Recognition

Complex events embody the patterns of business transactions. Low level RFID readings need to be recognized in terms of their time and location properties for the match of their high level business transactions. The first priority is to provide the key algorithms to deal with the complex events processing problems. Johnson et al. (2007) developed an efficient algorithm for finding out of order data in an IP packet data stream. Li et al. (2009) introduced a new architecture for out-of-order processing (OOP), that freed stream systems from the burden of order maintenance by using explicit stream progress indicators, such as punctuation or heartbeats, to unblock and purge operators. These out of order data processing did not take into account the changes in data over time. We consider using a sub-class of timed-automata called Event-clock Automata (ECA) (Alur et al. 1994). Unlike timed-automata, ECA does not control the reassignment of clocks, and, the value of a clock is fixed and associated with the time value of incoming events. ECA can have two types of clocks:

1. event-recording and
2. event-predicting clocks.

5.3.1 Event-Recording Clock

The value of this clock always equals to the time of the last occupancy of an incoming event relative to the current time. Figure 9 shows an example of using an ECA with event-recording clock for a simple query with a sequence of events (SEQ (a,b,c)). At the bottom, there is a stream of events and the numbers below the arrow are the timestamps. Unlike SASE the events here are not totally ordered. On top of that is an ECA with two event-recording clocks: $X_a < X_b$, $X_b < X_c$. This time-constraint can check the order of events and discard the unordered events which were ignored in SASE. As the output shows, only the two series of events recognized for our query and discarded the unordered events (e.g. a_9, b_6, c_7).

5.3.2 Event-predicting Clock

This clock is a prophecy variable whose value always equals the time of the next occurrence of an event relative to the current time. So the time-constraint in this clock will satisfy only if the incoming event happens after or within a certain period

Fig. 9 An example of ECA
with event timestamps

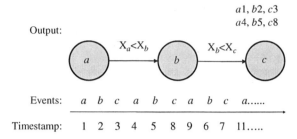

of time. For instance, in our example, event b is valid if only happens after 4 s from event a then we can simply change the time-constraint on ECA to $X_b \geq X_a + 4$ so then the output for this case will be a_1, b_5, c_8. In other cases which event b should happen within 4 s from event a then we have to change the time-constraint to $X_b \leq X_a + 4$.

Using ECA, we can filter and detect unordered events as well as discarding those events that are invalid with respect to the time constraints.

The event clock is essential to track the movement of an EPC in the FMCG supply chain. An example in NDP is the tracking of an EPC item `urn:epc:id:sgtin:49024300.24741.276`, which is a P&G Pantene Colour Revival Conditioner. Tracking of this item requires processing the clock data associated with each and every event on this item and shipment information from all available EPCIS.

In Table 2, an item was recorded in dispatch by an RFID reader on 11th April. It took almost a month before another time stamp was recorded on 9th May. By associating the reader location with the geo-layer information, the item was recognised as received at the Metcash receiving dock. The item was continuously read until 23rd May and then disappeared from that date onwards.

By filtering unnecessary event data, it was concluded that the item was picked and dispatched to the distribution centre on 23rd May. The case showed that there were significant delays from P&G supplying the goods to receiving at Metcash. The reason was unknown. The data also showed that it then took 2 weeks for the goods to stay in the warehouse before it was finally sent to the store.

Table 2 Sequence of events
on a product

Event	Date
P&G Dispatch	11-Apr-06
Metcash Receiving	9-May-06
Metcash Pick Face Initial Read	15-May-06
Metcash Pick Face Final Read	23-May-06
Metcash Pick Event	23-May-06

5.4 Real-Time Query Language

We need a general RFID data processing framework to process high-volume RFID data streams in real time and automatically transform physical RFID observations into virtual counterparts linked to business applications (Liu et al. 1998). RFID applications need a real-time query language that specifies how individual events are filtered and how multiple events are correlated via time- and value-based conditions. Based on our experiences with the NDPs, a real-time RFID query language should be able to query:

1. A sequence of complex events involving multiple readings from different local storage spaces and time ranges
2. Event patterns in given occurrence frequencies; and
3. Network data streams along with classical functions based on the static data stored in a central database.

Unfortunately, no such language currently exists that satisfies all these properties. To enable temporal event tracing, we proposed a new time notion, time-to-live (TTL) (Li et al. 2009), which represents the length of time an RFID event can legally live in an RFID system. TTL covers various complex temporal event patterns, including those that existing systems can process. In particular, we classify TTL into four categories to denote different RFID (primitive/complex) events:

- Absolute TTL, which specifies the time that an RFID tag can live in the physical world;
- Relative TTL, which specifies the time that an RFID tag can be used for a particular application (after which we can reassign the tag to other applications);
- Periodic TTL, which specifies the time between two successive events with the same event type — in other words, it controls the period in which similar events occur periodically; and
- Sequential TTL, which specifies the time between any two successive events.

TTL overcomes the current problem of event data management as described in Sect. 3. For example, in a sequence A B C , if A and B are more than 1 min apart, then B and C should be less than 2 min apart, but if A and B are more than 1 min apart, B and C can be more than 5 min apart. We can express the rules about time constraints:

- RULE 1: IF $[t(B) - t(A)] > 1$ Minute THEN $[t(B) - t(C)] < 2$ minutes
- RULE 2: IF $[t(A) - t(B)] > 1$ Minute THEN $[t(B) - t(C)] > 5$ minutes

Apparently these are two independent event patterns, one has initial event A (for RULE 1) and the other has the initial event B (for RULE 2). So the event sequence of ABC will be successfully identified whenever the time constraints of RULE 1 are satisfied, and BAC will be identified whenever RULE 2 are satisfied.

A TTL-enabled query language can capture a wide range of interesting RFID applications with temporal restrictions. A typical supply chain, such as the one the NDPs implemented, requires companies to pack and repack items due to redistribution requirements. Having full traceability at the geo-layer would generate global location data that is unified with temporal data and would provide a data-independant model for globally tracing objects. Using the TTL concept to maintain temporal information about items, especially returnable assets, is also critical in tracking items globally.

5.5 System Architecture and Prototyping

A complex event management system (CEMS) shown in Fig. 10 is designed to cope with the business requirement of FMCG supply-chain management. The architecture consists of:

1. Three databases: the Event Type Database, the Query Database, and the Central Database, and,
2. Six modules: the Filter and Cleaner, the Event Type Checker, the Query Checker, the Query Analyser, the Event Processor, and the Intermediate Results Updater.

CEMS takes the input of queries and an infinite RFID data stream and outputs the query results of events that match with the queries. For pattern recognition queries, the identified event sequences over a user specified frequency will be returned as the result.

There are three independent information flows in CEMS, namely the Database Control Flow, the Query Flow, and the Data Flow. The first one is for administrators to manage the three databases. The second one represents the process where CEMS handles queries submitted by users. Queries are sent to the Query Analyzer, which updates the Query Database, Event Type Database, and Intermediate Results Updater.

Data Flow in CEMS flows from receiving primitive events captured by RFID readers to publishing query results to the user. The raw RFID data are first filtered and cleaned, and then the events are sent to the Event Processor, waiting for being processed. When receiving the list of queries related to the events, the Event Processor starts to process the queries. First, queries that are not related to event sequences are processed, followed by the queries that are related to event sequences. Both query results will be sent to users, and the Central Database will be updated if the query results require updating the information.

EPCIS architecture provides services on data logic. Figure 10 shows how different parts of the business logic works. Hence, the system architecture for complex event management is at the application level by which complex events are detected in FMCG supply chain system.

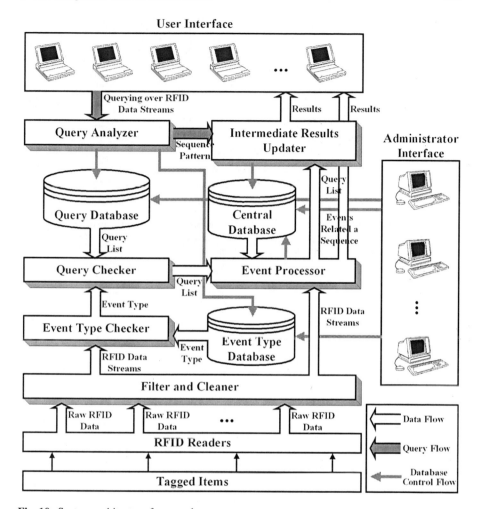

Fig. 10 System architecture for complex event management

6 Conclusions

The wide deployment of RFID devices is one of the key enablers for the FMCG supply chain. Our experience in the national projects showed that in order to ensure accuracy and validity of RFID generated information, it is necessary to have a good RFID event management process. This process should not only manage efficient capturing of RFID data. It should also manage the process in which RFID data captured have all related meta-data of the event such as context in which the data is captured. Hence, in FMCG supply chain management, complex event processing and pattern recognition is a key technology. Complex event processing offers significant benefits for fine-grind inventory control, moving object tracking, automated transition of ownerships, and many other applications.

The proposed model has a strong focus on intelligent RFID applications which involve pattern recognitions of moving objects. Based on the design and implementation experiences of the two national RFID demonstration projects, this chapter described a three level architecture of the complex event processing system. The software system architecture is also discussed. The core technology addressed in this chapter will have a direct application to supply chain management that involves frequent changes of the business environment. With technology playing such an integral role in society, the system ability to automatically adapt changes requires its real-time application can support the facilitation and monitoring of operations not only efficiently (in the form of timely decisions) but also effectively.

Acknowledgments The NDP and NDP Extension projects were supported by the Australian Government through two research grants from the Information Technology Online (ITOL) Program managed by the Department of Communications, Information Technology, and the Arts.

References

Alur R, Fix L, Henzinger TA (1994) Event-clock automata: a determinizable class of timed automata. In: Proceeding. of the 6th conference on computer aided verification (CAV'94). Lect Notes Comput Sci 818:1–13

Armenio F, Barthel H, Burstein L, Dietrich P, Duker J, Garrett J, Hogan B, Ryaboy O, Sarma S, Schmidt J, Suen KK, Traub K, Williams J (2007) The EPCglobal architecture framework. EPCglobal final version 1.2. http://www.epcglobalinc.org/home

Bai Y, Wang F, Liu P, Zaniolo C, Liu S (2007) RFID Data processing with a data stream query language. In: Proceedings of the 23rd international conference on data engineering (ICDE'07). Istanbul, Turkey, 15–20 April 2007, pp 1184–1193

Borriello G (2005) RFID: tagging the world. Commun ACM 48(9):34–37

Brenna L, Demers A, Gehrke J, Hong M, Ossher J, Panda B, Riedewald M, Thatte M, White W. (2007) Cayuga: a high-performance event processing engine. In: Proceedings of the ACM SIGMOD international conference on management of data (SIGMOD'07). Beijing, China 11–14 June 2007, pp 1100–1102

Chawathe SS, Krishnamurthy V, Ramachandran S, Sarma S (2004) Managing RFID Data. In: Proceedings of the 30th international conference on very large data bases (VLDB'04), Toronto, Canada 31 August–3 September 2006 vol 30, pp 1189–1195

Derakhshan R, Orlowska M, Li X (2007) RFID data management: challenges and opportunities. In: IEEE first international conference on RFID, Grapevine, TX, USA 27–28 March 2007, pp 175–182

Franklin MJ, Jeffery SR, Krishnamurthy S, Reiss F, Rizvi S, Wu E, Cooper O, Edakkunni A, Hong W (2005) Design considerations for high fan-in systems: the HiFi approach. In: Proceedings of the second Biennial conference on innovative data systems research (CIDR'05) Asilomar, CA, USA 7–10 January 2005, pp 290–304

Gershenfeld N, Krikorian R, Cohen D (2004) The internet of things. Sci Am 291(4):76–81

Gonzalez H, Han J, Li X, Klabjan D (2006) Warehousing and analysing massive RFID data sets. In: IEEE 22nd international conference on data engineering (ICDE'06), Atlanta, GA, USA April 2006, pp 83–93

GS1 Australia (2006) EPC Network™ Australia Demonstrator Project Report. Available on request from http://www.gs1au.org/. Accessed 24 June 2009

GS1 Australia (2007) National EPC Network™ Demonstrator Project Extension. Public report distributed in Smart Conference 2007. Sydney, Australia 20–21 June. Available on request from http://www.gs1au.org/products/epcglobal/australian_activities/demonstrator_project.asp

Gyllstrom D, Agrawal J, Diao Y, Immerman N (2008) On supporting Kleene closure over event streams data engineering. In: Proceedings of 24th IEEE international conference on data engineering (ICDE'08), Cancún, México 7–12 April, pp 1391–1393

Johnson T, Muthukrishnan S, Rozenbaum I (2007) Monitoring regular expressions on out-of-order streams. In: IEEE 23rd international conference on data engineering (ICDE 2007), Istanbul 15–20 April 2007, pp 1315–1319

Landt J (2005) The history of RFID. IEEE Potentials Oct/Nov 24(4):8–11

Lehtonen M, Michahelles F, Fleisch E (2007) Trust and security in RFID-based product authentication systems. IEEE Syst J, Special Issue on RFID Technology: Opportunities and Challenges. December 1(2):129–144

Li X, Liu J, Sheng QZ, Zeadally S, Zhong W (2009) TMS-RFID: temporal management of large-scale RFID applications. Int J Inf Syst Front. Springer. DOI:10.1007/s10796-009-9211-y

Liu G, Mok A, Konana P (1998) A unified approach for specifying timing constraints and composite events in active real-time database systems. In: Proceedings of the 4th IEEE real-time technology and applications symposium (RTAS'98), USA 3–5 June 1998, pp 199–209

Mansouri-Samani M, Sloman M (1997) GEM: a generalized event monitoring language for distributed systems. Dist Syst Eng 4(2):96–108

Mo JPT (2008) Development of a national EPC network for the tracking of fast moving consumer goods. Int J Enter Network Manag 2(1):25–46

Mo JPT, Gajzer S, Fane M, Wind G, Snioch T, Larnach K, Seitam D, Saito H, Brown S, Wilson F, Lerias G (2009) Process integration for paperless delivery using EPC compliance technology. J Manuf Tech Manag 20(6):866–886.

Mo JPT, Lorchirachoonkul W, Gajzer S (2009a) Enterprise systems design for RFID enabled supply chains from experience in two national projects. Int J Eng Bus Manag 1(2):49–54

Rizvi S, Jeffery SR, Krishnamurthy S, Franklin MJ, Burkhart N, Edakkunni A, Liang L (2005) Events on the Edge. In: Proceedings of the ACM SIGMOD international conference on management of data (SIGMOD'05), Baltimore, MD, USA 13–16 June 2005, pp 885–887.

Srivastava U, Widom J (2004) Flexible time management in data stream systems. In: Proceedings of the 23rd ACM symposium on principles of database systems (PODS'04), Paris, France 14–16 June 2004, pp 263–274

Wang F, Liu P (2005) Temporal management of RFID data. In: Proceedings of the 31st international conference on very large data bases (VLDB'05), Trondheim, Norway 30 August–2 September 2005, pp 1128–1139

Wang F, Liu S, Liu P, Bai Y (2006) Bridging physical and virtual worlds: complex event processing for RFID data streams. In: Proceedings of the 10th International Conference on extending database technology (EDBT'2006), Munich, Germany 26–31 March 2006, pp 588–607

Wu E, Diao Y, Rizvi S (2006) High-performance complex event processing over streams. In: Proceedings of the ACM international conference on management of data (SIGMOD'06). Chicago, IL, USA 26–29 June 2006, pp 407–418

Semantic-Based RFID Data Management

Roberto De Virgilio, Eugenio Di Sciascio, Michele Ruta, Floriano Scioscia, and Riccardo Torlone

Abstract Traditional Radio-Frequency IDentification (RFID) applications have been focused on replacing bar codes in supply chain management. Leveraging a ubiquitous computing architecture, the chapter presents a framework allowing both quick decentralized on-line item discovery and centralized off-line massive business logic analysis, according to needs and requirements of supply chain actors. A semantic-based environment, where tagged objects become resources exposing to an RFID reader not a trivial identification code but a semantic annotation, enables tagged objects to describe themselves *on the fly* without depending on a centralized infrastructure. On the other hand, facing on data management issues, a proposal is formulated for an effective *off-line* multidimensional analysis of huge amounts of RFID data generated and stored along the supply chain.

1 Introduction

A supply chain is a complex system composed by organizations and people with their activities involved in transferring a product or service from a supplier to a final customer. The key to make a successful supply chain relies on an extended collaboration, implying the integration among actors involved in the productive and logistic network. An integrated and flexible management of logistics (physical and information flows) has to be set-up both inside and outside factory boundaries. Specialized production and distribution processes suffer from the limited interactions allowed by rigid networks. As a result, nowadays a relevant component of competition in the market occurs among logistical chains. The supply chain can no longer be represented as static or linear, but it needs to be evaluated dynamically,

R. De Virgilio (✉)
Università Roma Tre, Rome, Italy
e-mail: dvr@dia.uniroma3.it

D.C. Ranasinghe et al. (eds.), *Unique Radio Innovation for the 21st Century*,
DOI 10.1007/978-3-642-03462-6_6, © Springer-Verlag Berlin Heidelberg 2010

as a complex system made of interactions and connections among actors operating along the chain.

Many empirical investigations have demonstrated there is a positive correlation between enterprise performances and its propensity and attitude to be integrated into larger systems. This is the reason why enterprises are more and more attentive to the opportunity offered by both coordination and cooperation among their internal functions and the other external actors contributing in different ways to the business. Hence, information has been an increasingly strategic asset in the last few years. It covers a determinant position in both logistics and marketing. The physical flow of raw materials, products and their related information is considered as a strategic element for quality standards of products/services, for business analysis and evaluation and finally to allow corrective actions. In particular, current trends in the consumer products market assign a growing significance to retailers in the governance of supply chains. The growing dimension of retail groups – sustained by the reached degree of concentration in that field – increases their power against producers and makes them privileged centers of value accumulation, acting as filters for the information flow for the whole chain. As a result, the main retailers are investing in new technology in order to boost the information exchange and are mandating the adoption of interoperable solutions to commercial partners (Smith 2005; Karkkainen 2003).

Radio-Frequency IDentification (RFID) is an AutoID (automatic identification) technology, interconnecting via radio two main components: (1) transponders (commonly named *tags*) carrying data, located on the objects to be identified; (2) interrogators (also known as *readers*) able to receive the transmitted data. Benefits introduced by RFID technology w.r.t. barcodes include: (i) unlike optical scan, alignment between reader and tag is not needed; (ii) longer read range (up to few meters); (iii) nearly simultaneous detection of multiple RFID tags; (iv) higher tag storage capacity (up to several kBs Ayoub Khan and Manoj 2009). Because of these features, RFID provides better levels of automation in the supply chain and helps prevent human errors (e.g., a reader can inventory an entire shipment in one pass while it is loaded into a warehouse, without having to scan each product manually). In latest years, industry is progressively rallying around few worldwide standards for RFID technologies. In this effort a leading role is played by the EPCglobal consortium EPC global. However, two main issues restrain a more advanced exploitation of RFID capabilities. Firstly, the original identification mechanism only enables a trivial string matching of ID codes, providing exclusively "yes/no" replies. Furthermore, RFID-based technology usually relies on a stable support infrastructure and fixed information servers where massive data analysis is quite difficult without the support of proper data management and aggregation schemes. Recall that an accurate evaluation of enhancements in RFID-based supply chains rely on global trend inspections over the supply chain itself which requires multidimensional analyses of huge amounts of RFID data generated and stored in central DBMSs. Serious data management issues are then inevitably inherited and they must be faced on.

In this chapter an innovative model for supply chain management is presented, aiming to overcome these limitations by adopting the Ubiquitous Computing (ubicomp) paradigm. As originally introduced by Mark Weiser (1999), ubicomp requires both information and computational capabilities to be deeply integrated into common objects and/or actions and the user will interact with many computational devices simultaneously, exploiting data automatically extracted from "smart objects" permeating the environment during his/her ordinary activities.

Leveraging a distributed architecture, the model provides a unified framework for both quick run-time analysis (with respect to a local fragment of the overall infrastructure) and stand-alone massive business logic elaborations (with respect to a centralized DBMS) following needs and requirements of the supply chain actors. An extension of current RFID technology supporting logic-based formalisms for knowledge representation is exploited. Semantic-based object/product annotations are stored into RFID tags, exploiting machine-understandable ontological languages originally created for the Semantic Web effort and based on Description Logics (DLs) (Bargide), such as RDF (Resource Description Framework)[1] and OWL (Web Ontology Language)[2] Semantically rich and unambiguous information is allowed to follow a product in each step of its life cycle. The model allows to trace and discover the information flow –associated to products thanks to their RFID tags– along the supply chain, and to formalize various supply chain analyses. Different perspectives can be so followed (e.g., product-centric, node-centric, path-oriented, time-oriented). Exploiting semantic-based queries, product and process information can be read, updated and integrated during manufacturing, packaging and supply chain management, thus allowing full traceability up to sales, as well as intelligent and de-localized interrogation of product data.

The chapter is structured as follows: Sect. 2 highlights the benefits of semantic-based approach for RFID data management in a typical scenario and Sect. 3 surveys relevant related work. Section 4 reports on basics of formalism and notation exploited in the detailed framework presentation of Sect. 5. Finally, Sect. 6 recalls some experimental evaluation corroborating the approach and Sect. 7 concludes the chapter.

2 Motivating Scenario

A simple reference example should clarify our approach, also highlighting its benefits. Let us suppose to monitor the life cycle of an apparel item, a *cotton shirt*, and to follow every production step surveying and extracting relevant product/process

---

[1]RDF Primer, W3C Recommendation, February 10th 2004, available at http://www.w3.org/TR/rdf-primer/.

[2]OWL Web Ontology Language, W3C Recommendation, February 10th 2004, available at http://www.w3.org/TR/owlfeatures/.

information. Each production stage will see the progressive joining of annotations to enhanced RFID tags attached – for instance – to cotton yarn containers shipped to the factory (first production stage), shirt pallets (logistic step) and single product packages (final sale phase). Because of traceability requirements, a tag will store: (1) quantitative data pertaining to the product besides the Electronic Product Code (EPC) identifier; (2) high-level qualitative information about production or delivery/logistics processes, expressed as semantic annotations w.r.t. a reference ontology of the specific industry domain. Information extracted via RFID can be used for a variety of purposes. First of all – at each stage of the product evolution – accurate verifications can be performed about expected quality requirements of the product/process. Moreover, intelligent deliveries can be routed from warehouse to different production departments according to their specific characteristics.

A product can inherit (relevant parts of) the semantically annotated description of its raw materials, through properties defined in the reference ontology for the relevant domain. Further product attributes can also be stored on the RFID tag, such as size, production date and (for perishable products) expiration date. Finally, location, entering and exiting times are stamped by each supply chain actor conveying the item, such that advanced applications can be enabled. Beyond rather basic features allowed by a traditional data-oriented usage of RFID, a semantic-based approach makes possible further interactions.

A relevant aspect of the approach is that the semantic-enhanced RFID technology allows to share information, so optimizing the supply chain and improving performance both in terms of logistics features and by providing innovative services available for all involved actors. The envisioned framework can support a range of use cases, involving different stakeholders along a product life cycle. Several tangible (economic) and intangible benefits are expected. During product manufacturing and distribution, a wide-area support network interconnecting commercial partners is not strictly needed. This is a significant innovation with respect to common RFID supply chain management solutions (De et al. 2004).

Semantic-enabled RFID tags contain a structured and detailed description of product features, endowed with unambiguous and machine-understandable semantics. Goods auto-expose their description to any RFID-enabled computing environment is reached. This favors decentralized approaches in order to offer context-aware application solutions, based on less expensive and more manageable mobile ad hoc networks. In addition to improved traceability, a semantic-based approach provides unique value-added capabilities. By combining standard and non-standard inference services devised in (Di Noia et al. 2004), several semantic-based matchmaking schemes can be designed to meet goals and requirements of specific applications. Adopting a logic-based approach, query flexibility and expressiveness are much greater than both keyword-based information retrieval and standard resource discovery protocols, which support code-based exact matches only. This enables an effective query refinement process and can increase user trust in the discovery facility. Semantic-enhanced RFID object discovery can be leveraged also for sales and post-sale services, by assisting customers in using their purchased products more effectively.

3 Related Work

In the latest few years automatic identification technologies are gaining more and more interest. This is mostly due to their possible use in many industrial applications. AutoID systems allow the exchange of information about moving people, animals or goods for tracking in real time. Now RFID is the fastest growing sector of the AutoID business (Raza 1999).

RFID benefits in supply chains have been widely acknowledged in both distribution and warehousing sectors. In latest years, they are becoming evident also in the retail and post-sales domains. RFID-derived benefits include timeliness, accuracy and completeness (Karkkainen 2003). Wal-Mart, the world leading retailer, has been at the forefront of RFID experimentation, leveraging its retail power to mandate RFID adoption amongst selected suppliers (Pepe and Risso 2008; Tajima 2007). Significant retailer cost savings associated with RFID-monitored short-shelf-life products are achieved and the introduction of RFID technology by retailers might improve their supply chain efficiency, accuracy and security (Jones et al. 2004; Angeles 2005). De et al. (2004) introduced a reference system for real time tracking of items in a ubiquitous context. That work constitutes a research prototype of current technological architectures for supply chain management, endorsed by worldwide special interest groups such as the EPCglobal consortium (Traub et al 2005). A further relevant example outlines a retail shopping scenario with RFID technology allowing the consumer to engage in a seamless shopping experience (Sellitto 2007).

The benefits of RFID have always been associated with the powerful and dynamic automated data acquiring capabilities of the technology (Smith 2005). Nevertheless, RFID has received relatively little investigation from the informative standpoint. RFID technology should actually be viewed as an information facilitator that can directly improve the decision-making capabilities of personnel within an organization and information sharing across boundaries of partner organizations. Currently, researchers appear to overlook the important process of using RFID-derived information, which is a significant factor in deriving benefits such as visibility of the supply chain, product traceability and retailer's inventory monitoring. In other words, supply chains should be considered not only as product flow networks, but also as information flow networks. Effective exploitation of the supply chain information infrastructure can increase business process awareness and thus improve both performance analysis and support to decision processes.

Innovative information technology solutions are required for RFID-based data storage and analysis in order to enable such a new kind of smart supply chain. With reference to business processes, for deep analysis with respect to great lapse and huge amounts of data, the most interesting research field is in defining new storage models for a more expressive and sophisticated description of RFID data. There are two main directions.

The first one concerns run-time processing of data streams (Bai et al. 2006; Bai et al. 2007 Wang et al. 2006; Jeffery et al. 2006). Most current approaches, however, track only very basic information, namely *raw data* produced by RFID readers.

Raw data consist in *(EPC, location, time)* triples, where EPC is a unique product identifier, while location and time mark each RFID reading event. More advanced information and knowledge representation techniques have not been significantly integrated yet within RFID technology in smarter supply chain management solutions able to support analyses with higher-level semantics.

The second research direction focuses on off-line computation and efficient data storage (Wang and Liu 2005; Ban et al. 2005; Gonzalez 2006) Wang et al. (2005) formalized some features and semantics of RFID events, proposing an extension of the Entity-Relationship model. Based on such extended conceptual model, data streams are analyzed w.r.t. temporal aspects. Bai et al. (2007) studied limits of using SQL to detect temporal events in a database and illustrated a SQL-like language to query such events in an efficient way. Ban et al. (2005) presented a location-oriented indexing which traces paths registered by RFID readers through a novel representation model. The exploited indexing schema is named *Time Parameterized Interval R-Tree*, as a variant of common R-Tree. Gonzalez et al. (2006) provided instead a new storage model borrowed from the datawarehouse literature. Its core idea is to group items moving together so that the multidimensional analysis can be based on dimensions and measurements like in a typical datawarehouse.

For quick chain interactions aiming at an object/group discovery given a semantically annotated request, we refer to the Semantic Web initiative and adapt techniques and technologies to RFID-based supply chains. The basic goal is to fully characterize products equipped with RFID tags by means of semantic languages such as RDF, OWL or DIG. Semantic Web technologies allow a formalization of annotations in a machine understandable way, so promoting interoperability. A range of tools can be used for information processing and analysis, including rule-based systems, logic-based inference engines and query engines based on declarative languages like SPARQL[3]. The validity of the Open World Assumption (OWA) enables meaningful analyses even in the presence of incomplete information. This feature allows to overcome shortcomings of widely adopted "closed world" paradigms –such as the relational model– that often arise when interfacing heterogeneous information systems of independent partner organizations. This is indeed the case of supply chain management architectures. By means of formal ontologies, knowledge about a specific domain can be modeled and exploited in order to derive new implied information from the one stated within metadata associated to each resource Di Noia (2004).

Few proposals for semantic-based annotation of physical products can be found in literature. A solution applied to ubiquitous commerce environments was introduced in Maass and Filter (2006). RFID tags, however, stored only a product code, which was used as a key to retrieve the corresponding RDF annotation from a central backend information system. This "virtual counterpart" approach Pömel et al. (2006), inherited from traditional RFID applications in supply chain management,

---

[3]SPARQL Query Language for RDF, W3C Recommendation 15 January 2008, http://www.w3.org/TR/rdf-sparql-query/

poses major architectural and organizational challenges for information sharing in complex multi-party supply chains. Conversely, our core idea is that, as physical products flow among supply chain partners, ipso facto relevant high-level information about them is conveyed and can be exploited for meaningful business analysis at different levels.

Finally the graph-based nature of semantic models produces relevant challenges. Although simple and general, these models cannot be used in their basic form as storage models. Many proposals can be found in literature concerning alternative logic organizations to efficiently analyze semantic data (Marcus et al. 2007). The contribution, however, is based on optimizations strongly dependent on the physical structures used (i.e., dependent on tools that support these structures). Our proposal is enriched by the provisioning of data models that describe semantic information at different levels of abstraction (not only at the physical one). Therefore the optimization is independent from any physical environment.

4 Preliminaries

In order to make the chapter self-contained, hereafter some details about adopted formalisms and languages will be provided.

4.1 Semantic-Based Matchmaking

Semantic-based object discovery is grounded on Description Logics (DLs) which provides the Ontology Web Language (OWL-DL) semantics. Furthermore, there is a strict correspondence between the OWL-DL syntax and the DIG (Description Logic Implementation Group) one which is exploited as interface for HTTP-based reasoners (Bechhofer et al. 2003). Particularly, when facing on implementation issues, DIG 2.0 formalism should be adopted to express both requests and resource descriptions, because it is less verbose and more compact, a mandatory requirement in mobile ad hoc applications. Anyway, here we formalize examples by adopting DL syntax for the sake of readability.

DLs are a family of logic formalisms for Knowledge Representation (Baader et al. 2002) whose basic syntax elements are *concept names, role names, individuals*. Intuitively, concepts stand for sets of objects in the domain, and roles link objects in different concepts. Individuals are used for special named elements belonging to concepts. Formally, concepts are interpreted as subsets of a domain of interpretation δ, and roles as binary relations (subsets of $\Delta \times \Delta$).

DL formulas give a semantics by defining the interpretation function $\cdot^{\mathcal{I}}$ over each construct. For example, if A and D are two generic concepts, their conjunction $A \sqcap D$ is interpreted as set intersection: $(A \sqcap D)^{\mathcal{I}} = A^{\mathcal{I}} \cap D^{\mathcal{I}}$, and also the other boolean connectives $\sqcup$ and $\neg$, when present, give the usual set-theoretic interpretation of union and complement.

Concepts can be used in *inclusion assertions* $O \sqsubseteq D$, and *definitions* $O \equiv D$, which impose restrictions on possible interpretations according to the knowledge elicited for a specific domain. A DL theory (a.k.a. *TBox* or *ontology*) is basically a set of inclusion assertions and definitions. A *model* of a TBox $\mathcal{T}$ is an interpretation satisfying all inclusions and definitions in $\mathcal{T}$. Many other constructs can be defined, so increasing the expressiveness of the DL. Nevertheless, this usually leads to a growth in computational complexity of inference services (Brachman and Levesque 1984). Hence a trade-off is necessary.

The core idea of the Semantic Web initiative (Berners-Lee et al. 2001) is to annotate information by means of markup languages, based on XML, such as RDF, RDFS and OWL. These languages have been conceived to allow machine understandable, unambiguous representation of Web contents through the creation of domain ontologies, increasing openness and interoperability in the WWW. The strong relationship between DLs and the above referenced languages (Baader et al. 2003) is also evident in the classification of the OWL sub-languages.

- *OWL-Lite*. It allows class hierarchy and simple constraints on relations between classes.
- *OWL-DL*. Based on DLs theoretical studies, it allows a great expressiveness keeping computational completeness and decidability.
- *OWL-Full*. It has a huge syntactic flexibility and expressiveness. This freedom is paid in terms of no computational guarantee.

In this chapter we will refer to the *Attributive Language with unqualified Number restrictions and Concrete Domains* $(\mathcal{ALN}(D))$ DL, a subset of OWL-DL, which has a polynomial complexity both for standard and non-standard inferences. Constructs of $\mathcal{ALN}(D)$ DL are reported in what follows:

- $\top$, *universal concept*. All the objects in the domain.
- $\bot$, *bottom concept*. The empty set.
- A, *atomic concepts*. All the objects belonging to the set A.
- $\neg A$, *atomic negation*. All the objects not belonging to the set A.
- $C \sqcap D$, *intersection*. The objects belonging to both C and D sets.
- $\forall R.C$, *universal restriction*. All the objects participating in the R relation whose range are all the objects belonging to C set.
- $\exists R$, *unqualified existential restriction*. At least one object participating in the relation R.
- $(\geq n R)$[4], $(\leq n R)$, $(= n R)$[5], *unqualified number restrictions*. Respectively the minimum, the maximum and the exact number of objects participating in the relation R.

---

[4]Notice that $\exists R$ is equivalent to $(\geq 1R)$

[5]We write $(= n R)$ for $(\geq n R \sqcap \leq n R)$

- *f, concrete features.* An extension to basic DLs that allows to link concepts to a concrete domain D (e.g., integers, reals, time and so on) through a set of unary predicates p. Each concrete feature f can be expressed as $p(f)$ with $p : \delta \Rightarrow D$, where δ is the feature domain. In this chapter only the concrete domain of integers and the following unary predicates will be considered: $(\geq_k g), (\leq_k g), (=_k g)$, with g a feature and k an integer value.

Knowledge Representation (KR) approaches to matchmaking usually exploit classical deductive services, namely *classification* (also known as *subsumption*) and *consistency* (i.e., *satisfiability*). Basically, given a request/resource pair annotated w.r.t. a common reference ontology, classification allows to check whether all request specifications are included within the resource description. Whereas consistency verifies whether some specifications in the request contradicts (some of) the ones within the resource annotation. In both cases, the response is then a binary *true/false* value. Although these inference services are very useful in the early phases of a discovery process, they are not sufficient to rank a set of resources with respect to a request.

Given R (for Request) and O (for Offer) both consistent with respect to an ontology T, logic-based approaches to matchmaking proposed in literature (Paolucci et al. 2002; Li and Horrocks 2004) use classification and consistency to grade match results in five categories:

- *Exact.* All the features requested in R are exactly provided by O and vice versa –in formulae $T \models R \Leftrightarrow O$.
- *Full-Subsumption.* All the features requested in R are contained in O –in formulae $T \models O \Rightarrow R$.
- *Plug-In.* All the features offered in O are contained in R –in formulae $T \models R \Rightarrow O$.
- *Potential-Intersection.* There is an intersection between the features offered in O and the ones requested in R –in formulae $T \not\models \neg(R \wedge O)$.
- *Partial-Disjoint.* Some features requested in R are conflicting with some offered in O –in formulae $T \models \neg(R \wedge O)$.

While exact and full matches seldom occur, a user may get several potential and partial matches. Then a logic-based matchmaker should provide a – logic – ranking of available resources w.r.t. the request, but what we get using classification and consistency is a boolean answer. Also partial matches might be just "near miss" (e.g., maybe just one requirement is in conflict), but a pure consistency check returns a hopeless *false* result, whereas it could be interesting to retrieve "not so bad" resources according to their similarity to the request.

Let us consider concepts R and O and an ontology T. If a partial match occurs, i.e., if they are not compatible with each other with respect to T, one may want to retract some specifications in R, G (for *Give up*), to obtain a concept K (for *Keep*) such that $K \sqcap O$ is satisfiable in T.

In Colucci et al. (2003) the Concept Contraction problem was first defined as:

– *Concept Contraction.* Let $\mathcal{L}$ be a DL, R, O be two concepts in $\mathcal{L}$ and $\mathcal{T}$ be a set of axioms in $\mathcal{L}$, where both R and O are satisfiable in $\mathcal{T}$. A *Concept Contraction Problem* (CCP), identified by $\langle \mathcal{L}, \mathcal{R}, \mathcal{O}, \mathcal{I} \rangle$, consists of finding a pair $\langle G, K \rangle \in \mathcal{L} \times \mathcal{L}$ such that $\mathcal{T} \models R \equiv G \sqcap K$, and $K \sqcap O$ is satisfiable in $\mathcal{T}$. Then K is a *contraction* of R according to O and $\mathcal{T}$.

If nothing can be kept in R during the contraction process, we get the worst solution –from a matchmaking standpoint– $\langle G, K \rangle = \langle R, \top \rangle$, that is give up everything of R. Conversely, if $R \sqcap O$ is satisfiable in $\mathcal{T}$, that is a potential match occurs, nothing has to be given up and the solution is $\langle \top, R \rangle$. Hence, a Concept Contraction problem amounts to an extension of a satisfiability one. Since usually one wants to give up as few things as possible, some minimality criteria in the contraction must be defined Gärdenfors (1988).

If the offered resource O is a potential match for R, it is necessary to assess what should be hypothesized H in O in order to completely satisfy R and then move to a full match. In Di Noia et al. (2003) the Concept Abduction problem was first defined as:

– *Concept Abduction.* Let $\mathcal{L}$ be a DL, R, O be two concepts in $\mathcal{L}$, and $\mathcal{T}$ be a set of axioms in $\mathcal{L}$, where both O and R are satisfiable w.r.t. $\mathcal{T}$. A *Concept Abduction Problem* (CAP), identified by $\langle \mathcal{L}, R, O, \mathcal{I} \rangle$, is to find a concept $H \in L$ such that $\mathcal{T} \models O \sqcap H \sqsubseteq R$, and moreover $O \sqcap H$ is satisfiable in $\mathcal{T}$. We define H a *hypothesis* about O according to R and $\mathcal{T}$.

If $O \sqsubseteq R$ then we have $H = \top$ as a solution to the related CAP. Hence, Concept Abduction amounts to extending subsumption. On the other hand, if $O \equiv \top$ then $H \sqsubseteq R$.

4.2 RFID Data Representation

Whatever RFID application usually generates a tuple stream in the form of a triple (E, l, t), where:

- E is an *EPC* (Electronic Product Code), i.e., a unique identifier stored in a tag and associated to each tagged object;
- l is the *location* where an RFID reader has scanned an object having the E EPC;
- t is the *time* when the reading took place.

As a single tag may have multiple readings at the same location – thus producing a great amount of raw data – cleaning techniques have to be applied. The most used compression converts raw data in *stay records* in the form: (E, l, t_{in}, t_{out}) where t_{in} is the time when the object enters the location l, and t_{out} is the time when the object leaves it. Although this basic solution reduces the amount of data to be stored (even if not considerably), previous data representation loses object transitions information. So an alternative representation of RFID data has been proposed in Lee and

Chung (2008). It involves *trace records* and has the form:

$$E : l_1[t_{in}^1; t_{out}^1] \Rightarrow \ldots \Rightarrow l_k[t_{in}^k; t_{out}^k]$$

where:

- $l_1, \ldots, l_k$ are locations along the path followed by the tag with E EPC ;
- t_{in}^i is the entering time at location l_i;
- t_{out}^i is the exiting time from the location l_i;
- the sequence is ordered by t_{in}^i.

The drawback of such data representations is that they are path-dependent, and therefore only path queries over objects moving together can benefit from them (e.g., *find the average time for jackets to go from tailor's to stores in Rome*). To overcome this limitation, the notion of *entry records* is introduced, which describes product information that can be used in multidimensional analysis. Entry records have the form:

$$E : [A_1, v_1], [A_2, v_2], \ldots, [A_n, v_n]$$

where:

- A_i describes an attribute representing the object with E EPC;
- v_i is the value associated to A_i for this object.

Notice that: (i) an entry record can be used to represent collections of RFID data at different level of details (e.g. raw data or stay records) and (ii) aggregates are based on different combinations of attributes A_i.

4.3 Supply Chain Indexing

In supply chain management, a basic need is to analyze object transitions. A product with an RFID tag can cross many locations in a chain. Tracing its movements, transitions can be expressed as a path $l_1, \ldots, l_n$ in a graph describing the supply chain itself. Different approaches and techniques have been proposed for supply chain indexing in order to effectively compute the path of a tag. Currently they are supported by DBMS physical optimizations. In Ban et al. (2005) and Gonzalez et al. (2006), EPC data features are exploited to group tags and arrange them through bitmap indexes. Lee and Chung (2008) devised an alternative encoding scheme that assigns to each path a pair (*Element List Encoding Number (ELEN) - Order Encoding Number (OEN)*). By assigning a prime number to each node in the chain, ELEN is obtained by multiplying path nodes (with related values) among them. In this way, the path followed by a tag is computed as factorization of the integer assigned to the path itself. Primality of assigned numbers guarantees the correctness of the result. Additionally, OEN is a value able to encode the arrangement among nodes in the path.

Both the above solutions present some non-negligible drawback. Approaches in Ban et al. (2005) and Gonzalez et al. (2006) are dependent on DBMS optimizations and on the similarity assumption between EPCs, whereas framework in Lee and chung (2008) does not cope with computational complexity of prime factorization of an integer. Since a supply chain can present several levels (typically from four to ten), to the best of our knowledge, no efficient algorithms are available for very large values.

To efficiently manage the object transitions, an elementary method was introduced in De Virgilio et al. (2009) for encoding each path $l_1, \ldots, l_n$. The basic idea is to model a supply chain s by a directed *sc-graph* G_s whose nodes are locations of s and there is an edge from a node l_1 to a node l_2 if there is some movement of objects from l_1 to l_2 in s. The *source nodes* of a sc-graph, i.e., the nodes having no incoming edge, usually represent the place where objects are produced, whereas *target nodes* of a sc-graph, i.e., nodes having no outgoing edge, are usually the final stores where products are sold. Then, a *token* is associated to each possible path from a source to a target node. Finally, the encoding is performed in each node n of a sc-graph by assigning to n the set of tokens representing each path from a source to a target node traversing n. The supply chain in Fig. 1 shows as an example the proposed path encoding.

4.3.1 Data Compression

In order to aggregate queries over a large amount of RFID data, it is useful to compress them with respect to different *aggregation factors*. For instance, considering location, data can be aggregated either at city, region or country levels. Similarly, a product can be grouped by brand, category or price.

Let us consider a set of attributes $S = \{A_1, A_2, \ldots, A_n\}$ describing available entry records. By borrowing a typical OLAP (on-line Analytical Processing) approach, attributes are grouped in S according to different *factors* such as location, time or product. Moreover, each factor allows to build a set $R = \{(x, r, y) : x, y \in S\}$,

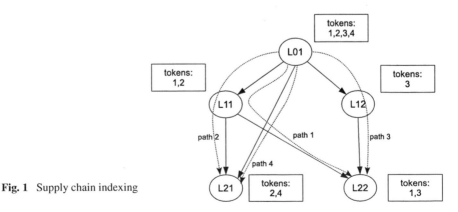

Fig. 1 Supply chain indexing

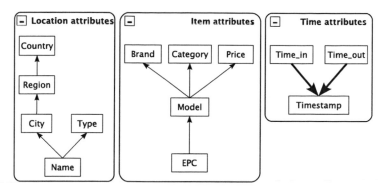

Fig. 2 A taxonomy example

where r is a binary relationship between attributes such as *is-a*, *part-of*, and so on. In this way *taxonomies* can be built starting from the original attribute set. Figure 2 shows possible taxonomies: thin arrows represent *part-of* relationships, thick ones represent *is-a* ones between attributes.

The set R, augmented with taxonomies defined over proper factors in R, suggests interesting attribute aggregations. An *aggregation factor* is defined as a propositional logic expression φ involving attributes occurring in $A_1, \ldots, A_n$. It can be easily noticed that φ depends on taxonomies defined over them. For instance, if $A_i \wedge A_j$ occurs in φ and $(A_i,$ part-of, $A_j)$ then the conjunction can be replaced by A_j, because A_j has a higher aggregation degree higher than A_i. Hence, a set of rules can be defined to simplify an aggregation factor using taxonomies. With respect to the set of attributes in Fig. 2, possible aggregation factors for that domain are: $\varphi_1 = (location \wedge time\_in \wedge time\_out)$ and $\varphi_2 = (model)$. Intuitively, φ_1 aggregates records for items entering and leaving the same location at the same time, while φ_2 aggregates records for items which are similar among them.

Given two entry records $r_1 = \{E_1; [A_1', v_1'], \ldots, [A_n', v_n']\}$ and $r_2 = \{E_2; [A_1'', v_1''], \ldots, [A_m'', v_m'']\}$, an *aggregated record* r_A has the form: $\{E_{r_A}; [r_1, r_2]\}$. Two entry records r_1 and r_2 are aggregated w.r.t. an aggregation factor φ if r_1 and r_2 *satisfy* φ as follows.

Definition 4.1 (φ-satisfiability) *Given an expression φ over a set of attributes $A_1, \ldots, A_n$, and two entry records $r_1 = \{E_1; [A_1, v_1'], \ldots, [A_j, v_j'], \ldots, [A_m, v_m']\}$ and $r_2 = \{E_2; [A_1, v_1''], \ldots, [A_j, v_j''], \ldots, [A_m, v_m'']\}$, r_1 and r_2 satisfy φ if, by replacing each A_i in φ with the atom $(v_i' = v_i'')$ for $1 \leq i \leq n$, the obtained formula is true.*

This notion can be used to introduce the one of aggregation based on a given aggregation factor.

Definition 4.2 (φ-aggregation) *Given an aggregated record r_A, r_A is a φ-aggregate if for each pair of entry records r_1, and r_2 occurring in r_A:r_1 and r_2 satisfy φ.*

Hence, an aggregation F can be obtained by a sequence of aggregation factors $\varphi_1, \varphi_2, \ldots, \varphi_n$. Basically, in a supply chain an aggregated record is an item *stock*.

Object movement can be seen as a collection of stocks that split moving from sources to targets. Under this assumption, a useful definition is reported hereafter.

Definition 4.3 (Subsumption) *Given two aggregated records* $r_{A_1} = \{E_{r_{A_1}};$ $[r_1, r_2, \ldots, r_n]\}$ *and* $r_{A_2} = \{E_{r_{A_2}}; [r'_1, r'_2, \ldots, r'_m]\}$, r_{A_1} *is subsumed by* r_{A_2}, *denoted by* $r_{A_1} \lhd r_{A_2}$, *if each* r_i *occurring in* r_{A_1} *also occurs in* r_{A_2}.

An object transition can be then seen as a movement from a stock to another one. Hence, an *observation* is defined as an object transition event and it can be represented by a pair (s_1, s_2), where s_1 is a stock including a set of items coming from the stock s_2. It ensues that $s_1 \lhd s_2$.

5 Framework and Approach

Analyses related to a supply chain could be basically intended as involving two different aspects: *on-product* and *batch* control. The former is performed on-line and aims at an object (product) discovery, given features explicitly stated in a request by either controller or customer. The latter happens off-line and refers to a higher level abstraction related to product batches flow. Both kinds of analysis help various actors involved in a supply chain (from producers to retailers to customers) in taking under control large amounts of goods with their related properties as well as to easily track and retrieve specific things or groups of things.

This section describes an integrated EPCglobal RFID-based framework where tagged goods can be either individually and automatically discovered in a pervasive computing scenario or aggregated for enabling centralized massive analysis involving off-line data storage. Both components will be examined in detail below.

5.1 Online Object Discovery

On-line object discovery requires an in-depth comparison between product characteristics and user requirements. In order to quickly produce useful results, advanced (possibly automated) matchmaking techniques are required where both user's request and resource characterization are expressed in a machine understandable format, which allows the needed expressiveness while keeping an acceptable computational complexity. Due to space, power and cost constraints, RFIDs are still currently endowed with low storage, no processing capability and short-range, low-throughput wireless links. Furthermore, each mobile reader in the field can access information only on micro devices in its range. As a consequence, approaches based on centralized control and unique information storage are impractical. On the other hand, when effective network infrastructures are lacking and exploited devices are resource-constrained, the discovery process can be strongly enhanced by exploiting KR techniques and technologies. Semantic-based resource annotation and matchmaking, as well as logic-based ranking and explanation services (Di Noia, et al. 2004), seriously improve resource retrieval experience. Furthermore, the enhancements to EPCglobal RFID standard protocol let users interact with the system without requiring dependable wired infrastructures while hiding technicalities from them.

In the research paper Di Noia (2008) we introduced a knowledge-based variant of EPCglobal RFID, whose primary goal was to keep a backward compatibility with the original technology as much as possible. Protocols to read/write tags were preserved, maintaining original code-based access, in order to ensure compatibility and smooth coexistence of new semantic-based object discovery applications and legacy identification and tracking ones. In our framework we refer to RFID transponders conforming to the EPCglobal standard for class I - second generation UHF tags (Traub 2005) (we assume the reader be familiar with basics of this technology). The proposal was tested in a simulation campaign and in application case studies, showing the benefits of the approach for stakeholders involved each stage of the product lifecycle (Ruta et al. 2007) – from raw materials to production, retail and post-sale services. Here an evolution of the approach is presented.

The practical feasibility of advanced semantic-based usage of RFID technologies must take into account some important constraints. First of all, the severe bandwidth and memory limitations of current RFID systems, in order to meet cost requirements for large-scale adoption. Due to technological advances and growing demand, passive RFID tags with greater memory amounts are expected to be available (Ayoub Khan and Manoj 2009). Nevertheless, XML-based ontological languages like OWL and DIG are far too verbose for a direct storage on RFID tags. Moreover, a mechanism is clearly required to distinguish semantic enabled tags from standard ones, so that semantic based applications can exploit the new features without interfering with legacy applications.

To enable the outlined enhancements, RFID tags and the air interface protocol must provide read/write capabilities for semantically annotated product descriptions with respect to a reference ontology, along with additional data-oriented attributes. Neither new commands nor modification to existing ones have been introduced. To accomplish that, we extend the memory organization of tags compliant with the above referenced standard.

Contents of TID memory area up to $1F_h$ bit are invariable. For tags having class identifier value $E2_h$ stored in the first byte of the TID bank, optional information could be stored in additional TID memory from 20_h address. There we store:

•
- a 16 bit word for optional protocol features, stored starting from 20_h address most significant bit first: currently only the most significant bit is used to indicate whether the tag is semantic-enabled or not; other bits are reserved for future uses;
- a 32-bit *Ontology Universally Unique Identifier* (OUUID) marking the ontology with respect to the description stored in the tag is expressed Ruta et al. (2006).

In this way, a reader can easily distinguish semantic based tags by means of a *Select* command with parameter values as in Table 1. Values for the triple ⟨*MemBank, Pointer, Length*⟩ identify the bit at 20_h address in the TID memory bank. The reader commands each tag in range to compare it with bit mask 1_2. The match

Table 1 *Select* command able to detect only semantic enabled tags

Parameter	Target	Action	MemBank	Pointer	Length	Mask
Value	100_2	000_2	10_2	00100000_2	00000001_2	1_2
Description	SL flag	Set (if match)	TID bank	Initial address	Bit to be compared	Bit mask

outcome will be positive for semantic enabled tags only. The *Target* and *Action* parameter values mean that in case of positive match the tag must set its *SL* flag and clear it otherwise. The following inventory step will skip tags having SL flag cleared, thus allowing a reader to identify only semantic enabled tags. Protocol commands belonging to the inventory step have not been described, because they are used in the standard fashion.

In order to retrieve the OUUID stored within a tag, a reader will exploit a *Read* command by adopting parameter values as in Table 2. *MemBank* parameter identifies the TID memory bank and the *WordPtr* value specifies that the reading must start from the third 16-bit memory word, i.e., from 20_h address. Finally, the *WordCount* parameter indicates that 32 bits (two 16-bit words) have to be read.

Contextual parameters (whose meaning may depend on the specific application) are stored within the *User memory bank* of the tag. There, we also store the product annotation. To overcome storage space limitations, it is encoded with a specialized compression algorithm designed for XML-based ontological languages. An RFID reader can perform extraction and storing of a description from/on a tag by means of one or more *Read* or *Write* commands, respectively. Both commands are obviously compliant with the RFID air interface protocol. Table 3 reports parameter values of the *Read* command for extracting the full contents of the User memory, comprising both contextual parameters and the compressed annotation.

The EPCglobal standard also provides a support infrastructure for RFID applications by means of the so called *Object Naming Service* (ONS) EPCglobal (2005). In our approach the ONS mechanism is considered as a supplementary system able to grant the *ontology support*. If a reader does not manage the ontology which provides terminology for the description within the tag, we may retrieve it exploiting the ONS service. The *EPCglobal Network Protocol Parameter Registry* maintains all the registered service suffixes (*ws* for a Web service, *epcis* for a EPCglobal Information Service (providing authoritative information about objects associated with an EPC code), *html* for a Web Page of the manufacturer). We hypothesize to register the novel *dig* suffix to indicate a provisioning service for ontologies with a specified OUUID value.

Table 2 *Read* command able to extract the OUUID from the TID memory bank

Parameter	MemBank	WordPtr	WordCount
Value	10_2	000000010_2	00000010_2
Description	TID memory bank	Starting address	Read 2 words (32 bits)

Table 3 READ command able to extract the semantically annotated description from the User memory bank

Parameter	MemBank	WordPtr	WordCount
Value	11_2	000000000_2	00000000_2
Description	User memory bank	Starting address	Read up to the end

5.2 Off-Line Batch Analysis

An off-line supply chain data analysis basically involves the following steps.

1. *Import* a reference supply chain. It takes as input the topology of a given supply chain in terms of a sc-graph $G(V, E)$ where nodes V correspond to locations and edges E to movements between locations. The graph is imported into a relational DBMS and indexed.
2. An expert user sets an attribute taxonomy and defines the aggregation factors according to queries suggested by stakeholders.
3. Raw data are collected from RFID readers in the field. In a preprocessing phase, raw data are compressed into stay records and ordered by t_{in}. Then aggregation factors defined by the user are incrementally applied to generate a set of aggregated entry records, which are materialized and indexed. In this way a compressed data set is imported into the RDBMS.
4. Finally, users can submit queries to a front-end interface and inspect the answers provided by the system.

In the following subsections we present in greater detail the aggregation process and illustrate the proposed storage schema for aggregated records able to guarantee an efficient query processing.

5.2.1 Data Aggregation

Given an aggregation $F = [\varphi_1, \varphi_2, \ldots, \varphi_n]$ and a stay records sequence SR ordered by t_{in}, F is applied to SR to generate a set of aggregated entry records ARS. Notice that an aggregation makes sense if it is locally applied. In other words, only records having the same location are aggregated. Since information details are contained within tagged items, there is no need for a centralized backend for the whole supply chain. Each supply chain partner has visibility of the relevant attributes and entry records of items passing through its premises and can use its own support infrastructure for off-line data analysis. For each node, aggregation can be then performed according to the most relevant factors for business analysis, selected according to specific management goals. This approach also solves the problem of change of item "ownership" when items flow among different partners (and even across country borders) or they are redistributed in different packages.

Table 4 Entry records aggregation

Record	EPC	Model	Price	Count	Loc	t_{in}	t_{out}
r_1	clt01	Polo	160	1	L12	4	6
r_2	clt02	Polo	160	1	L12	4	6
r_3	clt03	Suit	400	1	L12	4	6

Hereafter a straightforward aggregation strategy is outlined.

1. Records characterized by the same location are grouped in a set L.
2. The aggregation factors φ_i are sequentially applied to elements in L: if there exists an aggregated record rd_A such that current element R' of L and rd_A satisfy υ, then R' can be inserted into rd_A; otherwise a new aggregated record from R' is created.

For example, let us consider three entry records representing clothes (*i.e.*, items) moving into the supply chain shown in Figure 1, as in Table 4.

Now, let us consider an aggregation $F = [\varphi_1, \varphi_2]$ resulting from factors $\varphi_1 = (time\_in \wedge time\_out)$ and $\varphi_2 = (model)$. With respect to φ_2 two aggregated records are produced: $r_{A_1} = \{r_1, r_2\}$ and $r_{A_2} = \{r_3\}$. Finally, applying φ_1, the aggregated record $r_{A_3} = \{r_{A_1}, r_{A_2}\}$ is obtained.

The final step involves building observations, assigning a token to each of them. The following algorithm illustrates that process. It takes as input: a sequence of aggregated records ARS increasingly ordered by $time\_in$, the chain C in terms of a sc-graph $C(V_C, E_C)$ and a map LM where a list of tokens is assigned to each location. The output is a map OM whose observations have a given token.

1. Starting from the last record AR of ARS, go back to find the aggregated record that subsumes AR.
2. If it does not exist, a new observation is generated, the list of tokens corresponding to AR location is extracted from LM and the first token is assigned to the new observation.
3. Otherwise, a new observation is generated where $AR \lhd AR'$. The token to be assigned results from the intersection between the lists of tokens of both locations in AR and AR' excluding the ones coming from other paths reaching AR' location.

5.2.2 Data Storage

Figure 3 shows the relational schema to contain RFID data processed as described above. `PATHMAP` and `LOCATION` store chain information. Each entry in `PATHMAP` has an external reference `LOC_ID` to `LOCATION` with an associated *token*. If more tokens are associated to a given location, then `PATHMAP` will have more entries for that location. The `DEPTH` attribute measures the current path length up to location identified by `LOC_ID`. `ITEM` stores information about products, in terms of EPC

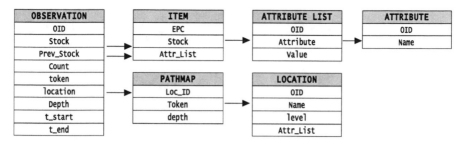

Fig. 3 Storage schema

and last stock where the item was aggregated. ATTRIBUTE gathers attributes featuring both ITEM and LOCATION and they are related to an attribute list through ATTRIBUTE_LIST which associates a value to each attribute. OBSERVATION traces all transitions of aggregates. There, the STOCK attribute is the current aggregate, containing items coming from PREV_STOCK, observed in a given LOCATION. Each observation stores stock items number as well as traversed locations (COUNT and DEPTH attributes), assigned token and time when the stock entered and left the location respectively (T_START and T_END attributes).

As resulting tables could be very large, relational DBMS capabilities have been deeply exploited tuning the above schema to obtain better performances. The first optimization is the horizontal partitioning of tables into smaller physical ones, namely *partitions*. Each partition inherits the same structure (e.g., columns and datatypes) of the original table, namely *master*. Partitioning is performed row by row and it is based on column value (*range*). For instance, in OBSERVATION table, the column LOCATION is exploited as range (see Fig. 4). If needed, the resulting smaller tables can be further partitioned with respect to another range, as for example the *time* one. Recall that this step is totally hidden from the user. STOCK, OID and LOC_ID are set as ranges for ITEM, ATTRIBUTE_LIST and PATHMAP, respectively. Moreover, tree indexes B$^+$ are defined on the whole set of tables. Obviously, an unclustered index will be defined for each range column used for partitioning and other indexes on the single partition. In particular, clustered indexes will be used on STOCK in each OBSERVATION partition.

5.2.3 Query Processing

The supply chain analysis is based on querying object transitions. Many approaches (Gonzalez et al. 2006) only allow to find the movement history of a given tag, by means of path selection queries, also known as *tracking queries*. Other query templates are proposed in Lee and Chung (2008) in order to enable a more flexible analysis of supply chains: besides tracking queries, *path oriented queries* are considered, divided into *path oriented retrieval queries* and *path oriented aggregate queries*. The former finds tags that satisfy given conditions about path and time, the

OBSERVATION

oid	stock	prevstock	count	token	depth	location	t_start	t_end

oid	stock	prevstock	count	token	depth	location	t_start	t_end
obs1	stock01	stock01	8	1	1	L01	2	3
obs3	stock03	stock01	2	3	2	L12	4	6
obs4	stock04	stock01	3	1	2	L11	5	7
obs6	stock06	stock01	1	1	2	L11	7	8
obs2	stock02	stock01	1	4	2	L21	4	5
obs5	stock05	stock01	1	4	2	L21	5	6
obs9	stock09	stock04	1	2	3	L21	13	16
obs10	stock10	stock06	1	2	3	L21	14	18
obs7	stock07	stock03	1	3	3	L22	7	8
obs8	stock08	stock04	2	1	3	L22	8	9

Fig. 4 Horizontal partitioning

latter compute aggregate values for tags satisfying specifications. The authors formalize query templates and express queries using an XPath-like language. Here, we will use the same formalization and we will describe a method to process tracking queries and path oriented queries. Since we use an RDBMS, queries are translated into SQL (Structured Query Language).

In a tracking query, given a tag_id, a tag flow is returned (in terms of a list of locations). There is the need to find the stock containing the item as well as the token assigned to the observation corresponding to the stock itself. In the relational schema presented above, the table ITEM will be queried to extract the stock_id and the OBSERVATION table to extract the token assigned to the related observation. Finally PATHMAP will be queried to return all locations annotated with the extracted token. The list of locations is ordered by the attribute DEPTH. The corresponding SQL code ensues.

```
SELECT P.LOC_ID
FROM PATHMAP P, OBSERVATION O, ITEM I
WHERE I.EPC=<tag_id> AND O.STOCK=I.STOCK
   AND P.TOKEN=O.TOKEN
ORDER BY P.DEPTH
```

In a path oriented retrieval query, given a path expressed in terms of an XPath-like syntax, the list of tags that followed it is returned. Also in this case both stock and token information into OBSERVATION are exploited to select items. Moreover, ancestor and/or parent relationships between locations have to be processed. Given two locations l1 and l2, if they are in a parent-child relationship (i.e., l1/l2), then PATHMAP is queried and tokens are selected such that the depth of l2 is the depth of 11 incremented by 1. Otherwise, if locations are in the ancestor-descendant relationship (i.e., 11//12), tokens such that the depth of 12 is greater than the depth of 11 are selected. Let us consider the query //L1/L2. In this case we translate it into SQL as reported hereafter:

```
SELECT I.EPC
FROM PATHMAP P1, PATHMAP P2, OBSERVATION O, ITEM I
WHERE P1.LOC_ID=MD5('L1') AND P2.LOC_ID=MD5('L2')
  AND P2.DEPTH=(P1.DEPTH+1) AND O.LOCATION='L2'
  AND O.TOKEN=P2.TOKEN AND I.STOCK=O.STOCK
```

When translating //11//12 in SQL, the condition P2.depth > P1.depth has been considered. Anyway, path oriented retrieval queries may present different conditions and, in that case, selection conditions has to be added to ITEM and OBSERVATION. For instance, we can enrich the previous query with time conditions such as //11/12[t_end < 200]. With respect to the previous SQL code, condition O.t_end < 200 must be added to the WHERE clause. Path oriented aggregate queries present aggregate functions which can be considered in the SELECT clause of the SQL query such as COUNT, SUM, MAX and so on.

6 Experimental Evaluation

As stated in the above Sect. 5, the proposed framework enables analyses at two different stages. In what follows, an on-line object discovery toy example is presented exploiting semantic-based object annotation in a given u-commerce context; furthermore a massive data examination conducted off-line on large amount of data is outlined. The benchmarking system is a dual quad core 2.66 GHz Intel Xeon PC, running Linux Gentoo OS, with 8 GB RAM memory, 6 MB cache memory, and a 2-disk 1Tbyte striped RAID array. We carried out our experiments using PostgreSQL 8.3, because it has been proved (see Beckmann et al. in 2006) that it is significantly more efficient with respect to commercial database tools. Two different supply chains in the field of clothing production and distribution (namely $Chain_{20}$ and $Chain_{100}$) have been defined with 20 and 100 nodes, respectively. Both chains present seven levels, i.e., in both cases a path has a maximum length equal to 7.

6.1 On-line Resource Discovery

The on-the-fly discovery basically involves the end part of the supply chain, that is customers and retailers, but it can be also applied to different stages in product manufacturing and distribution, when real-time knowledge-based decision support in needed. Our matchmaking framework, leveraging the inference services described in Sect. 4, can be adapted to a wide range of resource discovery use cases. The example scenario reported here refers to the apparel product domain. A typical supply chain interaction sequence is outlined hereafter, but for the sake of brevity only the last step will be described in detail.

1. *Selection of raw materials.* Several yarn deliveries are directed to a jacket factory. Each one is identified and described by means of semantic-enabled RFID tags attached to containers, which are scanned upon arrival at the factory warehouse. Due to traceability requirements, a tag must store quantitative data such as volume, weight and a production time stamp besides the EPC code. Qualitative information about yarn and its production process are also stored within the tag, expressed as a semantic annotation with reference to an apparel industry ontology. Information extracted via RFID readers at warehouse gate can be used for several purposes by workers equipped with handheld devices:

   - product/process requirements established in the contract with supplier can be immediately verified by means of a semantic query upon yarn characteristics;
   - each yarn type can be routed to a different warehouse area, according to specific storage requirements (e.g., temperature, humidity) (Ruta et al. 2008);

2. Manufacturing and packaging. Each yarn type can be routed from the warehouse to a different production department according to its properties. Subsequently, during product packaging operations, RFID semantic-enabled tags are written and then attached to each item before shipping. Relevant portions of the semantically annotated description of raw materials can be inherited by the final product; in our scenario, a jacket can inherit information from yarn such as type of cloth and color. It is important to note that a centralized backend infrastructure is not required for this operation: just like production takes materials as input and returns products as output, information can be transferred from tags of materials to tags of products by a mobile RFID enabled ad hoc network within the factory.

3. Sales. A customer enters an outlet, where a ubiquitous commerce system is able to assist her in discovering additional available items. Details follow.

Shop items are tagged with RFID transponders containing the EPC code, OUUID, compressed semantic description and data-oriented attributes. For our case study, we adopted a simplified apparel ontology; price and body area are used as data-oriented attributes. Body area parameter is referred to a specific part of the body, as shown in Table 5.

A request is performed on behalf of the user when she enters a dressing room to try a product on. The request is based on the semantically annotated description of the product selected by the user, which is read by an RFID interrogator installed

Table 5 Correspondence between human body and values of contextual attribute

Value	1	2	3	4	5	6
Body area	Hands	Head	Chest (outer layer)	Chest (inner layer)	Legs	Feet

into the dressing room. In order to satisfy the request, a local shop hotspot retrieves resource descriptions referred to the specified ontology from the enterprise back-end database and performs a semantic-based matchmaking. RFID readers and wireless devices deployed in the outlet cooperate through an agent-based, message-oriented middleware infrastructure. A prototype has been built upon *IBM WebSphere RFID Tracking Kit* (Chamberlain et al. 2006), which provides a general framework for the orchestration of several heterogeneous devices in mobile applications. A semantic-based layer has been integrated in order to support features specific to our approach.

Let us consider the following case study: *a man enters the outlet to buy some elegant clothing. He notices a fine dark green jacket and decides to try it on.* Sensors detect the customer entering the dressing room. The RFID reader is triggered and reads data stored within the tag attached to the jacket, then it is deactivated again. Annotated data correspond to an elegant, large-sized, dark green jacket, in mostly linen cloth, suitable for young adult men and fall season, with buttons and five pockets. The corresponding DL expression w.r.t. the reference ontology (not reported here for the sake of conciseness) is:

- *Jacket* $\sqcap = 1$ *hasColors* $\sqcap \exists$ *hasFastenings* $\sqcap \forall$ *hasFastenings.Buttons* $\sqcap$
  $\forall$ *hasMainColor.DarkGreen* $\sqcap \exists$ *hasMaterials* $\sqcap \forall$ *hasMainMaterial.Linen* $\sqcap$
  $\forall$ *hasPattern.Plain* $\sqcap = 5$ *hasPockets* $\sqcap \forall$ *hasSize.Large* $\sqcap$
  $\forall$ *hasSleeves.LongSleeves* $\sqcap \forall$ *hasStyle.Elegant* $\sqcap$
  $\forall$ *suitableForAge.YoungAdult* $\sqcap \forall$ *suitableForGender.Male* $\sqcap$
  $\forall$ *suitableForSeason.Fall*

The equivalent DIG representation of this fairly simple annotation is 1592 B long. Realistic object descriptions can be a few times larger. Compression brings the annotation to 210 B, so that it can be stored on the RFID tag, along with the item EPC, ontology identifier and its data-oriented parameters. In this case price is $195 and body area value is 3.

The customer is trying on the jacket, as usual in a dressing room. Meanwhile, an embedded touchscreen is activated and item details are shown. Those elements will be used by the customer for building his semantic request. Let us note that the request is a DL conjunctive query, where each concept represents a desired feature. User can customize it through a graphical user interface.

Customer likes his jacket, so he would like to look for similar ones. He sets a target price of $200 and only removes the constraint on season from the system recommendation. The DL expression for user request thus becomes:

- **D:** *Jacket* $\sqcap = 1$ *hasColors* $\sqcap \exists$ *hasFastenings* $\sqcap$
  $\forall$ *hasFastenings.Buttons* $\sqcap \forall$ *hasMainColor.DarkGreen* $\sqcap \exists$ *hasMaterials* $\sqcap$
  $\forall$ *hasMainMaterial.Linen* $\sqcap \forall$ *hasPattern.Plain* $\sqcap = 5$ *hasPockets* $\sqcap$
  $\forall$ *hasSize.Large* $\sqcap \forall$ *hasSleeves.LongSleeves* $\sqcap \forall$ *hasStyle.Elegant* $\sqcap$
  $\forall$ *suitableForAge.YoungAdult* $\sqcap \forall$ *suitableForGender.Male*

Customer confirms his request. It is now translated into a compressed DIG description, in the same way as above, and associated with user-supplied target values for data-oriented attributes. Let us suppose the following products are available in the apparel store knowledge base:

- **S1:** an elegant, large-sized, gray suit, in mostly linen cloth, suitable for adult men and spring climate, with two button fastenings and ten pockets. Price is $678; body area is 3:
  Suit ⊓ ∀ *hasMainColor.LightGray* ⊓ ∀ *hasMainMaterial.Cotton* ⊓ = 10 *hasPockets* ⊓ ≥ 2 *hasFastenings* ⊓ ∀ *hasFastenings.Buttons* ⊓ ∀ *hasPattern.Plain* ⊓ ∀ *hasSize.Large* ⊓ = 2 *hasLegs* ⊓ ∀ *hasLegs.PipeLegs* ⊓ ∀ *suitableForAge.Adult* ⊓ ∀ *suitableForGender.Male* ⊓ ∀ *suitableForSeason.Spring* ⊓ = 2 *hasMaterials* ⊓ ∃ *hasColors* ⊓ ≤ 1 *hasColors*

- **S2:** medium-sized blue jeans, suitable for casual young adult men and spring climate, with pipe legs and five pockets. Price is $38; body area is 5:
  Trousers ⊓ = 1 *hasColors* ⊓ ∀ *hasMainColor.MediumBlue* ⊓ ≥ 2 *hasFastenings* ⊓ ≤ 2 *hasFastenings* ⊓ ∀ *hasLegs.PipeLegs* ⊓ ∀ *hasMainMaterial.Jeans* ⊓ = 1 *hasMaterials* ⊓ ∀ *hasPattern.Plain* ⊓ = 5 *hasPockets* ⊓ ∀ *hasSize.Medium* ⊓ ∀ *hasStyle.Casual* ⊓ ∀ *suitableForAge.YoungAdult* ⊓ ∀ *suitableForGender.Male* ⊓ ∀ *suitableForSeason.Spring*

- **S3:** an elegant, large-sized, midnight blue jacket, in mostly linen cloth, suitable for young adult men and fall climate, with buttons and five pockets. Price is $190; body area is 3:
  Jacket ⊓ = 1 *hasColors* ⊓ ∃ *hasFastenings* ⊓ ∀ *hasFastenings.Buttons* ⊓ ∀ *hasMainColor.MidnightBlue* ⊓ ∃ *hasMaterials* ⊓ ∀ *hasMainMaterial.Linen* ⊓ ∀ *hasPattern.Plain* ⊓ = 5 *hasPockets* ⊓ ∀ *hasSize.Large* ⊓ ∀ *hasSleeves.LongSleeves* ⊓ ∀ *hasStyle.Elegant* ⊓ ∀ *suitableForAge.YoungAdult* ⊓ ∀ *suitableForGender.Male* ⊓ ∀ *suitableForSeason.Fall*

- **S4:** an elegant, medium-sized striped lavender jacket, in mostly synthetic material, suitable for adult women and spring climate, with buttons and two pockets. Price is $194; body area is 3:
  Jacket ⊓ = 3 *hasColors* ⊓ ∃ *hasFacings* ⊓ = 1 *hasFastenings* ⊓ ∀ *hasFastenings.Buttons* ⊓ ∀ *hasMainColor.Lavender* ⊓ = 3 *hasMaterials* ⊓ ∀ *hasMainMaterial.Synthetic* ⊓ ∀ *hasPattern.Striped* ⊓ = 2 *hasPockets* ⊓ ∀ *hasSize.Medium* ⊓ ∀ *hasSleeves.LongSleeves* ⊓ ∀ *hasStyle.Elegant* ⊓ ∀ *suitableForAge.Adult* ⊓ ∀ *suitableForGender.Female* ⊓ ∀ *suitableForSeason.Spring*

The shop server performs the matchmaking. Results are reported in Table 6 where scores are shown in the last column; being a distance measure, a lower value implies a better match. *S3* is by far the best supply for similarity search. Among

Table 6 Matchmaking results

Supply	Compatibility (Y/N)	score(·)
S1: Men gray suit	Y	4.458
S2: Men blue jeans	N	4.618
S3: Men midnight blue jacket	Y	0.97
S4: Women lavender jacket	N	8.974

others, *S4* is incompatible with *D*, as they both represent jackets but *S4* is a women's garment.

Customer can reserve one or more items. Reservation will then be notified to the local server, so that products could be prepared in advance. *Otherwise, if customer is not satisfied with the results, he can refine his request and issue it again. Eventually he exits the dressing room to finalize his purchase.* Sensor detects the exit event and the dressing room becomes ready for another customer.

A thorough experimental evaluation of system performance is ongoing with our software-simulated RFID platform. Two kinds of experimental results can be reported so far:

1. compression rates for semantically annotated products descriptions and queries in DIG;
2. performance of reading and decoding compressed semantic resource annotations from simulated RFID tags.

Compression rate was tested with 70 DIG documents of various size (from 609 B to 793 kB). Our aim was to evaluate compression performance for both smaller instance descriptions and larger ontologies. Overall average compression rate is $92.58 \pm 3.58\%$. Higher compression rates were obviously achieved for larger documents, but even for DIG files shorter than 2 kB the result is $87.05 \pm 2.80\%$, which is surely satisfactory for our purposes.

Reading and decoding times referred to compressed semantic resource annotations from simulated RFID tags was evaluated to the aim of providing insight into the possible impact of our approach on RFID system performance. In preliminary tests using the above-mentioned RFID simulation platform, a read rate of nearly 500 tags/s was obtained, whereas independent sources estimated read rates ranging from 7 to approximately 100 tags/s with standard Class 1 Generation 2 UHF RFID systems in typical conditions (Kawakita and Mistvgi 2006). This is an early evidence that semantic-based RFID applications can have comparable performance with respect to traditional ones. The latter, in turn, will not suffer any direct performance degradation from the newly introduced features, as they will read the EPC only.

6.2 Off-line Data Analysis

Off-line batch data analysis can be performed at each node of the supply chain. Let us consider the above apparel scenario. A retail outlet can monitor sales performance by aggregating records according to attributes and performing queries in order to extract relevant information for its business goals.

Stay records stored in the outlet database contain timestamps of arrival and departure (i.e., sale) for sold items, along with EPC codes. They are aggregated by model and then by arrival time and departure time (like in the example in Sect. 5.2.1). The outlet management wants to check performance of sales of jackets in the current year. In particular, there is the need to compare sales figures among different suppliers, in order to see what jackets are most successful.

Let us suppose that the outlet is labeled by $L0$ in the supply chain location graph and that it has three suppliers $L1$, $L2$ and $L3$. In that case, for each supplier Lx, the number of sold jackets since January 1st 2010 can be retrieved through a path oriented aggregate query. It corresponds to the following expression in our adopted XPath-like language:

$//Lx/L0[t_{end} >'$ 2010-01-01 00:00:00$'$ and model='Jacket']

This information can be retrieved from the relational database described in the previous section by means of the following SQL query:

```
SELECT COUNT(I.EPC)
FROM PATHMAP P1, PATHMAP P2, OBSERVATION O, ITEM I,
ATTRIBUTE_LIST L, ATTRIBUTE A
WHERE P1.LOC_ID=MD5(Lx) AND P2.LOC_ID=MD5(L0)
AND P2.DEPTH=(P1.DEPTH+1) AND O.LOCATION=L0
AND O.TOKEN=P2.TOKEN AND I.STOCK=O.STOCK
AND I.T_END>'2010-01-01 00:00:00'
AND I.STOCK=L.OID AND L.ATTRIBUTE=A.OID
AND A.NAME='Jacket'
```

Note that COUNT aggregate operator has been used in order to evaluate sold item number. In a very similar way other interesting information can be extracted. For example, the management could be interested in knowing the average time trousers from each supplier stay in the outlet before a sale occurs, so that ordered quantities can be adjusted. It is important to point out that the selection of data aggregation attributes has a direct impact on applicable queries, therefore it must descend from KPI (Key Performance Indicators) decided by managers. Of course, the approach is general and can be applied by whatever actor in the supply chain to evaluate appropriate performance metrics with respect to its own business goals, by exploiting attribute-based data aggregation and a combination of tracking queries and path oriented queries.

An experimental campaign has been carried out to evaluate framework performance also from the massive data analysis standpoint. Here, we report main results. Performance of the proposed approach has been compared with the one provided by

Lee and Chung in (2008), from now on named *Path*. The authors devise a relational schema to store RFID data as in what follows:

```
PATH_TABLE(PATH_ID, ELEN, OEN)
TAG_TABLE(TAG_ID, PATH_ID, START, END, TYPE)
TIME_TABLE(START, END, LOC, START_TIME, END_TIME)
INFO_TABLE(TYPE, MODEL, BRAND, CATEGORY, PRICE)
```

In this schema, `TAG_TABLE` contains information related to the trace records representation. `PATH_TABLE` stores information about path movements, according to the ELEN and OEN coefficients (recalled in Sect. 4), and `TIME_TABLE` is related to temporal information (i.e., t_{in} and t_{out} for each location). Finally, `INFO_TABLE` describes product information (e.g., model, brand, category and price). The attribute taxonomy described in Sect. 4 has been exploited synthetic stay records have been generated by setting movement and time information randomly along each chain. For both chains 5 x 10^5 stay records were created. Then two aggregations were used: $F_1 = [(t_{in} \wedge t_{out}), (EPC)]$ and $F_2 = [(t_{in} \wedge t_{out}), (model)]$. Note that F_1 corresponds to a low aggregation whereas F_2 compresses records considerably. Using templates provided in Lee and Chung (2008), we formulate eight queries. They are shown in Table 7. $Q1$ is a tracking query, $Q2$, $Q3$, $Q4$ and $Q5$ are path oriented retrieval queries and finally $Q6$, $Q7$ and $Q8$ are path oriented aggregate queries. Figure 5 shows performance test results. First of all, it has to be noticed that times in $Chain_{20}$ are higher than in $Chain_{100}$. Object transitions are more widely spread in the latter, due to a greater number of nodes.

For the tracking query $Q1$, *Path* is faster than our approach with respect to F_1. However, in that system the query returns ELEN and OEN numbers, hence another application has to compute the factorization of ELEN ordering numbers with OEN.

Table 7 Test queries

Q1	7515
	17281
Q2	//L45
	//L628
Q3	/L01/L12/L22/L33/L45
	/L03/L16/L29/L321/L422/L628
Q4	//L12//L34
	//L16//L524
Q5	/L01/L12//L33/L45[t_start = 336501 AND t_end = 514515]
	/L03/L16//L321/L422/L628[t_start = 919250 AND t_end = 1023517]
Q6	COUNT():: //L45
	COUNT():: //L628
Q7	COUNT():: //L12//L34
	COUNT():: //L16//L524
Q8	COUNT():/L01/L12//L33/L45[t_start = 336501 AND t_end = 514515]
	COUNT():/L03/L16//L321/L422/L628[t_start = 919250 AND t_end = 1023517]

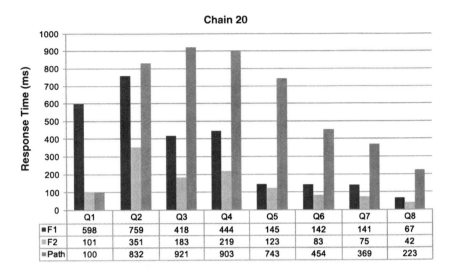

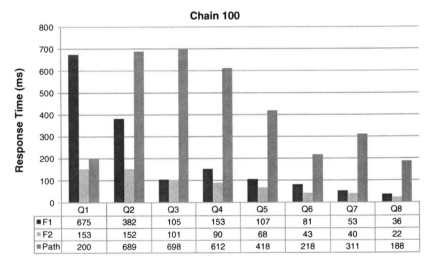

Fig. 5 Execution times

This is a high complexity task because the computation for large numbers is not efficient. Instead, the proposed approach directly returns the ordered list of traversed locations. Moreover performances are comparable if we use the aggregation F_2. This proves the usefulness of our aggregation mechanism.

For the path oriented retrieval queries $Q2$, $Q3$, $Q4$, $Q5$, our system exhibits better performance than *Path*. In all queries, *Path* has to execute a heavy join between TAG_TABLE, containing information of all items, and PATH_TABLE, containing ELEN and OEN numbers. Our system accesses OBSERVATION to extract all stocks following the input path, and then ITEM is accessed to return the EPC of corresponding items in those stocks. Furthermore, the partitioning technique supports the query process. Each OBSERVATION partition corresponds to a given location, therefore a direct access to interesting items is possible. Also in this case, the added value of aggregation is straightforward. With respect to F_1, the two approaches are comparable, whereas our system provides a better performance for the aggregation F_2. In $Q5$ time conditions lighten the processing of results. While *Path* has to introduce a new join with TIME_TABLE, containing timing information of transitions, our system adds a selection condition to WHERE clause; therefore also for the aggregation F_1 it has better performances. The path oriented aggregate queries $Q6$, $Q7$ and $Q8$ produce a similar behavior. In this case $Chain_{100}$ significantly exploits the partitioning technique, making aggregations F_1 and F_2 comparable.

7 Conclusions

This chapter presented an innovative approach for data management in supply chains based on RFID identification technology. Both *on-line* semantic-based object discovery and *off-line* analyses involving large amounts of RFID data are enabled. Distinguishing features are: (i) definite modifications to the EPCglobal standards allowing to exploit ontology-based data as well as to support non standard inference services, while keeping backward compatibility, (ii) advanced compression techniques enabling a significant space saving also maintaining a logical representation of data aggregation.

Such an approach may provide several benefits. Information about a product is structured and complete; it accurately follows the product history within the supply chain, being progressively built or updated during object lifecycle. This improves traceability of production and distribution, facilitates sales and post-sale services thanks to an advanced and selective discovery infrastructure. Indexing techniques that guarantees an efficient data access have been also proposed in a tool implementing the proposed approach.

Some experimental results are presented to show the feasibility of the proposed framework also evidencing its effectiveness. The coherent development of the approach allows a strengthening of the information to be shared between the actors involved in supply chains, reducing the costs of adoption of RFID in business. Furthermore, an increase in transparency and trust is achieved not only between supply chain partners, but also between retailers and customers. This may be a direct competitive advantage for companies that adopt the technology.

References

Angeles R (2005) RFID technologies: supply-chain applications and implementation issues. In Journal of Information System Management, **22**(1): 51–65

Ayoub Khan MBR, Manoj S (2009) A Survey of RFID Tags. In International Journal of Recent Trends in Engineering, **1**(4): 68–71

Baader F, Calvanese D, Mc Guinness D, Nardi D, Patel-Schneider P (2002) The description logic handbook. Cambridge University Press, Cambridge

Baader F, Horrocks I, Sattler U (2003) Description logics as ontology languages for the semantic web. Festschrift in honor of Jorg Siekmann, Lect Notes Artif Intell. Springer Heidelberg

Bai Y, Wang F, Liu P (2006) Efficiently filtering RFID data streams. CleanDB Workshop, 50–57, Seoul, Korea

Bai Y, Wang F, Liu P, Zaniolo C, Liu S (2007) RFID data processing with a data stream query language. Proceedings of the 23nd international conference on data engineering, ICDE 1184–1193, Istanbul, Turkey

Ban C, Hong B, Kim D (2005) Time parameterized interval r-tree for tracing tags in rfid systems. Database and Expert Systems Applications, 16th International Conference, DEXA 2005, Copenhagen, Denmark, 22-26, August 2005, Proceedings, vol. 3588 of Lecture Notes in Computer Science, pp. 503–513, Springer

Bechhofer S, Möller, R, Crowther P (2003) The DIG description logic interface. Proceedings of the 16th International workshop on description logics (DL'03), Rome, Italy, September, vol. 81 of CEUR Workshop Proceedings

Beckmann JL, Halverson A, Krishnamurthy R, Naughton JF (2006) Extending rdbmss to support sparse datasets using an interpreted attribute storage format. Proceedings of the 22nd international conference on data engineering, ICDE 2006, 3–8 April 2006, Atlanta, GA, USA, p. 58, IEEE Computer Society

Berners-Lee T, Hendler J, Lassila O (2001) The semantic Web. Sci Am **284**: 28–37

Borgida A Description logics in data management. IEEE Trans Knowledge Data Eng **7**: 671–682

Brachman R, Levesque, H (1984) The tractability of Subsumption in Frame-based description languages. 4th National Conference on Artificial Intelligence (AAAI-84), Morgan Kaufmann, Massachusetts, 34–37

Chamberlain J, Blanchard C, Burlingame S, Forestier SC, Griffith G, Mazzara M, Musti S, S.-Y, Son, Stump G, Weiss C (2006) IBM WebSphere RFID Handbook: A Solution Guide. IBM International Technical Support Organization

Colucci S, Di Noia T, Di Sciascio E, Donini F, Mongiello M (2003) Concept abduction and contraction in description logics. 16th International Workshop on Description Logics (DL'03), September, vol. 81 of CEUR Workshop Proceedings, Rome, Italy

De P, Basu K, Das S (2004) An ubiquitous architectural framework and protocol for object tracking using RFID tags. 1st International Conference on Mobile and Ubiquitous Systems:Networking and Services (MOBIQUITOUS 2004), 174–182, Boston, Massachusetts, USA

De Virgilio R, Sugamiele P, Torlone R (2009) Incremental aggregation of RFID data. Proceedings of the 2009 International Database Engineering & Applications Symposium, 194–205, ACM, Cetraro, Italy

Di Noia T, Di Sciascio E, Donini F, Mongiello M (2003) Abductive matchmaking using description logics. IJCAI '03, 337–342

Di Noia T, Di Sciascio E, Donini F, Mongiello M (2004) A system for principled matchmaking in an electronic marketplace. Int J Electron Commer, **8**: 9–37

Di Noia T, Di Sciascio E, Donini F, Ruta M, Scioscia F, Tinelli E (2008) Semantic-based Bluetooth-RFID interaction for advanced resource discovery in pervasive contexts. Int J Semant Web Inf Syst (IJSWIS), **4**: 50–74

EPCglobal Ratified Specification (October, 4, 2005) Object Naming Service (ONS - ver. 1.0). http://www.epcglobalinc.org

Gärdenfors P (1988) Knowledge in flux: modeling the dynamics of epistemic States. Bradford Books, MIT Press, Cambridge

Gonzalez H, Han J, Li X, Klabjan D (2006a) Warehousing and analyzing massive RFID data sets. Proceedings of the 22nd International Conference on Data Engineering, IEEE Computer Society Washington, DC P83

Gonzalez H, Han J, Li X, Klabjan D (2006) Warehousing and analyzing massive rfid data sets. Proceedings of the 22nd International Conference on Data Engineering, ICDE 2006, 3–8 April 2006, Atlanta, GA p. 83, IEEE Computer Society

Jeffery S, Garofalakis M, Franklin M (2006) Adaptive cleaning for RFID data streams. Proceedings of the 32nd international conference on Very large data bases, 163–174, VLDB Endowment, Seoul, Korea

Jones P, Clarke-Hill C, Shears P, Comfort D, Hillier D (2004) Radio frequency identification in the UK: opportunities and challenges. Int J Retail & Distribution Manag, 32: 164–171

Karkkainen M (2003) Increasing efficiency in the supply chain for short shelf life goods using RFID tagging. Int J Retail Distribution Management 31: 529–536

Kawakita Y, Mistugi J (2006) Anti-collision performance of Gen2 air protocol in random error communication link. Proceedings of the International Symposium on Applications and the Internet Workshops - SAINT 2006, 68–71, Phoenix, Arizona, USA

Lee C H, Chung C-W. (2008) Efficient storage scheme and query processing for supply chain management using rfid. Proceedings of the ACM SIGMOD international conference on management of Data, SIGMOD 2008, Vancouver, BC, Canada, 291–302 June 10–12, ACM

Li L, Horrocks I (2004) A software framework for matchmaking based on semantic web technology. Int J Elect Com 8: 39–60

Maass W, Filler A (2006) Towards an infrastructure for semantically annotated physical products. Informatik, 94: 544–549

Marcus DAA, Madden S, Hollenbach K (2007) Scalable semantic web data management using vertical partitioning. VLDB '07: Proceedings of the 33rd international conference on very large data bases, 411–422, VLDB Endowment, Vienna, Austria

Paolucci M, Kawamura T, Payne T, Sycara K (2002) Semantic matching of web services capabilities. The Semantic Web - ISWC 2002, pp. 333–347, no. 2342, Sardinia, Italy

Pepe MFC, Risso, M (2008) SME food suppliers versus large retailers: perspectives in the international supply chains. 15th Conference of European Institute of Retailing and Services Studies (EIRASS), 411–422, Zagreb, Croatia

Römer K, Schoch T, Mattern F, Dübendorfer T (2004) Smart identification frameworks for ubiquitous computing applications. Wire Networks, 10: 689–700

Raza N, Bradshaw V, Hague M, (1999) Applications of RFID technology. IEE Colloquium on RFID Technology, 1–5

Ruta M, Di Noia T, Di Sciascio E, Scioscia, F, Piscitelli G (2007) If objects could talk: A novel resource discovery approach for pervasive environments. Int J Internet Protoc Technol (IJIPT) 2: 199–217

Ruta M, Di Noia T, Di Sciascio E, Piscitelli G, Scioscia F. (2008) Semantic-based mobile registry for dynamic RFID-based logistics support. 10th International Conference on Electronic Commerce, ICEC 08, ACM Press, Pittsburgh, USA

Ruta M, Di Noia T, Di Sciascio E, Donini F (2006) Semantic-enhanced bluetooth discovery protocol for M-commerce applications. Int J Web Grid Serv 2: 424–452

Sellitto C, Burgess S, Hawking P (2007) Information quality attributes associated with RFID-derived benefits in the retail supply chain. Int J Retail & Distrib Manag, 35: 69–87

Smith A (2005) Exploring radio frequency identification technology and its impact on business systems. Inf Manag Compu Secur 13: 16–28

Tajima M (2007) Strategic value of RFID in supply chain management. J Purch Manag 13: 261–273

Traub K, et al. (2005) EPCglobal Architecture Framework. Tech. rep., EPCglobal

Wang F, Liu P (2005) Temporal management of RFID data. Proceedings of the 31st international conference on very large data bases, 1128–1139, VLDB Endowment, Trondheim, Norway

Wang F, Liu S, Liu P, Bai Y, (2006) Bridging physical and virtual worlds: complex event processing for RFID data streams. Lect Notes Comput Sci, 3896: 588, EDBT, Munich, Germany

Weiser M (1999) The computer for the 21st century. SIGMOBILE Mob. Comput Commun Rev 3: 3–11

RFID Data Cleaning for Shop Floor Applications

Holger Ziekow, Lenka Ivantysynova, and Oliver Günter

Abstract In several case studies we found that shop-floor applications in manufacturing pose special challenges to cleaning RFID data. The underlying problem in many scenarios is the uncertainty about the exact location of observed RFID tags. Simple filters provided in common middleware solutions do not cope well with these challenges. Therefore we have developed an approach based on maximum-likelihood estimation to infer a tag's location within the reader range. This enables improved RFID data cleaning in a number of application scenarios. We stress the benefits of our approach along exemplary application scenarios that we found in manufacturing. In simulations and experiments with real world data we show that our approach outperforms existing solutions. Our approach can extend RFID middleware or reader firmware, to improve the use of RFID in a range of shop-floor applications.

1 Introduction

Over the past few years manufacturers have increasingly shown interest in using RFID on their shop floors. We conducted several case studies in the manufacturing domain and identified numerous RFID applications on the shop floors of manufacturing plants (Ivantysynova et al. 2008). We found that many of these shop-floor applications pose special challenges for cleaning captured RFID data. Characteristics of radio-frequency signals lead to an intrinsic uncertainty in RFID read events (i.e., data captured by RFID readers). Consequently RFID applications must deal with this uncertainty.

Generally, a reader can capture every RFID tag in the signal field. Thus, a read event only indicates the presence of a tag within the reader's range. Yet, the signal field is not a sharply defined region. Readers may capture some tags only with close

H. Ziekow (✉)
Institute of Information Systems, Humboldt-Universität zu Berlin, Berlin, Germany
e-mail: ziekow@wiwi.hu-berlin.de

D.C. Ranasinghe et al. (eds.), *Unique Radio Innovation for the 21st Century*,
DOI 10.1007/978-3-642-03462-6_7, © Springer-Verlag Berlin Heidelberg 2010

proximity and intercept others from a great distance. Various factors like humidity or the tag's antenna alignment influence the success rate of tag reads. Read events may get lost because of collisions at the air interface level or shielding effects. Also, RFID readers may accidentally capture tags nearby which are not supposed to be registered. This is in particular the case for tags with high read ranges, like UHF tags. Consequently, reading RFID tags is a probabilistic process. This poses challenges to applications which require accurate data on a tag's proximity to the reader.

Existing middleware solutions typically provide low pass filters to suppress flickering and to smooth the read signal. This approach is sufficient for accurately detecting the presence of tags, e.g., in a shelf or to notice when a tag passes trough a gate. Yet, the effectiveness of such simple filters is limited when precise information about a tag's prosition is required. Simple filters do not allow reasoning about a tags location within the read range, which can lead to faulty detections. One example is a stack reader of a production robot that must identify the topmost item in the stack (Brusey et al. 2003). Another example is determining the order of items on a conveor that pass by a reader. Here, the problem is that items may not be detected in the same order in which they pass by.

In this paper we present a probabilistic data cleaning approach that considers basic properties of radio-frequency signals. This approach improves the accuracy in detecting the proximity of tags and thereby improves RFID-data capture in a number of application scenarios. Our solution outperforms existing alternatives in scenarios that require positioning of moving tags in the reader field. We are furthermore able to predict confidence intervals for the accuracy of tag detection. With this approach we improve data cleaning in a range of RFID applications, particularly in the manufacturing domain.

The remainder of the chapter is structured as follows: In Sect. 2 we provide details on the problem background and targeted application scenarios. In Sect. 3 we review related work. In Sect. 4 we present our probabilistic solution for cleaning RFID data. In Sect. 5 we report on experimental results and conclude in Sect. 6.

2 Application Scenarios and Problem Background

Due to the noisy nature of raw RFID reads it is necessary to clean the data before using them in higher level applications (e.g. applications for tracking items or warehouse management). This usually involves suppression of flickering and elimination of double reads. Subsequently, one may also be able to use context knowledge to filter out false positives (i.e., events from tags that did not purposely enter the reader field). During our case studies in manufacturing we found several application scenarios that challenged existing data-cleaning approaches. Figure 1 provides an overview of selected scenarios. These scenarios illustrate basic principles that occur in variations of different manufacturing processes.

In the first scenario, a reader monitors items on a conveyor (Fig. 1 left). Here, RFID is used to record the items that pass-by. For example conveyors are used to

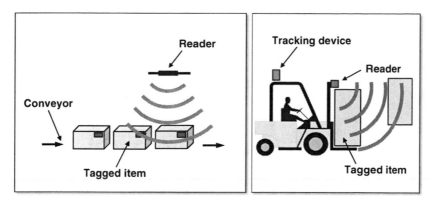

Fig. 1 Scenarios for RFID- data cleaning in manufacturing

load machines or to fill items into transportation units. The data-cleaning challenge here is that the order of RFID reads may not match the order of the items on the conveyor. That is, the reader may detect an item at greater distance before a closer one. This can result in erroneous process documentation and a mix-up of items in later processing.

In the second scenario, RFID is used to track movements of items in an asset tracking scenario (Fig. 1 right). A forklift tracks its own position, for example, via WLAN-positioning. It further uses a mobile RFID reader to detect items in its vicinity. Here the challenge for data cleaning is to infer from the RFID reads where items are picked or dropped. This supports accurate tracking of asset positions and can reduce search times. Another scenario that can benefit from our method is picking items from the top of a stack as described in in Brusey et al. (2003).

The underlying challenge in all the described scenarios is that raw RFID reads are noisy. Particularly, the read reliability decreases with the read distance. This is because the radio-frequency signal strength decreases with the square of the distance. The success rate of RFID read attempts is therefore a function of the distance. In proximity of the reader – where the signal strength is sufficient – the success rate is usually close to 100%. From this point on, the reliability drops with the signal strength. Thus the resulting curve shape of read reliability over read distance has typically a constant part followed by a hyperbolic drop (see Fig. 3 for an illustration). Some deriviations from the described curve shape may occur due to signal reflections and overlaps in the application environment. Note that our solution does not rely on a particular curve shape. Instead, we use the reliability function as an input to our algorithm.

Figure 2 depicts the reliability function acquired in tests with a Nordic ID PL3000 (UHF) handheld reader. The measurements show the typical curve shape with some distortions at about 70 cm distance, presumably due to some close by metal frame in our laboratoy setting. All tests were conducted with the same tag. This clearly shows the noisy nature of RFID reads.

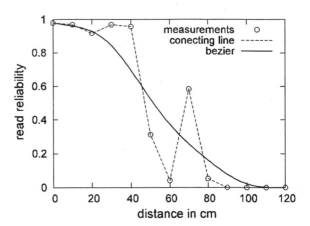

Fig. 2 Measured read reliability over distance

A technique to reliably infer when a tag is in a certain position relative to the reader could solve the problems in each of the discussed scenarios. In the first scenario one could correctly order the items if one knows when the tags passed a common reference point (e.g., right under the reader). In the second scenario one could accurately determine when items are picked up or dropped. This would be the case if you could infer when an item is right next to the reader (i.e., on the forklift). In the application described by Brusey et al. (2003), proximity to the reader allows inferring which item is on top of a stack.

Thus, in all these scenarios you need to detect this specific read event when the tag is physically closest to the reader. This event needs to be identified from the input stream of RFID read events. We subsequently present a method to infer such an event of interest and show the advantages compared to existing approaches in experiments.

3 Related Work

The noise and errors in RFID data pose various challenges to RFID-based applications. Thus, data cleaning is a well recognized requirement for RFID middleware (Floerkemeier and Lampe 2004). Existing approaches to RFID-data cleaning mainly target the filtering of false positives and negatives. False positives occur when readers can capture distant items outside the intended scope. False negatives occur if the reader fails to read out a tag in its scope. This can be because of collisions on the air interface or because of shielding effects (Finkenzeller 2003). Basic filters for reducing such errors are included in existing RFID middleware solutions (Bornhövd et al. 2004; Wang and Liu 2005).

A commonly used solution for supressing errors involve using low pass filters, also called smoothing filters. Such a filter can be realized by counting read events in a window over the read sequence and comparing them with a threshold. Unlike our solution such filters cannot assess the proximity of an item to the reader. Thus, detections may not preserve the oder of the tagged items.

Jeffery et al. (2006a) smooth RFID data in temporal sliding windows. They also group several readers in spatial granules to correct for missed readings and remove outliers. Additionally they provide support for specifying filter predicates. Jeffery et al. (2006b) use sampling estimators to automatically adapt the window size of their filters. Brusey et al. (2003) provide a solution with a weighted low-pass filter to order items on a stack. However, their solution is tailored to a specific application setup and is not directly applicable to all scenarios that we target. Bai et al. (2006) employ smoothing filters to eliminate noise and duplicates. They also provide means to preserve the order of detections in concurrent filter windows. Another set of approaches addresses the redundant reader problem. These approaches aim to associate a tag with the correct reader in scenarios with overlapping signal fields. For example Pupunwiwat and Stantic (2007) use set theory to associate tags in shelves with the correct reader. However, such approaches are designed for relatively static environments and are not applicable to our targeted problem of monitoring running production processes.

No cleaning approach known to us can infer the position of a moving tag within a reader's signal field. That is, only presence of tags in the field is detected but the position within this field is unknown. However, there are some approaches from the domain of asset tracking that infer tag position within the reader range. Bekkali et al. (2007) use multiple readers and the signal strength of tag responses to determine tag positions. Yet not every RFID reader interface provides measurements of the received signal strength (that is, it is not part of standard protocols) and the setup with multiple readers is more costly but at least not hard to apply for our second scenario. Other approaches use mobile devices, for example, devices mounted on robots, to scan RFID tags in an area, e.g., Song et al. (2007) and Hähnel et al. (2004). Such approaches derive accurate positions of the RFID tags from multiple reads and the position of the reader. However, such approaches are designed for relatively static environments (e.g., a warehouse) and are not applicable to monitoring processes in real time.

In contrast to existing solutions, our approach is able to infer the distance at which an observed item passes a reader. This enables improvements for applications that need highly accurate registration of objects at certain locations (e.g., to detect the order of items that pass a reader). Unlike existing approaches we are furthermore able to estimate confidence values for the accuracy of our filter.

4 The Cleaning Approach

In this section we present our approach to cleaning RFID data. We first present the basic idea. We then explain the concrete application of the approach along a simple and a more complex scenario. Finally we show how to calculate approximate confidence intervals for the accuracy of each detection.

4.1 Basic Idea

Our data cleaning approach uses maximum likelihood estimation to detect an event of interest for a certain RFID tag (e.g., when is the tag right at the reader?). Simplified, the idea is as follows: We first split the read stream into substreams such that each event in a substream R corresponds to the same RFID tag. We consider each read event e from the stream of raw read events R as a candidate for the event of interest.

For our approach we assume an event model in which each event e is defined by a tuple (r, l, t). Here, $r \in \{0, 1\}$ denotes the read success;1 being a success. The value l records relative physical movements of the reader. In the conveyor scenario this could be a record of how much the conveyor has moved so far. We subsequently refer to the value l as the location of event e. The value t is a timestamp. The cleaning algorithm then calculates the probability of seeing the event stream R under the assumption that a candidate event e is the event of interest. It then returns the event $e \in R$ where the calculated probability is maximized. Formally we need to calculate the maximum-likelihood estimation given in (1).

$$e_{\text{interest}} = \arg \max_e L(e) = \arg \max_e f(\{e_1, ..., e_n\}|\, e) \tag{1}$$

Here, f returns the probability of observing the read stream $R = \{e_1, \ldots, e_n\}$ under the assumption that e is the event of interest. The calculation of f depends on the particular process. We will present examples in the following two sections. In any case the calculation follows the same principle.

Under the assumption that all events in R are statistically independent[1] we get:

$$f(R|e) = \prod_{e_i \in R} f(\{\, e_i\} \,|\, e) \tag{2}$$

Calculating $f(e_i\,|e)$ requires knowledge of the reliability function f_{rel} for RFID reads which are dependent on the relative tag-reader position. This function depends on the used RFID hardware and should be measured before configuring the data-cleaning algorithm. It is likely that f_{rel} varies in each installation and therefore in each installation a configuration is necessary. Note that our experiments in Sect. 5 show that using rough approximations for this function still yield good results. Thus guessing the function can be an alternative if measurements are infeasible.

Each installation can result in different reliability functions, including asymetric signal spreads. For simplicity we subsequently assume a symmetric signal spread. The function f_{rel} then only depends on the distance between tag and reader. However, the real tag-reader distance for events in the stream is not known and must be inferred by the maximum likelihood estimation. The raw stream only contains

---

[1]In many cases, this assumption may not hold. Our experiments, however, show that the statistical dependencies have a limited influence and that the resulting error is negligible.

abstract coordinates for the relative reader position. For each event e as a parameter in the likelihood function, L, we can calculate the tag-reader distance that follows for the events in the stream. That is, from the assumption that e is the event of interest follows an absolute position form e and all other events in R accordingly. We use the notation $D(e', e)$ to denote this calculation of the location for an event $e' \in R$ under the assumption that $e = e_{\text{interest}}$.

For illustration consider the events $e_1 = (1, 10, 5)$ and $e_2 = (1, -20, 6)$. Assume that e_2 was captured directly at the reader position – the position of interest. Then the tag reader distance for e_1 must have been 30 length units.

The input stream R includes events that were successfully read and events where the read failed. Thus $f(e_i | e)$ equals $f_{\text{rel}}(e_i)$ for successful reads and the complementary probability $1 - f_{\text{rel}}(e_i)$ for failed attempts. For the maximum likelihood function this yields the following :

$$L(e) = f(R|e)$$
$$= \left(\prod_{(r_i,l_i,t_i)\in R:r_i=1} f_{\text{rel}}(D((r_i, l_i, t_i), e)) \right) \left(\prod_{(r_i,l_i,t_i)\in R:r_i=0} 1 - f_{\text{rel}}(D((r_i, l_i, t_i), e)) \right)$$
(3)

We show several realisations of the generic approach along sample scenarios in Sects. 4.2 and 4.3. However, to solve (3) we must consider that sequence R is potentially of infinite length. We therefore define a window over the stream R to create a limited sequence R'. We choose the window such that R' contains the event of interest. Furthermore it must be large enough to calculate (1) with sufficient accuracy.

In our tests we chose the window as follows.

- As the lower bound we use the first successful read event.
- To determine the upper bound we use the relative tag-reader movement. Starting from the lower window bound we know that the tag is out of range once the movement is more than twice the maximal reader distance. The last event that is possibly in range marks the upper window bound.

Note that there may be different approaches to windowing streams. However, we got good results using the described approach in our experiments.

It remains to find the event e that maximizes (2). This is trivial since we compute the equation with R' instead of R. In the limited window R' we can compute (2) for every event in R' and select the maximum.

4.2 Realizing a Simple Scenario

In this section we show the realisation of the above approach along a simple but relevant scenario. In the scenario, RFID equipped items move on a conveyor (see scenario two in Sect. 2). An RFID reader that is installed over the conveyor monitors the movement with periodic read requests (see Fig. 1, left). The data-cleaning task is to infer the order of items on the conveyor. We reduce this problem to detecting the RFID read event during which the corresponding item was directly under the reader.

For determining the window R' from the stream R we use the first observation of the corresponding item as a lower bound. The upper bound is when the conveyor has moved twice the maximal read distance from this point on. This results in a fixed number of events in the window since the conveyor speed and the frequency of read attempts is constant.

The function f_{rel} depends on the reader installation. In the sample scenario we assume that the reader signal spreads symmetrically along the conveyor. (This signal spread is likely for the chosen reader installation.) Thus we can model f_{rel} as a function that solely depends on the distance from the reader. (Measurement in the application setting would determine the concrete shape of the function.)

It remains to determine the function $D(e_i, e)$ for the given scenario to fully define (3). The function $D((r_i, \ l_i, t_i), (r, \ l, \ t))$ is $|l_i - l|$ because items on the conveyor move only in one dimension and the event of interest should be directly under the reader. This yields the following formula of the likelihood function:

$$L((r, l, t)) = f(R'|(r, l, t))$$
$$= \left(\prod_{(r_i, l_i, t_i) \in R': r_i = 1} f_{rel}(|l_i - l|) \right) \left(\prod_{(r_i, l_i, t_i) \in R': r_i = 0} 1 - f_{rel}(|l_i - l|) \right) \quad (4)$$

4.3 Realizing a Complex Scenario

In a more complex scenario a mobile RFID reader is used for asset tracking (see scenario two in Sect. 2). Therefore, a mobile reader is mounted to a transporter (see Fig. 1 right). The data-cleaning task is to detect the exact positions where items are picked up or dropped off. We limit the focus of the subsequent discussion to detecting picking up of items. The case for dropping items off is analogous.

For determining the window R' from the stream R we again use the first observation of the corresponding item as the lower bound. However, the upper bound is not at a fixed offset from this event. Instead, we extend the window until we know that the items must have been picked up. This is at the first successful read after the forklift has moved twice the maximum read range from the first read[2]. Since this is only possible if the item was picked up in between, we know that the event of interest is in the window.

---

[2]After a timeout, we drop streams where no successful read occurs after the forklift has moved twice the maximum read range from first read. These are streams for items where the forklift passed by without picking up the item.

For f_{rel} we again assume a radial spread of the reader signal. Thus we can model f_{rel} again as function that solely depends on the distance from the reader (For other signal shapes one needs a multi dimensional version of f_{rel}.).

It remains to determine the function $D(e_i, e)$ for the given scenario. Herein lies the main difference from the previous example. We must consider that the item is picked up at a certain point and remains right next to the reader afterwards. We must therefore distinguish between events e_i that are captured before e and those that are captured afterwards. This yields the likelihood function given by (5).

$$L((r, l, t) = f(e_1, ..., e_n | (r, l, t)) = f_1 f_2 f_3 f_4 \tag{5}$$

where f_1, f_2, f_3 and f_4 are given as below.

$$f_1 = \left(\prod_{(r_i, l_i, t_i) \in R' : r=1 \wedge t_i < t} f_{rel}(|l_i - l|) \right)$$

$$f_2 = \left(\prod_{(r_i, l_i, t_i) \in R' : r=0 \wedge t_i < t} 1 - f_{rel}(|l_i - l|) \right)$$

$$f_3 = \left(\prod_{(r_i, l_i, t_i) \in R' : r=1 \wedge t_i >= t} f_{rel}(0) \right)$$

$$f_4 = \left(\prod_{(r_i, l_i, t_i) \in R' : r=0 \wedge t_i >= t} 1 - f_{rel}(0) \right)$$

4.4 Calculating Confidence Intervals

In this section we show how to compute confidence intervals for each detection of $e_{interest}$ – the event of interest. Specifically we provide an algorithm that returns a spatial error *dist* given a confidence value *confVal* where, the correct event of interest is not more than *dist* length units away from the detected event $e_{interest}$ with the probability *confVal*.

During the calculation of (3) we determined a probability for each event $e_i \in R'$. This value is the probability of observing the sequence R' given that e_i is the event of interest ($e_{interest}$). The basic idea is to use these values for deriving a confidence interval for each detection of $e_{interest}$. Since we know that $e_{interest}$ is in R' we can normalize the probability values for all $e_i \in R'$ such that their sum equals 1. This yields an approximate density function for which e is $e_{interest}$ given the observed R' (note that this solution is approximate because (2) has a bias for limited samples). We can then integrate this function over an spatial interval around the location of the detected $e_{interest}$. That is, the summed up normalized probability values in this interval yield the confidence that the correct event of interest is in this interval. This approach can be implemented as follows:

In the first step we use R' to build a set H of events with corresponding normalized probability values:

$$H = \{(r,l,t,p)|(r,l,t) \in R' \wedge p = \frac{f((r,l,t)|\mathrm{R})'}{\displaystyle\sum_{(r,l,t)\in R'} f((r,l,t)|\mathrm{R})'} \} \tag{6}$$

We can then obtain the confidence estimation using Algorithm 1 by evaluting H based on detected event (r, l, t, p) using (6).

Algorithm 1 *confCalc(confVal, H, (t, r, l, p))*

1. $\mathrm{H_{SORTED}}$ = sortByDistanceToEvent(H, (t, r, l, p))
2. probabilitySum = 0
3. distance = 0
4. **for** i = 0 to $|\mathrm{H_{SORTED}}|$ **do**
5.   **if** probabilitySum $\geq$ confVal **then**
6.   **break**
7.   **end if**
8. $(t_i, r_i, l_i, p_i) = \mathrm{H_{SORTED}}[i]$
9.   probabilitySum = probabilitySum + p_i
10.   distance = $|l_i - l|$
11. **end for**
12. **return** distance

The confidence estimation can be used by downstream applications for reasoning about information from our filter. For instance, models for probabilistic data management may use the confidence for storing the inferred location data accordingly (e.g., Garofalakis et al. 2006). Furthermore, in critical situations the application may ask the worker for manual verification if the corresponding event was detected with too low confidence.

5 Experiments

In this section we present experimental results for the evaluation of our data-cleaning approach. We compared our solution against existing filters and scenario-specific trivial solutions. We further evaluated the impact of read frequency on the accuracy of data cleaning and predicted confidence values.

For the evaluation we used simulations of the sample scenarios that we described in Sect. 2. Using simulations allowed us to test the approach with a large number of

Fig. 3 Reliability function in the simulation and the approximation used for the probabilistic operator

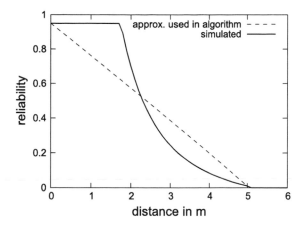

test runs in various settings. In addtion, to stress that our solution also works with real data we conducted an experiment using real RFID hardware.

In the simulations we reconstructed the analyzed processes with virtual items and readers. We generated read events randomly using a distance-dependent reliability function to model the likelihood of a successful read attempt. Figure 3 depicts the functions used and an approximation of the reliability function. In testing our algorithm we intentionally used a very rough reliability function approximation for our calculations. This is to account for the fact that the reliability function may not always be known accurately. Using the very rough approximation we evaluated our approach with a worst case assumption.

5.1 Evaluation with a Simple Scenario

In scenario one, RFID is used to monitor the order of items on a conveyor. To evaluate this scenario we simulated a conveyor that moves with walking speed (0.83 m/s). We simulated a reader that is installed directly over the conveyor. For the simulated read reliability we used the function depicted in Fig. 3. Here, the distance refers to positions on the conveyor, i.e. zero refers to the position closest to the reader.Note that we assumed a symmetric read signal.

Our algorithm detects when an item is exactly under the reader. With this information one can deduct the order of items on the conveyor. In our experiments we analyzed the following.

1. We compared the spatial error in detection with low-pass filters and trivial solutions.
2. We compared how many items were detected in the wrong order – that is, how many items were swapped.
3. Finally, we tested the accuracy of our confidence prediction for the spatial errors.

Table 1 Error variances for different cleaning approaches

Max. L. Opt. 14 cm	Max. L. Approx. 20 cm	Window Middle 64 cm
Reads Middle: 42 cm	Low Pass 2: 35 cm	Low Pass 3: 28 cm
Low Pass 4: 33 cm	Low Pass 5: 40 cm	Low Pass 6: 45 cm

Table 2 Average spatial error for Different Cleaning Approaches

Max. L. Opt. 29 cm	Max. L. Approx. 35 cm	Window Middle 66 cm
Reads Middle: 53 cm	Low Pass 2: 44 cm	Low Pass 3: 39 cm
Low Pass 4: 42 cm	Low Pass 5: 46 cm	Low Pass 6: 48 cm

Table 1 shows the variance of spatial errors for the tested approaches. The values in the table are determined using 1000 test runs and a simulated read frequency of two read attempts per second. Table 2 lists the average spatial error for the same experiment (note that we shifted each detected event by the median error to align the tests).

In both tables *Max. L. Opt.* and *Max. L. Approx.* refer to our optimization approach. The implementation for *Max. L. Opt.* used the exact reliability function in the algorithm. For the implementation of *Max. L. Approx.* we used the rough linear approximation. *Window Middle* and *Reads Middle* are trivial cleaning implementations and can be used as a baseline. *Window Middle* simply returns the event in the middle of the detection window R'. *Reads Middle* returns the event in the middle of the first and the last detection of an observed item. *Low Pass 2–6* are low-pass filters commonly used for RFID- data cleaning. They apply a tuple based sliding window over the stream and return an event once the whole window is filled with successful read attempts. The numbers 2–6 denote the window size in tuples.

The numbers in the tables already indicate the superiority of our cleaning approach over existing approaches. However, the number that really matters in the targeted scenario is how many items on the conveyor are swapped due to false detections. Figure 4 displays the number of swapped items for two test setups. In the first setup, the items were spaced one meter apart; in the second, 1.5 m apart.

The results show that our approach clearly outperforms the alternatives. Furthermore, they show that our approach copes well with errors in the assumed reliability function. That is, we achieved good results even for the implementation with the rough linear approximation. With the exact function only 3 out of 1000 items were swapped when they were 1.5 m apart and 31 when they were 1 m apart. Using the approximation 8 items where swapped with 1.5 m apart; 55 were swapped with 1 m apart.

In both cases the best performing alternative was a low-pass filter with a window size of three. This filter resulted in 25 swaps for a distance of 1.5 m and 87 swaps for a distance of 1 m.

In further tests we analyzed the accuracy of our predictions of confidence intervals. In the experiments we calculated a 95% confidence interval for each run. We then counted how many detections were outside this interval. Table 3 lists the results for 1000 test runs with different read frequencies. The numbers obtained show that the confidence prediction overstates the confidence particularly for the low read frequency of 1 Hz. However, the predicted values are close to the correct results. To our knowledge no other approach predicts confidence values for the detection. Thus a comparison of alternatives was not possible.

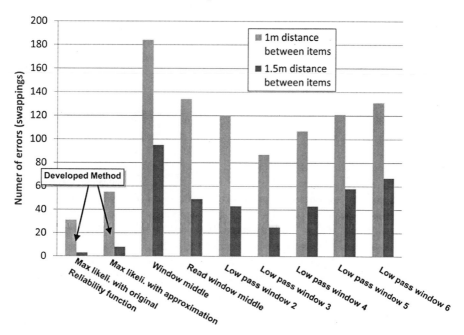

Fig. 4 Impact of the read frequency on swapped items in the conveyor simulation

Table 3 Average spatial error for different cleaning approaches

Read frequency	Events outside the confidence interval
1 Hz	12.7%
2 Hz	6.9%
4 Hz	6.7%
6 Hz	7.16%
8 Hz	7%

In additional test runs we analysed the impact of the read frequency on the accuracy of the cleaning approach. We compared our approach with the best performing low-pass filter for different read frequencies. Figure 5 displays the results. In the figure, *Max. L. Opt* refers to our approach using the exact reliability function for f_{rel} in the implementation. *Max. L. Approx.* refers to the implementation of our approach that uses the linear approximation of the reliability function for f_{rel}. *LowPass3* refers to the best performing alternative approach, which was a low-pass filter with a

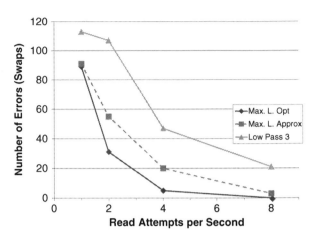

Fig. 5 Impact of the read frequency on swapped items in the conveyor simulation

window size of three. All other alternatives performed worse and are not displayed for simplicity. Again we used 1000 simulations for the generation of each data point. Figure 5 shows a reduction of errors with increasing read frequency. Starting from the low frequency of 1 Hz, the error rate first decreases drastically and then converges against small values. The curve shape is similar for all approaches but our solution outperforms the alternatives for each frequency. We have no errors using the high frequency of 8 Hz and *Max. L. Opt.* For *Max. L. Approx.* we had only three errors in the experiments. Using *Low Pass 3* still resulted in 21 errors at this point. Besides the superiority of our approach for all frequencies, the experiments verify that our approach copes well with errors in f_{rel}. For very low (1 Hz) and very high frequencies (8 Hz) *Max. L. Approx.* performed very similarly to *Max. L. Opt.* For the tested frequencies *Max. L. Opt.* clearly outperforms *Max. L. Approx.* but *Max. L. Approx.* also shows low error rates and is clearly better than all existing alternatives.

5.2 Evaluation with a Complex Scenario

To evaluate our approach in a complex scenario we simulated an asset tracking scenario (scenario two in Sect. 2). In the scenario a carrier (forklift or trolley) moves toward an item, picks it up, and continues moving in a random direction. The simulated speed of movement was walking speed (0.83 m/s). We tested how well our approach could detect from the RFID-read stream where the item was picked up. We thereby tested how our approach works for monitoring complex movements.

Figure 3 depicts the reliability function that we used in the simulation as well as the approximation that we used in our algorithm. We simulated 1000 test runs of picking events for different sample rates (see Fig. 6). The generated events were processed with our approach for detecting picking events using the approximate reliability function (here labeled *prob. pick*). We compared our approach with

Fig. 6 Experimental results

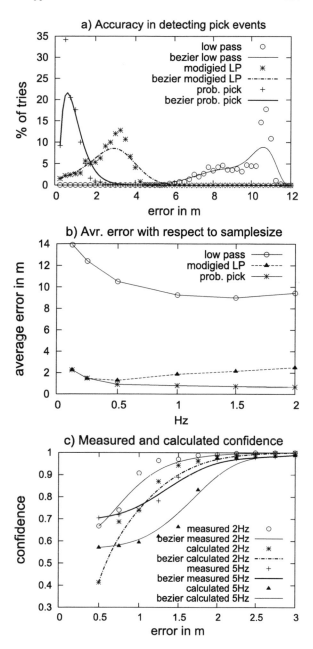

trivial solutions to this problem. Note that the typically used low-pass filters are not applicable in this scenario. For example a tuple based low-pass filter would detect picking up of an item if the carrier resides in its proximity for a while. We therefore used a low-pass filter with a location based window (*lowpass*) and a modification of it (*modifiedLP*). The filter *lowpass* takes the upper bound of the spatially defined window that we used in our probabilistic approach (see (5)).

Figure 6a shows the percentage of detections with errors in a certain range. Each point refers to a scope of 0.25 m and is positioned at the upper bound of the range. The sample rate was 2 Hz. We found that *probPick* outperforms the other operators in terms of accuracy with an average error of about 0.7 m. Figure 6b shows the effect of the sample rate on the operators' accuracy. Figure 6c shows the predicted and measured confidence that errors are below a given distance. We obtained this data using sample rates of 5 Hz and 2 Hz. The results show very different shapes for the measured confidence curves. We see that the predicted curves match the measured curves with some understatement. Overall the experiments stress the applicability of our approach to complex scenarios. Note that no existing approach, to our knowledge, is directly applicable to this scenario and by our algorithm clearly outperforms trivial solutions to the problem.

5.3 Tests with Real Data

So far, we have presented simulations. Below we describe an experiment we conducted using real RFID hardware. The tested scenario is a slight modification of the conveyor scenario (scenario two in Sect. 2). In our experiments, we moved an item straight to a reader and let it remain next to it. This process might be used, for example, to monitor the order in which a conveyor packs items into a box. In this case the reader would be installed right at the box. The cleaning task is to determine the order in which the items were packed into the box.

For the experiments we used a Nordic ID PL3000 handheld reader. Figure 2 shows the reliability function that we measured for this reader. We did not use this function for f_{rel} in the implementation for our test. This is because we had a small sample set and did not want to use the same data for determining f_{rel} and testing the approach. Instead we used a rough linear approximation with 95% read reliability at a distance of zero and 0% read realiability at a distance of 1.1 m. In the experiments we issued read requests such that we had one read attempt every five centimeters. To count the swapped items we assumed a distance of 10 cm between each RFID tag.

Figure 7 shows the results of 200 test runs where the number of errors (swapped items) with our approach, low-pass filters, and trivial alternative implementations are depicted (these are the same algorithms as tested in the simulation of the first conveyor scenario). The results show that our solution clearly outperforms the alternatives. We achieve these result despite the fact that we used a very rough approximation for f_{rel} in the algorithm and despite the irregularity at 70 cm distance (see Fig. 2). For real world applications one would use measurements to determine f_{rel} and likely achieve even better results. Overall, the tests with real data also demonstrated that our approach is superior to other alternatives.

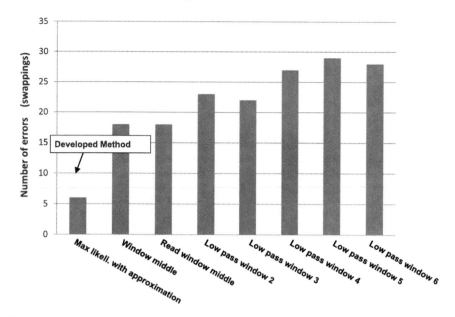

Fig. 7 Swapped items in tests with real RFID hardware

6 Conclusion

In our case studies we found a range of scenarios that pose special challenges to RFID data cleaning. A number of scenarios required the determination of the relative position of tags and readers with high accuracy. However, noise and uncertainty in RFID reads makes this positioning challenging. Therefore we developed a probabilistic cleaning approach that can solve these problems and extend current middleware solutions or reader firmware. A strength of our solution is that it allows precise reasoning about tag locations within the read range. This enables more accurate RFID reads in many scenarios, e.g., capturing the order of items on a conveyor.

Experiments with simulations as well as real data showed that our solution improves the state of the art by several orders of magnitude. Furthermore, our approach can provide confidence values for its results. Downstream applications may use these confidence values to reason about their input. For example approaches that use probabilistic data models for RFID data would benefit (e.g., Garofalakis et al. 2006).

Particularly for monitoring the order of items on a conveyor our approach shows very good results. Compared to existing solutions, manufacturers can increase the throughput of items on a conveyor without increasing the risk for data errors. This can make RFID installations feasible in more scenarios or improve the performance of an installation. A possible extension to further improve our work is using additional sensor data – such as light barriers – and integrating them in our algorithm.

Addressing this extension and implementing our approach in RFID middleware soultions is the subject of future work.

References

Bai Y, Wang F, Liu P (2006) Efficiently filtering RFID data streams. In: proceedings of report on the first international VLDB workshop on clean databases (CleanDB 2006), Seoul, Korea, 11 September, pp 50–57

Bekkali A, Sanson H, Matsumoto M (2007) RFID indoor positioning based on probabilistic RFID map and Kalman Filtering. In: proceedings of the third IEEE international conference on wireless and mobile computing, networking and communications, Washington, DC, p 21

Bornhövd C, Lin T, Haller S, Schaper J (2004) Integrating automatic data acquisition with business processes – experiences with SAP's Auto-ID infrastructure. In: proceedings of the thirtieth international conference on very large data bases, Toronto, Canada, August 31 – September 3, pp 1182–1188

Brusey J, Harrison M, Floerkemeier C, Fletcher M (2003) Reasoning about uncertainty in location identification with RFID. In: proceedings of the workshop on reasoning with uncertainty in robotics at IJCAI-2003, Acapulco, Mexico.

Finkenzeller K (2003) RFID handbook: fundamentals and applications in contactless smart cards and identification. Wiley, New York

Floerkemeier C, Lampe M (2004) Issues with RFID usage in ubiquitous computing applications. In: proceedings of pervasive computing, Springer, Linz,Vienna, pp 188–193

Garofalakis MN, Brown KP, Franklin MJ, Hellerstein JM, Wang DZ, Michelakis E, Tancau L, Wu E, Jeffery SR, Aipperspach R (2006) Probabilistic data management for pervasive computing: the ata furnace project. IEEE Data Eng Bull 29(1):57–63

Hähnel D, Burgard W, Fox D, Fishkin K (2004) Mapping and localization with RFID technology. In: proceedings of the IEEE international conference on robotics and automation, New Orleans, USA, 26 April – 1 May

Ivantysynova L, Ziekow H, Günther O, Kletti W, Kubach U (2008) Case studies. In: Günther O, Kletti W, Kubach U (eds) RFID in manufacturing. Springer, Heidelberg, pp 61–111

Jeffery S, Garofalakis M, Franklin M (2006a) Adaptive cleaning for RFID data streams. In: proceedings of the 32nd international conference on Very Large Data Bases, Seoul, Korea, 12–15 September, pp 163–174

Jeffery S, Alonso G, Franklin M, Hong W, Widom J (2006b) A pipelined framework for online cleaning of sensor data streams. In: proceedings of the 22nd international conference on data engineering, Washington, DC, p 140

Pupunwiwat P, Stantic B (2007) Location filtering and duplication elimination for RFID data streams. Int J Prin Appl Inf Sci Tech 1(1): 29–43

Song J, Haas CT, Caldas CH (2007) A proximity-based method for locating RFID tagged objects. Adv Eng Inform 21:367–376

Wang F, Liu P (2005) Temporal management of RFID data. In: proceedings of the 31st international conference on Very Large Data Bases, Trondheim, Norway, 30 August-2 September, pp 1128–1139

Part III
Global Information Architectures and Systems

Autonomous Control and the Internet of Things: Increasing Robustness, Scalability and Agility in Logistic Networks

Dieter Uckelmann, M.-A. Isenberg, M. Teucke, H. Halfar,
and B. Scholz-Reiter

Abstract The Internet of Things and Autonomous Control are considered key concepts to enhance logistic processes in supply networks. Here the question that often comes up is about how both concepts relate to each other. The principal aim of this article is to evaluate, whether the Internet of Things and the paradigm of Autonomous Control in logistics can complement each other's capabilities. There are numerous different architectural approaches towards an Internet of Things. In this article, the EPCglobal Network is chosen, as it provides a standardised and well accepted approach based on open interfaces. A state of the art analysis is performed, concerning the existing technologies and research on Autonomous Control and the EPCglobal Network. In an integrative approach both concepts are merged. In this context, the EPCglobal Network is used as an information broker for Autonomous Logistic Objects. One possible application of this integrative approach is described based on an intelligent truck to illustrate the potential of the extended EPCglobal framework through means of Autonomous Control.

1 Introduction

The vision of an Internet of Things is becoming more realistic. A growing number of scenarios are realized, where electronic (portable) devices interconnect (e.g. home environment) or embedded sensors report their measurements. These emerging applications are made available by using communication technologies such as wireless local area network (WLAN) or Universal Mobile Telecommunications System (UMTS) for ubiquitous Internet connections. They make life more comfortable, and will probably lead to a trend of increasing networking within our society but they do not yet provide an overall concept that considers all future impacts of

D. Uckelmann (✉)
Planning and Control of Production Systems, University of Bremen, Hochschulring 20, D 28359
Bremen, Germany
e-mail: uck@biba.uni-bremen.de

D.C. Ranasinghe et al. (eds.), *Unique Radio Innovation for the 21st Century*,
DOI 10.1007/978-3-642-03462-6_8, © Springer-Verlag Berlin Heidelberg 2010

an increasing number of communication partners in the Internet. Their high number will pose new challenges to the current Internet architecture in terms of scalability and robustness.

There are several opportunities for scaling down the communication effort within the Internet of Things. One possible scaling mechanism is to integrate intelligent objects. The intelligence of these objects consists of capabilities for a decentralised coordination, such as measuring environmental influences on the objects by sensors (temperature, pressure, acceleration, etc.) as well as contextual and autonomous decision making, thus providing these objects with a high level of autonomy. The paradigm of a centralised authority for information processing and control would be replaced by an autonomy paradigm, which allows objects to act on their own. In particular, logistical objects would benefit from this new approach, since transport processes include diverse locations, limited online access, and changing environmental parameters including humidity. The Autonomous Control approach transfers the responsibility for the object from centralised authorities to the object itself. While autonomous objects do not require an online connection they still have the capability to communicate to other nearby objects. Therefore, it is possible to cluster Autonomous Logistic Objects for building contextual and self-administrating networks in the Internet of Things. These groups can be formed by adjacent autonomous objects (on the lowest level) or Group Leaders (on higher levels).

The aim of this paper is to evaluate whether the Internet of Things and the paradigm of Autonomous Control in logistics can complement each other in a sensible way. Another goal is to figure out which modifications are necessary to combine both concepts. This contribution is structured as follows: Sect. 2 describes a state of the art analysis, focused on the Internet of Things, autonomous concepts, the required technical components and the EPC global Network as a possible platform that may be extended to combine both concepts. In Sect. 3 the concepts, technologies and current EPCglobal Network implementation are merged, to develop a new design of Internet of Things enabled autonomous objects. A possible application focussed on an intelligent truck is discussed in Sect. 4. Finally, in Sect. 5 the results are analyzed and an outlook to further required research is provided.

2 State-of-the-Art Analysis and Technical Background

2.1 The Internet of Things

The term Internet of Things has widely been used by researchers and practitioners to describe the combination of the real world of physical objects with the virtual world of information technology (Fleisch and Mattern 2005; Bullinger and ten Hompel 2007; Floerkemeier et al. 2008). The continued miniaturisation of information processing technology and its combination with new technologies, such as radio frequency identification (RFID), real-time locating systems (RTLS), sensor technologies and actuators, are usually seen as core technologies for the Internet of Things.

There are multiple approaches towards the Internet of Things. Some consider a pure Internet approach, such as IPv6 as sufficient, requiring that every object has an IP-address and supports the corresponding protocols. In the case of other approaches, more holistic architectures have been developed as part of projects, such as PROMISE[1], BRIDGE[2] or EURIDICE[3]. Additionally, projects and activities focussed on embedded and wireless sensor networks or actuators, such as COBIS[4], SENSEI[5] and the Open Geospatial Consortium (OGC) Sensor Web Enablement (SWE), have researched certain components of the Internet of Things (Botts et al. 2008).

One of the most accepted approaches towards the Internet of Things today, is the EPCglobal Network. It is based on a simple identifier, attached to a physical object that links related information in an Internet-alike structure, which includes discovery services and information services. The objects themselves are not intelligent – they do not hold relevant business data, nor do they offer processing capability. Intelligence in the EPCglobal Network can only be achieved in a combination with business databases and processing capability, thus requiring ubiquitous online capability.

The EPCglobal Network does not yet consider the inclusion of sensor values on the streams of data, but there are some activities on the EPC Sensor Network at the Auto-ID Labs Korea to incorporate Wireless Sensor Networks (WSN) and sensor data into the EPCglobal Network architecture and standards (Sung et al. 2007).

One reason for the success of the EPCglobal Network is that it defines the interfaces required between applications or products, rather than the applications or products themselves. This ensures that applications may be easily integrated with the Internet of Things, as long as they comply with the interface specifications. There are several commercial and free implementations of the relevant software components for the EPCglobal Network available. Large software companies as well as integrators are offering corresponding services and help for implementation. Therefore, in this paper's approaches, the EPCglobal Network will be used as the basic infrastructure, providing a certain level of reliability, stability and user acceptance as well as the flexibility to be extended with missing functionalities, such as sensor and actuator integration.

The vision of the future Internet of Things is still a much broader vision than the current scope of the EPCglobal Network and the proposed enhancements. It includes business-to-consumer interactions (e.g. using Near Field Communication (NFC) tags and mobile readers), location-based context-aware services and even non-commercial interactions between citizens for social purposes, such as the geo-tagging of photos, video clips and commentary. As this paper focuses on business

---

[1] See www.promise.no

[2] See www.bridge-project.eu

[3] See www.euridice-project.eu

[4] See www.cobis-online.de

[5] See www.sensei-project.eu

aspects in logistic networks, these issues are not addressed. Yet, the concept of extending the EPCglobal Network towards a more holistic Internet of Things may be utilized for the above mentioned features as well.

The Internet of Things offers a wide range of potential benefits. These include:

- better supply chain visibility,
- improved life-cycle knowledge,
- enhanced customer relationship management and
- smart products and services.

The first three topics are derived from cost reduction, improved speed, process quality and security, whereas smart products and services are aiming for revenue growth (Fleisch et al. 2005).

Besides requiring anytime online capability, scalability is another main obstacle to the Internet of Things. If every object is connected to the Internet of Things, the current available infrastructure will be overloaded. Other approaches, such as autonomous strategies, do not suffer the problem of scalability, nor do they require always online connections.

2.2 Autonomous Control

The concept of Autonomous Control is less technically focused than the Internet of Things and has been influenced more by fundamental than applied research. The concept of Autonomous Control is being researched in detail in one of the core domains of the Internet of Things – logistics. The goal is to achieve a higher level of agility in logistics. The Collaborative Research Centre (CRC) 637[6] 'Autonomous Cooperating Logistic Processes – A paradigm Shift and its Limitations' at the University of Bremen, which has been analysing Autonomous Control in logistics since 2004, defines the characteristics and objectives of Autonomous Control as follows:

> Autonomous Control describes processes of decentralised decision-making in heterarchical structures. It presumes interacting elements in non-deterministic systems, which possess the capability and possibility to render decisions independently. The objective of Autonomous Control is the achievement of increased robustness and positive emergence of the total system due to distributed and flexible coping with dynamics and complexity. (Hülsmann and Windt 2007)

The mentioned characteristics and objectives of Autonomous Control have been explained in detail by the CRC 637. The discussions of these terms by Hülsmann, Windt and Böse (Hülsmann and Windt 2007; Böse and Windt 2007) will be briefly summarised as follows. Additionally, this paper considers *agility* as an important

---

[6]See www.sfb637.uni-bremen.de

Fig. 1 Structure of autonomous control and its objectives leading to higher agility

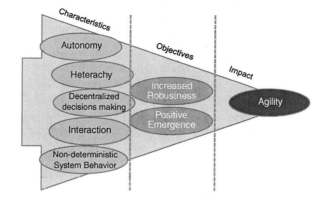

impact of Autonomous Control. Figure 1 illustrates the relations between structure, objectives and impact of Autonomous Control.

The concepts of *autonomy* and *heterarchy* are closely related to each other. Autonomy, in this context, refers to properties of an object, while heterarchy refers to those of a system:

• An object is considered autonomous if it can make decisions independent of influences by external entities. Such external entities may be central planning and control instances, or other objects within the same system. A system is considered heterarchical, if no entity within that system exerts a permanently dominant influence on the system elements (Probst 1992). In a heterarchical system the system components act in such a way that no component is excluded from the decision processes on an organisational basis. Every component has the same opportunity to participate in the interactions of the system and hence the shaping of its actions (Hejl 1990).

Both, autonomy and heterarchy, are in turn related to *decentralised decision-making,* which means that decisions are not taken by a central planning and co-ordinating instance, but instead are delegated to individual system elements. Decision-making is understood as the selection between different (and normally conflicting) action alternatives, based on the processing of necessary information and in pursuit of certain goals (Laux 1998).

Implementation of Autonomous Control implies both, the existence of decision alternatives within a logistic network, and the presence of decision competence at logistic objects in form of methods and algorithms.

Interaction: Interaction describes the influence that system elements can have on each other during decision making processes. Though autonomous objects should not be dominated by other system elements, their decision making processes are not performed in isolation. Instead, decision relevant data or information is exchanged between the involved elements. Several different forms of interaction are possible. These are allocation of data, which other objects can access, bi-directional data exchange between objects, and co-ordination, which is the ability of Autonomous Logistic Objects to cooperate with and co-ordinate activities of other objects.

Non-deterministic system behaviour: In a non-deterministic system, it is not possible to predict a definite system output for a given input to the system. The behaviour cannot be predicted despite knowledge about the laws governing the system, the input to the system and the states of the system and its elements. Thus, for exactly identical inputs and beginning states, the resulting output of the system may be different (Pugachev and Sinitsy 2002).

The main application goals for Autonomous Control of the CRC 637 are *increased robustness* and *positive emergence* of the logistic system.

Increased robustness: The CRC 637 postulates that a system will be more robust against outer disturbances if its elements are autonomously controlled. The main argument supporting this assumption says that each system element can react more immediately to unforeseeable, dynamic influences, updating its own decision making to the modified local situation with no need to wait for new instructions from a central control instance.

Positive emergence means that the interplay of the decisions taken by the system elements (individual logistic objects) results in a better achievement of objectives of the system as a whole, than can be explained from the combined achievement of individual objectives (e.g. lower delivery times and higher adherence to delivery dates) of each single logistic object.

Finally, increased robustness and positive emergence will yield an impact in achieving system agility.

Agility: Beyond the findings of the collaborative research centre 637, the term agility is used in this paper to describe the business impact of Autonomous Control. Agility is known as a concept to increase the ability of companies to respond effectively to either expected, or unexpected changes in market demand (Sharifi and Zhang 2001; Brown and Bessant 2003). Based on increased robustness and positive emergence, Autonomous Control enables an increase in agility. This does not imply that agility may not be achieved without Autonomous Control. But centralised approaches suffer from scalability issues and demand online access, as mentioned before.

Positive emergence and agility depend on achievement of logistic targets such as delivery times or service levels by the logistic system. Logistic target achievement is assumed to vary for different levels of system complexity and different levels of Autonomous Control. The shape of the resulting curve, which is shown in Fig. 2, reflects the following assumptions: In complex systems, the achievement of logistic targets will increase with the level of Autonomous Control until Autonomous Control reaches a specific level. Systems with a low level of Autonomous Control (e.g. using conventional planning or control methods) will yield high logistic target achievement only when the system's complexity will remain low, but drop

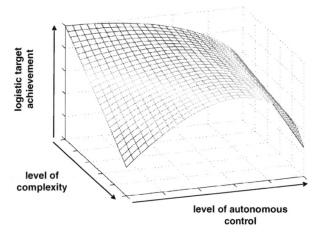

Fig. 2 Correlation between logistic target achievement, level of complexity and level of Autonomous Control of a logistic system (compare to Philipp et al. 2007)

markedly for higher system complexity. However, a very high level of Autonomous Control will result in low achievement of logistic targets, as system behaviour will increasingly resemble a state of anarchy (Philipp et al. 2007).

In the field of production logistics simulation studies of several decentralised production control paradigms indicate an increased logistic target achievement, compared to the conventional configuration (Scholz-Reiter et al. 2005a, b, 2006). For example, a simulation of the job shop floor of a metal working medium-sized enterprise, regarding adherence to delivery dates and throughput times, resulted in average delivery delays over a 3 year period, reduced by up to 20% compared to the real life operations data. A confirmation of the efficiency of Autonomous Control in a transport scenario is made by Rekersbrink et al. They have developed the *Distributed Logistics Routing Protocol* (DLRP) and built a simulation scenario, which is based on a real world network and includes the German autobahn-net, 18 cities and 35 undirected edges (Rekersbrink et al. 2009). Rekersbrink et al. compared the DLRP solution to an optimal solution as lower bound, the tabu search algorithm and a random-like solution as upper bound. Within small scenarios, the tabu search algorithm dominated the DLRP. However, in more extensive scenarios, the DLRP started to dominate the tabu search algorithm – that shows, that Autonomous Control fits very well to complex situations, like the example above.

2.3 Technical Basis for Implementation of Autonomous Control in Logistics

In the last section, the key concepts and objectives of Autonomous Control have been described. This section, in contrast, will discuss how Autonomous Control can be realized technically, by using current or emerging future information and communication technologies to make objects autonomous.

As mentioned in the last section, the ability for decentralised decision made by autonomous objects is a key aspect of Autonomous Control concepts. This is also emphasised in a short definition, which has been provided by the CRC 637 to summarise the most important aspects of Autonomous Control:

> Autonomous Control in logistics systems is characterized by the ability of logistic objects to process information, to render and to execute decisions on their own. (Hülsmann and Windt 2007)

Autonomous Logistic Objects may be physical (material) objects or immaterial (virtual) objects (Böse and Windt 2007).This paper focuses on physical objects that can hold a certain degree of intelligence.

To achieve Autonomous Control, different intelligent logistic objects of a logistic system have to be given capabilities to store and process relevant, locally stored data, communicate and interact with other objects and get information about the surroundings and interact with it. In the field of logistic processes, these intelligent objects should be able to collect and process information of their environments and to identify and evaluate alternative process executions (e.g. alternative transport routes within a logistic network), according to their individual evaluation system (Böse and Windt 2007; Hülsmann and Windt 2007).

Scholz-Reiter et al. (2008) define five levels of technical capabilities, which are relevant to the implementation of Autonomous Control. These levels, listed below, range from simple identification of parts via storage of dynamic data, decentralised data processing and communication, to intelligent integrated information based material handling, allowing autonomous co-operating logistic processes:

- *Level 1*: In the simplest form, RFID and other Auto-ID technologies are used for identifying logistic objects and storing static data, which does not change over a products' life-cycle.
- *Level 2*: In the second step Auto-ID technology is enhanced by storing dynamic data. This includes time and location recording and storing of data created by sensors on transponders, e.g. temperature or humidity data for shelf life prediction in the food sector.
- *Level 3*: More enhanced intelligent logistic objects are assumed to process data in order to create new, pre-processed data and solve tasks. Pre-processing labels, for example, are combinations of passive transponders and a micro-processor with limited processing capacity (Overmeyer et al. 2006). Another approach to decentralised data processing is characterised by usage of software agents, which are hosted on intelligent Auto-ID readers or Personal Digital Assistants (PDA). Software agents are software entities, which can act autonomously, according to a set of stated goals (Bussmann et al. 2004).
- *Level 4*: Logistic objects are given the capability to exchange data with other objects in order to interact with each other. Classical radio frequency identification technology is mostly limited to data exchange between tags and special reading and writing devices. Still, there are other RF communication protocols for ad-hoc networks, such as ZIGBEE, allowing communication between

objects. For the communication of data between more advanced devices other technologies are used, such as WLAN or UMTS.

- *Level 5*: Finally, intelligent information based material handling is needed to initiate actions, using flexible material handling systems. The exchange of dynamic data, such as production tolerances, between product components and an assembly robot, may be an example.

3 Merging the Concept of Autonomous Logistic Objects with the Internet of Things

3.1 Identifying Possible Synergies Among Both Concepts

The EPCglobal Network and Autonomous Control are both concepts for collecting, storing and providing information. Collection and distribution of information is the domain of the EPCglobal Network, whereas Autonomous Control allows automated decision rendering. These decisions are taken based on intelligent processing of available real-time information. In order to combine both concepts, it seems logical to consider the well standardized and accepted EPCglobal Network as the basic information and communication structure for autonomous objects and extend it to a more holistic Internet of Things infrastructure, including Autonomous Control capabilities. For identifying possible synergies between the concepts, the following questions, concerning the two concepts and their technological basis, have to be answered:

3.1.1 Where Is Data Generated and Needed?

In the EPC-Network, raw data is generated through *things*, filtered and enriched through business data, thus delivering valuable information that can be accessed anywhere through the network. In autonomous controlled systems, data needs to be filtered and accessed for decision rendering at the physical place, where the decision taking agent is operating independently, without the need for an online connection. Autonomous Logistic Objects will have to interact with each other, even if they are geographically separated. As an example, communication of waiting transport goods (at a transport station) with their scheduled but unpunctual means of transport may be mentioned. The EPCglobal Network could be a communication medium for spatially separated software agents. Additionally, the EPCglobal Network can offer software agents a direct access to relevant business information systems. Due to its standardized and robust network structure, the EPCglobal Network is well suited to act as an information broker between objects, business information systems and software agents. The EPCglobal Network provides information transparency, thus enabling location-independent data filtering, storing, business content enriching, and accessing capabilities for autonomous decision making.

3.1.2 How Can a *Capacity to Act* Be Assured in Offline Situations?

Logistical objects are not always connected to the Internet. The *capacity to act* in the EPCglobal Network is based on real-time online access to the logistical objects. The EPCglobal Network can benefit from software agents as used in Autonomous Control to enable acting without a connection to the Internet. This requires synchronisation capabilities between autonomous objects and the EPCglobal Network to update data, decisions and preference-sets for software agents.

3.1.3 How Can Authorisation and Security Be Ensured?

The EPCglobal Network offers mechanisms for secure and authorized information access. In an autonomous controlled system, the decision rendering agents need to support these mechanisms. This has to be taken into account while designing the information structures and formats as well as the interfaces of the integrated approach. Automated, secure and authorized data access has to be guaranteed for software agents.

3.1.4 Where Are Decisions Rendered?

Using the concept of Autonomous Control, decisions are rendered in a decentralised way by processing power directly attached at the objects or by geographically distant agents or persons associated with the objects.

Autonomous decision processes through software agents may become an integral part of the EPCglobal Network. This requires that the EPCglobal Network is extended, to store status information about software agents associated with the object or the software agent itself. Additionally, Autonomous Logistic Objects need important background information, necessary for adequate decision making, and they need to report the results of these decisions back to higher planning and controlling instances. In turn, higher authorities will provide updates of background information (e.g. new possibilities of transport), and they will give feedback to Autonomous Logistic Objects urging them to modify their plans once their reported decisions are not in line with objectives set by these higher authorities (e.g. overstress of the decided route of transport). The latter will trigger new decision making processes by the Autonomous Logistic Objects.

3.1.5 How Are Decisions Executed?

Decisions need to be communicated to actuators of material handling systems. This can be done locally, for example if a parcel triggers a sorter. While the EPCglobal Network is fast enough to forward information in most applications, it may be too slow to handle real-time requirements for machine-to-machine communication. In instances that do not require sub-second responses, the EPCglobal Network with its otherwise excellent communication infrastructure may be utilized to communicate the decision from the software agent through its network to an actuator.

This would require a corresponding actuator interface extension to the Internet of Things. There should be a clear differentiation between sensors and actuators in this context. While sensors provide information *to* the Internet of Things and the concept or Autonomous Control, actuators execute decisions *from* these infrastructures. The protocols will most likely be very different. Especially when considering numeric controlled machines, the necessary interface would need to respect existing approaches, such as the OPC Unified Architecture (Mahnke et al. 2009), or define new lightweight solutions that allow a similar functionality.

3.2 Integrative Approach Towards Internet of Things Enabled Autonomous Objects

The concept of Internet of Things and the concept of Autonomous Logistic Objects share basic technologies. Both approaches can complement each other. The approach suggested in this contribution, is to use the Internet of Things as an 'information broker' for autonomous objects, while using software agents for decentralised information processing and interaction capabilities to reduce data traffic within the network and make objects less dependent on network access. In other words, the Internet of Things will provide the macro structure (linking to a central server, communication channels etc.), while the approach of Autonomous Logistic Objects defines the object behaviour.

Table 1 lists the technologies used for the EPCglobal Network as it is today, autonomous logistics and the integrative approach on Internet of Things enabled autonomous objects.

Building on the five levels to autonomous logistics as defined by Scholz-Reiter et al. (2008), the model is extended by distinguishing between traditional data processing (3a) versus autonomous data processing and decision rendering (3b). The table shows the different technology steps and their usage in logistical applications. Additionally, it compares their utilisation in the EPCglobal Network, the concept of Autonomous Logistics, and the integrated approach developed in this paper. While the EPCglobal Network is extended by dynamic data capturing, autonomous data processing and integration of intelligent material handling systems, the concept of Autonomous Control benefits from the integration of traditional data processing and the standardised, well accepted structure of the EPCglobal Network.

Each of the technologies, such as sensors and data processors, are affected by the trend towards miniaturisation, which has been a constant from the beginning of microprocessor technology, and is manifest in the emergence of micro- and nanotechnology. However, since the prices of available miniaturized microprocessors are still high (proportional to the price of the object), the technical feasibility of process capability on the object itself still needs some development time. Hence, it is reasonable that first only high value objects in closed loop applications and grouped objects will be equipped with powerful microprocessors. Low value objects and

Table 1 Comparing logistical automation levels (• → supported and ○ → not supported)

No.	Logistical automation level and technologies	Logistical applications	EPCglobal Network (today)	Autonomous logistics	Internet of Things enabled autonomous objects
1	Identification: data carrier, decoder/encoder	identify items in real time, link to related information	•	•	•
2	Dynamic data collection: integrated sensors, wireless sensor networks	collect data about physical status of things and their environment, raise awareness about context	○	•	•
3a	Traditional data processing: ERP, SCM	centralised data processing	•	○	•
3b	Autonomous data processing and decision rendering: (miniaturized) distributed data processing technology, software agents	data pre-processing, independent decisions rendered by things, reaction of things to external stimuli, smart characteristics	○	•	•
4	Communication capability: Bluetooth, WLAN, UMTS, etc.	interact with infrastructures	•	•	•
5	Intelligent handling systems: integrated actuators, robotics	industrial usage, personal usage	○	•	•

open loop applications will follow step by step when the prices decrease. To realize each object's individual processing capability, in this early stage of the Internet of Things, the process capability can be transferred, partially or totally, into the Internet of Things resources, dependent on the automation level (compare Table 1) of the object. Figure 3 illustrates possible distributions of process capability among the object and the Internet of Things.

With the objective of a highly agile logistical system, the most suitable distribution is the implementation of the process capability on the object itself. These fully capable intelligent objects are able to act independently for long phases of their existence. Because objects will rarely have the possibility for energy-intensive long

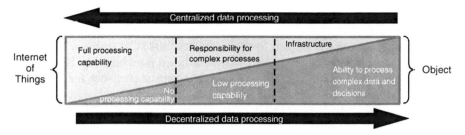

Fig. 3 Possible distributions of process capability by using the integrated approach

range communication (e.g. UMTS, satellite communication), the biggest dependency is in the need of an infrastructural connection to the Internet of Things backbone for getting information updates. Additionally, synchronisation capabilities for decisions and preference-sets for software agents are needed.

This paper suggests to utilize, and to extend, the EPCglobal Network to realize Internet of Things enabled autonomous objects as shown in Fig. 3. Fig. 4

The architecture considers Smart Objects (Lopez et al. 2010) (e.g. parcels), Group Leaders (e.g. trucks) and the Internet of Things. Smart Objects contain static data, such as unique identifier and dynamic data, through sensors. The architecture on Group Leader level includes the Internet of Things Information Server (IoT IS), the Query Interface with data synchronisation capabilities, the Repository that stores

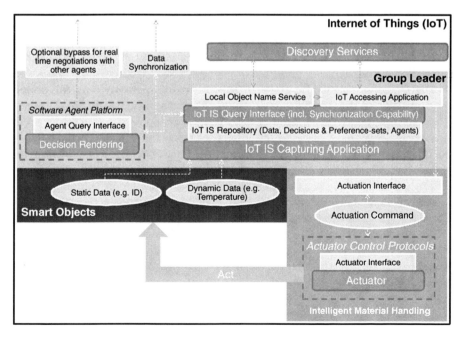

Fig. 4 Combining the internet of things and autonomous concepts

data as well as decisions, software-agents and preference sets, and the Capturing Application. The IoT IS ensures the correct and secure data exchange with the Internet of Things that will connect to traditional ERP and SCM systems. The Group Leader has the ability of contextual and attribute-based team building. By communicating with Smart Objects, it will find sufficient similar attributes for building a group. The Group Leader acts as an interface to the next higher hierarchical level and negotiates with other representatives of similar groups for building a team on a higher hierarchical level (cluster of clusters). The Group Leader architecture consists of several components known from the EPCglobal Network. The Capturing Application connects to Auto-ID infrastructures. The EPCglobal Network offers this functionality for EPC compliant RFID today. Capturing Application can be extended to handle sensor data as well.

The Query Interface should be extended with offline capabilities, requiring synchronisation with the Internet of Things, and it should be enabled to communicate through a standardized interface to a software agent platform. Each Smart Object, even with no or low processing capability (see Fig. 2), may have its own product-centric software agent, running on this platform that is a part of the Group Leader infrastructure. In further stages of technical developments, the Group Leader infrastructure and the software agent will be able to run on the objects themselves, but due to the current development status and cost of technical components, this paper considers that both functionalities have to run separately from the object. The advantages of this solution are that objects can be already transformed into intelligent objects without the necessary process capability and that the representing software agent is geographically near. If objects change the locality, the software agent will be transferred to the next appropriate software agent platform, e.g. based on proximity to the object or group interests. Additionally, the software agent platform negotiates in real-time with other software agents through the Internet of Things, if an online connection is available.

The Repository should be extended as well to hold not only data, but preference sets for a software agent infrastructure, different software agent templates and store decisions for logging reasons as well. The Accessing Application finally needs to support actuators. A standard interface for connecting actuators would enable the integration of handling systems to act on the Smart Objects, thus providing the highest level of autonomy.

The capabilities of the contextual and attribute-based group building as well as self-organizing and self-hierarchisation, can help to give the architecture of the Internet of Things a more flexible topology. Having Billions of addressable objects in the Internet of Things would increase the data throughput rapidly by producing trillions of queries and commands (Commission 2009). The clustering approach would have a direct impact on the scalability and robustness of the logistical network, depicted by the Internet of Things. The communication effort and energy expenditure for contacting each object separately could be significantly reduced by using the Group Leader as the only possible contact to the group members. As an interface, the Group Leader would forward information to corresponding objects or send aggregated group data to the next higher hierarchical level. Additionally, groups would be able to act autonomously when they are not connected to the Internet of Things (offline usage), thus offering a higher level of robustness and agility.

Using the integrated technical approach as well as the clustering concept control commands, requests for resources and status queries will not necessarily go across the Internet and put a strain on the technical infrastructure, partly these things can be managed locally by communication between the intelligent objects. Furthermore, this approach enables a faster reaction to unforeseen events, it significantly reduces the communication effort for informing and controlling and it increases the energy efficiency within the Internet of Things (by less communication) as well as on the level of objects (by clustering).

4 Field of Application

To explain the suggested architecture, this paper considers an application in transport logistics, based on an intelligent truck. Grouping of products with identical or similar destinations are an efficient cost-effective way to optimise transport capacities and routing distances. Additionally, the reduction of emissions and energy consumption caused by the transport processes are seen as common goals (Gudehus 2005). Today, these goals are addressed through central planning and optimisation. Large logistics networks consist of numerous different destinations and vehicles. Due to this complexity and scale of transportation processes, an optimal solution cannot be computed centrally for the complex logistic network in a reasonable time. Although heuristic algorithms offer a practical way to approximate an optimal solution, dynamic changes and offline situations to central planning and control facilities lead to major deviations. Dynamic changes include for example unforeseen events, such as traffic jams, breakdowns, changed quantities, and spontaneous orders.

The integration of the presented approach in transportation logistics is demonstrated by the following scenario, illustrated in Fig. 5.

Dozens of pallets and thousands of items may be carried by one truck; often those goods contain perishable goods, such as fresh vegetables, fish or frozen foods, which require refrigerated vehicles. In a prototypic scenario at the LogDynamics Lab at the University of Bremen each item as well as the refrigerated truck is identified through a unique RFID-tag and fixed sensors are installed in the truck to measure the temperature of the different climate zones. The installed RFID-reader utilizes the EPCglobal Generation 2 UHF protocol and is currently connected to a PC, holding the software agent platform based on the Java agent development framework JADE. Additionally, there have been tests with sensor networks in the intelligent truck as part of the Collaborative Research Centre 637. The next necessary steps to extend this prototypic scenario towards Autonomous Control and the EPCglobal Network would be:

- implementation of the EPC Information Service (EPCIS) on the intelligent truck,
- integration of offline data synchronisation capabilities (mobile EPCIS),

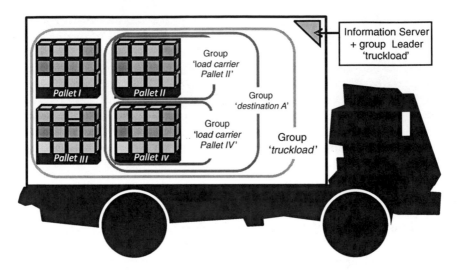

Fig. 5 Groups and subgroups of autonomous controlled objects

- integration of an actuator interface with the EPCIS to activate the cooling unit,
- development of a sensor interface between the EPCIS and the sensors,
- integration of the JADE and the EPCIS.

Based on this enhanced infrastructure it would be possible to implement the grouping algorithms. Pallets with an identical destination form a group (see group 'destination A', which includes pallets II and IV in Fig. 5) by themselves, supported by a software agent running on the software agent platform that decides on behalf of the associated group. The group destination A is a member (subgroup) of the group truckload, which group leader would be the intelligent truck, represented by the mobile EPCIS (IoT IS, see Fig. 3). Since the truck itself represents an intelligent object in the EPCglobal Network, instructions for several packages just have to be forwarded to it once. The truck software agent splits the instructions and transfers the relevant subsets to the corresponding groups, respectively its software agent, to reduce the data throughput and increase the scalability within the Internet.

In the case of an unforeseen event, such as increasing temperature during the transport of perishable goods, due to a broken cooling unit, objects equipped with temperature sensors can communicate the temperature measurements to the truck (Chill-on 2009). As modern refrigerated trucks offer multi-temperature refrigeration units (Nuhn 2009) with separately controllable cooling sections, the software agent of the intelligent truck can use its Internet of Things Accessing Application and triggers the cooling actuators for adjusting the temperature of the compartments exactly to the needs of the refrigerated goods. The target values for the temperature and other dynamic values, the quality model for the truckload and the software agents and instruction settings may be communicated through the EPCglobal Network. If the cooling device is broken, the truck's software agent will be able to use the EPCglobal Network to find the next cold storage warehouse or send a request for a replacement truck with free capacities and a similar route. Hence, the decentralised intelligence and the connection to the Internet enable an autonomous, efficient and preference-based operation to critical incidents and help to strengthen the robustness of the logistic processes and the overall agility.

This solution utilizes autonomy for low level automation while instruction sets that are communicated to the agents through the EPCglobal Network are used to balance non-deterministic behaviour of autonomous systems and overall business targets.

5 Conclusion and Outlook

This article examined how the concept of autonomous objects in logistics can be integrated with the existing EPCglobal Network to contribute to a more holistic Internet of Things.

The Internet of Things combines physical objects with entities of information technology, including identification, real-time location, sensors, actuators and information processing technologies. It requires connectivity of objects and location based context aware services for business and social applications.

The concept of autonomous objects is focused on autonomy, heterarchy and decentralised decision-making. It aims for increased robustness, positive emergence and agility of the system. Technical implementation is based on object identification, dynamic data storing, data processing and data exchange as well as intelligent information based material handling, using flexible material based systems coordinated by intelligent objects.

To combine both concepts, the article examined the EPCglobal Network as an information source for autonomous objects and tried to answer how the decision processes are implemented, where data or information are stored, which technologies are used for the concepts and what potential applications both concepts have.

To form an integrative approach, the article proposed using the Internet of Things as an information broker for Autonomous Logistic Objects, while Autonomous Logistic Objects can reduce network dependency of intelligent objects and reduce network traffic load. The corresponding architecture, based on an extended EPCglobal Network, was designed and explained. The architecture shows several synergies and advantages. Standardized interfaces, as known from the EPCglobal Network, were suggested for the integration of sensors, actuators and software agents. Clustering mechanisms were proposed to reduce communication needs. Autonomous decision making that does not rely on continuous online connection and provides a higher degree of stability was combined with the concept of the Internet of Things.

Based on an example of an intelligent truck, a practical use-case was evaluated for the new concept. Overall, the integrated approach offers immense potentials for logistic applications and beyond.

Still, a combination of the Internet of Things and autonomous logistic processes leaves many open research issues to be solved. These include the guaranteeing that preference-sets and decisions are not outdated and conflicting decisions between centralised and decentralised infrastructures are resolved as well as determining different objects, either close or remote, as their appropriate interaction partners, and ensuring data integrity in case of replication conflicts.

References

Bullinger HJ, ten Hompel M (ed) (2007) Internet der Dinge. Springer, Berlin

Botts M, Percivall G, Reed C, Davidson J (2008) OGC® Sensor web enablement: overview and high level architecture. In: Nittel S, Labrinidis A, Stefanidis A (eds) GeoSensor networks 2006. Springer, Berlin

Bussmann S, Jennings N, Wooldridge M (2004) Multiagent systems for manufacturing control: a design methodology. Springer, Berlin

Böse F, Windt K (2007) Catalogue of criteria for autonomous control in logistics. In: Hülsmann M, Windt K (eds) Understanding autonomous cooperation and control in logistics. Springer, Berlin

Brown S, Bessant J (2003) The manufacturing strategy-capabilities links in mass customisation and agile manufacturing – an exploratory study. IJOPM. doi: 10.1108/01443570310481522

Chill-On Project (2009) Novel technologies for a safe and transparent food supply. Chill-On. http://www.chill-on.com/. Accessed 27 Aug 2009

Commission of the European Communities (2009) Internet of things – An action plan for Europe. European Commission. http://ec.europa.eu/information_society/policy/rfid/documents/commiot 2009.pdf. Accessed 26 Aug 2009

Fleisch E, Mattern F (eds) (2005) Das Internet der Dinge. Ubiquitous computing und RFID in der Praxis. Springer, Berlin

Fleisch E, Christ O, Dierkes M (2005) Die betriebswirtschaftliche Vision des Internets der Dinge. In: Fleisch E, Mattern F (eds.) Das Internet der Dinge. Ubiquitous computing und RFID in der Praxis. Springer, Berlin

Floerkemeier C, Fleisch E, Langheinrich M, Mattern F, Sarma SE (eds) (2008) The internet of things.In: Proceedings of the first international conference, IOT 2008. Lecture notes in computer science, Issue: 4952. Springer, Berlin

Gudehus T (2005) Logistik . Grundlagen – Strategien – Anwendungen, 3rd edn. Springer, Berlin

Hejl P (1990) Self-regulation in social systems. In: Krohn W, Küppers G, Nowotny H (eds) Selforganization: portrait of a scientific revolution. Kluver, Dordrecht

Hülsmann M, Windt K (eds) (2007) Understanding autonomous cooperation and control in logistics. Springer, Berlin

Laux H. (1998) Entscheidungstheorie. 4th ed. Springer, Berlin

Nuhn H (2009) Lastkraftfahrzeuge. Herbert Nuhn. http://www.kaelte-nuhn.com/transportkueh-lung/lkw.htm. Accessed 11 Mar 2009

Lopez TS, Ranasinghe DC, Patkai B, McFarlane M (2010) Taxonomy technology and applications of smart objects. Information systems frontiers, special issue on RFID. doi: 10.1007/s10796-009-9218-4

Mahnke W, Leitner S, Damm M (2009) OPC Unified architecture. Springer, Berlin

Overmeyer L, Nyhuis P, Höhn R, Fischer A (2006) Controlling in der Intralogistik mit Hilfe von pre processing labels. In: Pfohl H, Wimmer T (eds) Steuerung von Logistiksystemen – auf dem Weg zur Selbststeuerung. Dt. Verkehrs-Verlag, Hamburg

Pugachev V, Sinitsy, I (2002) Stochastic systems: theory and applications. World Scientific, Singapore

Philipp T, de Beer C, Windt K, Scholz-Reiter B (2007) Evaluation of autonomous logistic Processes – analysis of the influence of structural complexity. In: Hülsmann M, Windt K (eds) Understanding autonomous cooperation and control in logistics – the impact on management, information and communication and material flow. Springer, Berlin

Probst GJ (1992) Organisation. Strukturen, Lenkungsinstrumente und Entwicklungsperspektiven. Landsberg, verlag moderne industrie

Rekersbrink H, Makuschewitz T, Scholz-Reiter B (2009) A distributed routing concept for vehicle routing problems. In: Logist Res doi: 10.1007/s12159-008-0003-4

Sharifi H, Zhang Z (2001) Agile manufacturing in practice. Application of a methodology. IJOPM. doi: 10.1108/01443570110390462

Scholz-Reiter B, Delhoum S, Zschintzsch M, Jagalski T, Freitag M (2006) Inventory control in shop floors, production networks and supply chains using system dynamics. In: Tagungsband der 12. ASIM Fachtagung "Simulation in Produktion und Logistik, SCS Publishing House, Erlangen, pp 273–282

Scholz-Reiter B, Freitag M, de Beer C, Jagalski T (2005a) Modelling and analysis of autonomous shop floor control. In: Proceedings of the 38th CIRP international seminar on manufacturing systems. Florianopolis 2005 (on CD-ROM)

Scholz-Reiter, B, Freitag, M, de Beer, Ch, Jagalski, T (2005b) Modelling dynamics of autonomous logistic processes: discrete-event versus continuous approaches. Ann CIRP 55(1):413–417

Scholz-Reiter B, Uckelmann D, Gorldt C, Hinrichs U, Tervo, JT (2008) Moving from RFID to autonomous cooperating logistic processes. In: Miles SB, Sarma SE, Williams JR (eds) RFID technology and applications. Cambridge University Press, Cambridge

Sung J, Sanchez Lopez T, Kim D (2007) The EPC sensor network for RFID and WSN Integration infrastructure. PerComW'07. doi: 10.1109/PERCOMW.2007.113

A Design for Secure Discovery Services in the EPCglobal Architecture

José J. Cantero, Miguel A. Guijarro, Antonio Plaza, Guillermo Arrebola, and Janie Baños

Abstract The EPCglobal Network architecture defines a functional component, named Discovery Service (DS), whose role is to enable the gathering of complete information from multiple information providers across an object's lifecycle. The DS has great potential for improving business processes and providing new services to customers. However, at present, despite the progress in ratification of EPCglobal standards for other interfaces such as the EPC Information Service (EPCIS) or Object Naming Service (ONS), a technical standard for DS interfaces is still under development. In this early stage of the standardisation process, this chapter presents an analysis of potential models for DS design. We present a detailed design based on a model providing greater security. We have both implemented and deployed Discovery Service modules based on our selected model. Finally, we present solutions implemented to provide secure communications between different entities and fine-grained access control to DS records in a design that is scalable.

1 Introduction

The main motivation for RFID-based applications is to provide item-level visibility in business and logistic processes. By tagging items using a globally unique numbering scheme, such as the EPC (Electronic Product Code), and deploying RFID infrastructure throughout the supply chain, trace information can be collected and processed in an automated and efficient way. The item traces are relevant from a business perspective as they represent footprints of business processes, such as logistical or commercial operations.

J.J. Cantero (✉)
AT4 wireless. S.A. C/ Severo Ochoa, nº 2, Parque Tecnológico de Andalucía, 29590 Campanillas, Málaga, Spain
e-mail: jjcantero@at4wireless.com

D.C. Ranasinghe et al. (eds.), *Unique Radio Innovation for the 21st Century*, DOI 10.1007/978-3-642-03462-6_9, © Springer-Verlag Berlin Heidelberg 2010

But in order to take advantage of event data by developing new services and applications to improve business processes (e-Pedigree, track & trace, anti-counterfeiting, etc.), information from multiple partners across the supply chain must be shared. While organisations realize that information sharing is the key to achieve new business benefits, they are naturally reluctant to share information in an uncontrolled way. We assume the following framework for information sharing between organisations:

• Multiple organisations are involved during an item's life-cycle (from manufacturer to the point-of-sale and beyond) and therefore generate their own item traces with their RFID readers.
• Organisations want to retain control over their item traces, having local data storage for which they control access.
• Information sharing will only be done in purposeful arrangements of selected and known partners.

These assumptions imply the existence of distributed information repositories which are under the control of corresponding local organisations. According to the EPCglobal architecture framework (also called the EPC Network) (EPCglobal 2009), these distributed information repositories are named EPC Information Services (EPCIS) (EPCglobal 2007), and they provide an accessing interface to authorised partners. As supply chains consist of complex relationships, multiple EPCIS systems exist. In order to enable efficient information sharing, the main role of Discovery Services (DS) is to identify the information sources that actually contain information related to a certain EPC. By providing this functionality, a Discovery Service retains the local data governance in the EPCIS, while providing the basis for sharing information between supply chain companies in an efficient and secure way.

The EPC Network already provides some capabilities for tracking goods: given a certain EPC, the Object Name Service (ONS) (EPCglobal 2008) provides a URL (Uniform Resource Locator) of the company responsible for allocating the EPC, typically the brand owner. However, DS provides a list of URLs to multiple companies across the supply chain where information about a certain EPC is available (usually a list of links to companies' EPCIS). Figure 1 shows the complementary role of ONS and DS in relation to multiple EPCIS.

Unlike ONS, clients querying a DS will be required to provide authentication credentials and the amount of information returned in response to their query will be subject to filtering by an access control system based on policies. These policies will be defined by each information provider (publisher) that registers records in a DS, according to its business relationships with potential DS querying clients.

DS will need to be designed to accept updates in close to real time from multiple publishers across the supply chain or the lifecycle of an object (including organisations that handle the object beyond the point of sale or delivery, e.g. for repair purposes, maintenance, returns and reverse logistics, as well as recycling, remanufacturing and other end-of-life processes). Because DS store serial-level records,

Fig. 1 EPC Network lookup
services: ONS and DS

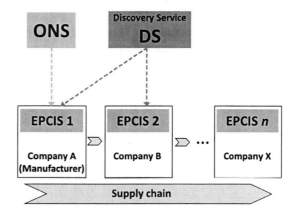

they will need to be sufficiently scalable to store large volumes of data, possibly up
to trillions of records per year. DS will also need to provide a secure environment
for authentication of both publishers and clients making queries. Policies to enforce
access control need to be managed in a scalable manner.

2 Analysis of Potential Models for Discovery Services

Key aspects to be taken into account in the DS design process (BT Research et al.
2007a) are:

- Minimise the information to be stored in DS databases (due to the potentially
  large volumes of data).
- Minimise query response time.
- Keep external interfaces independent of the underlying implementation technol-
  ogy, especially the database technology.
- Allow companies to control the visibility of their data.
- Obtain a scalable service from the point of view of data volumes, security, access
  control and performance.
- Be easily integrated into the EPC Network, facilitating the integration with
  EPCIS which comply with current standard (EPCglobal 2007).

Different models can be selected to design a DS considering the key aspects
identified. According to how query responses are obtained from the DS, different
design models may be considered. These models can be more broadly categorised
into two families (BT Research et al. 2007b):

- *Directory of Resources* model: DS clients receive a synchronous answer directly
  from the requested DS.

- *Query Relay* model: DS propagates the client query to resources, and client receives an asynchronous answer from resources (typically the EPCIS).

In this section both potential models are described and analysed.

2.1 Directory of Resources Model

The figure below shows the operation of a DS based on the Directory of Resources model.

The main processes of this model are the following:

- *Publishing to the DS*: The resources, consisting typically of EPCIS, publish selected information to the DS. In this model, the minimal information published to a DS must be the EPC, the unique identification number associated with the RFID tag, along with a link that can be used to communicate with the resource (usually an EPCIS). For an EPCIS with a web-service interface, this link would consist of the location of the web service interface description.
- *Querying DS (one-off query)*: The client queries the DS providing filtering conditions (at least one EPC or an EPC list/range). The query is answered by the DS without recourse to additional parties, providing a list of EPC numbers and associated links to EPCIS instances to the client. The EPCIS resources do not have to be available at the instant of the client query, although the EPCIS must be available at a later time to respond to the client's EPCIS query.
- *Querying EPCIS*: Once the list of addresses to EPC Information Services are received (step 5 in Fig. 2), the client directly contacts one or multiple of the returned EPCIS instances. This process can be a burden for the client when many EPCIS need to be contacted, but the client is in full control of its requests and may choose to alter the query in subsequent requests or may choose to abort the requests altogether.
- *Standing Queries*: The client may perform a subscription to the DS with query parameters. In this case, any new publication of records to the DS will be compared against the list of current subscriptions. Where client subscriptions match (and subject to security constraints) the subscribers are notified (or delegated to message queues) of the new records.
- *Advanced Queries*: Although in the 'pure' Directory of Resources model only references to resources must be published (as in step 5 in Fig. 2), optionally additional information may be included to provide business context (business step, state, etc.). Such information can facilitate more advanced client queries.

DS based on the Directory of Resources model must maintain a number of security policies related to:

- *Publications*: Specifying who can publish information to the DS, and what information they can/must publish.

- *Client Queries*: It is expected that the publishers of the DS records will have the ability to define security policies about who can see published records, maintaining fine-grained control over the visibility of the published information.

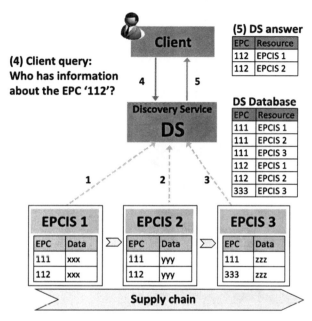

Fig. 2 Directory of resources model. Steps 1, 2 and 3: Publishing to the DS. Step 4: Query DS. Step 5: Response from DS

2.2 Query Relay Model

The key idea of this design is not to reveal any of the resources' data to clients (i.e., EPC numbers and resource references stored in the DS) but instead to propagate client queries to resources. DS forwards client queries to waiting resources (i.e., typically EPCIS instances), which can then respond directly to the client. Alternatively, this model also supports Standing Queries (see Sect. 2.1) for which a subscription is maintained (and stored) at the DS. Standing Queries are forwarded to resources as soon as they register new events matching the query.

Figure 3 shows the basic operation of the Query Relay model. EPCIS resources register at DS to receive client queries. This registration specifies the EPC numbers for which the EPCIS holds information and a reference to the EPCIS resource. This may optionally include further routing data (i.e., secondary keys), for example, the EPC event type (i.e., business step) to allow more selective routing of client requests.

The main processes of this model are the following:

Fig. 3 Query relay model

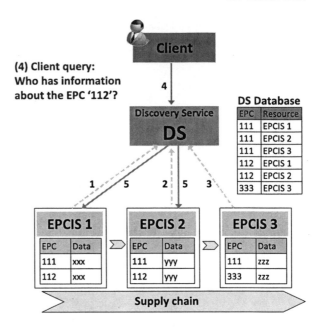

- *Relaying Queries*: Relaying the client's query has the advantage that EPCIS resources retain full data ownership, that is:

  - Resources keep full control of which data is released to whom because they maintain the access control to their data. No data from resources is revealed to clients directly.
  - Resources can log all (successful and failed) attempts to access their data.
  - Resources can deny access to certain clients unawares to a client because access control policies are typically considered sensitive data.

  Relaying queries on behalf of the client also releases the client from having to connect to and access the relevant resources directly. Without increasing complexity on the client side, the DS could implement dynamic strategies for parallel querying multiple resources concurrently and to cope with the idiosyncrasies of the network, intermitted disconnects, as well as with slow and unresponsive clients.

- *Routing Replies*: There are two principal alternatives for routing the query response back to the client:

  - Queried EPCIS can establish direct connection to the client in order to convey the query response, bypassing the DS.
  - Query response is routed through the DS, allowing the consolidation of the responses from multiple EPCIS resources. It also allows decoupling client

from resources, for example, to hide the client's network address from EPCIS instances and vice versa. In this case, the DS must be specially trusted to manage the confidentiality of the client requests.

- *Receiving Replies*: If query responses are returned to the client without consolidation at the DS, clients need to receive and combine, potentially, multiple replies. However, if the client has no indication about the number of replies to expect, it cannot decide when to terminate the query (i.e if and when all replies from resources willing to reply have been received by the client or if particularly "slow" replies are still underway).

2.3 Comparison and Conclusion

A key requirement is to protect the information stored in the DS from unauthorized client access. According to the Directory of Resources model, any DS client will have to pass two access controls: once at the DS (to obtain the list of EPCIS links) and once again at the subsequent query to the EPCIS to obtain detailed event information.

In the Query Relay model, security policies will determine which EPCIS resources are allowed to register with the DS for which EPCs and other associated routing data. They will also determine which clients are allowed to submit queries to the DS, and may optionally constrain the parameters of the query or the response messages that are allowed.

Access control policies for the Query Relay model serve a different function compared to the Directory of Resources model. In the latter they are used to protect the confidentiality of the resource, that is, to protect access to resources' sensitive data (EPCs and resource references) from unauthorized and potentially harmful clients. To serve this function they may need to be very fine-grained. In the Query Relay model, access control policies are intended to protect the systems from denial-of-service attacks and to reduce the query traffic reaching an EPCIS. This may be achieved with relatively lightweight access control mechanisms

The Table 1 summarizes a comparison between both models, according to different criteria.

Table 1 Main characteristics of different models

Characteristics	Directory of resources	Query relay
Client trust	Good	Concerns
Resources trust	Concerns	Good
Response latency	Good	Concerns
Standing queries	Yes	Yes
Access control	Complex	Simple

3 Discovery Service Design

Within the scope of the BRIDGE project, the Directory of Resources model was selected for the implementation and deployment of a running DS prototype, according to its characteristics related to security and response latency.

3.1 Discovery Service Architecture

Following the Directory of Resources model, the resultant DS architecture (Guijjarro et al. 2008) consists of a data repository (database), proxy interfaces and external interfaces to interact with external entities (see Fig. 4):

- EPCIS to publish event information
- Clients or user applications to make queries and receive replies.

The objective of the designed DS is to return a time-ordered list of links to multiple EPCIS instances that hold information related to a specific EPC (or a list/range of EPCs). Therefore, the DS is designed not to duplicate or aggregate information stored within each individual EPCIS repository, but to store only relevant information to be able to create the list of links. It is worth pointing out that this list of EPCIS links is constructed considering the identity of the client making the query and the corresponding access control policies defined by the DS publishers, generally according to the business relationship and level of trust with clients.

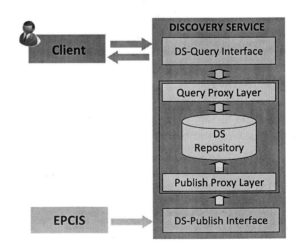

Fig. 4 Discovery service architecture

3.2 Information Storage Technology

Due to the large information volumes to be potentially managed by Discovery Services (information related to millions or even billions of individual objects), scalability (i.e. the ability to handle growing amounts of data and operations between them, including the capacity of the system to evolve in order to manage that growth) is the key criteria to select the data storage technology. Other criteria (Cantero et al. 2008) that must also be considered are: potential bottlenecks, response time, capacity to support multiple updates simultaneously, internal organization of data, robustness, etc.

Several technologies, such as LDAP (Lightweight Directory Access Protocol), DNS (Domain Name Server), DHT (Distributed Hash Tables) and Search Engines were compared (BT Research et al. 2007b). If we consider scalability then DHT and LDAP are the more suitable technology options for data storage. One of the most attractive characteristics of LDAP, which was the technology selected for the DS prototype, is an optimal response time to queries, due to the hierarchical organization of data storage. A 5-level hierarchical structure was selected for the development on the DS prototype (see Fig. 5).

However, DHT is clearly more promising in a large-scale real-life scenario, due to its decentralized model. DHT is designed to scale to a large number of nodes, allowing continuous node arrivals and failures by constructing a structured overlay network.

To allow interfacing to alternative data storage technologies, the DS architecture makes use of Proxy Interfaces (Publish Proxy Interface and Query Proxy Interface), which are dependent on the database technology. Their role is to interface between the external interfaces and the underlying database technology, providing flexibility to use any underlying database technology while keeping the same external interfaces.

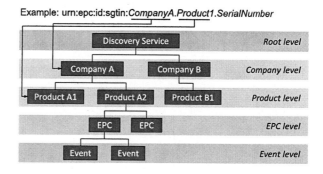

Fig. 5 DS prototype database structure based on LDAP

3.3 External Interfaces

The DS will provide two external interfaces to interact with other components of the network:

- DS-Publish Interface: to capture information from multiple information providers (mainly EPCIS).
- DS-Query Interface: to allow user applications to make queries and receive replies.

Web Services technology was selected for the external interfaces. The detailed design of the different interfaces has been made following, as much as possible, already existing methods. As an example, the description of one of the DS-Publish Interface methods is described below.

Method: publishRecord
Parameter: ds_publishRecord
Return: ds_publishRecordResponse
Parameter fields:
Return fields:

Field	Description
recordID (former uniqueID)	Unique identifier of a concrete record within the DS

Mandatory fields:

Field	Description
publisher ProfileID	Identifies which publisher provided this record
epcList	List of EPCs (Pure-identity URN) affected by the event. It SHALL NOT be present in the case of an aggregation event. This particular situation will be overcome by the definition of optional record fields related with aggregation. (eg. parentId and childID see below)
	Field Description
	epc EPC (Pure-identity URN, that will be useful to filter information)

Optional fields:

Field	Provided by	Description
action	Publisher	A string from an enumerated list. This is used to indicate how the record corresponds to a particular stage in the lifecycle of the object. Allowed values are "LINK" (default), "CREATE", "CLOSE", "DESTROY" In case of aggregation event, the action values can be: "ADD", "OBSERVE", "DELETE" Within this field all kind of events are unified
businessStep	Publisher	A vocabulary whose elements denote steps in the business process. E.g. an identifier that denotes "shipping"
disposition	Publisher	A vocabulary whose elements denote a business state of the object after the event happened. E.g. an identifier that denotes "available for sale" or "received"
eventTime	Publisher	The timestamp asserted by the publisher for this record. The timestamp should be expressed with a resolution of 1 second and be time zone qualified, relative to UTC. i.e. YYYY-MM-DDThh:mm:ssTZD (e.g. "1997-07-16T19:20:30+01:00") The Data Base implementation, in addition, will store a copy of the UTC time, in the interest of comparison possibilities. See http://www.w3.org/TR/NOTE-datetime E.g. YYYY-MM-DDThh:mm:ssTZD (i.e. "1997-07-16T19:20:30+01:00")
parentID	Publisher	Present only in case of Aggregation events. It represents the id of the parent (as EPC pure identity or URI)
childEPCs	Publisher	Present only in case of Aggregation events. It represents the list of EPCs added, observed or deleted from the parentID

Field	Description
epc	EPC (Pure-identity URN)

Example of publishing a basic record:

```xml
<?xml version="1.0" encoding="UTF-8"?>
<ds_publishrecord>
  <ds_header>
    <TTL>P1Y2M3DT10H30M<TTL>
  </ds_header>
  <ds_record>
     <publisherProfileID>
        7D4BA819CE274985F36F
     </publisherProfileID>
     <action>LINK</action>
     <eventTime>2005-04-03T20:33:31.116-06:00</eventTime>
```

```
   <epcList>
       <epc>urn:epc:id:sgtin:0614141.107340.1</epc>
       <epc>urn:epc:id:sgtin:0614141.107340.2</epc>
       <epc>urn:epc:id:sgtin:0614141.107340.7</epc>
       <epc>urn:epc:id:sgtin:0614141.107340.8</epc>
   </epcList>
   <bizStep>urn:epcglobal:cbv:bizstep:shipping</bizStep>
   </ds_record>
</ds_publishrecord>
```

3.4 Integration with EPCIS

The current EPCIS standard does not include a specific communication mechanism between an EPCIS and a DS to publish event information. For that reason, an integration module named IS-DS Module has been developed to be integrated with any software that complies with the EPCIS Capture interface standard (Fig. 6).

The IS-DS module's mission is to filter relevant EPCIS event information and to send it to the DS, using the DS-Publish Interface. For example, a filter might be set to trigger the publishing of a record to the DS when an object is first received by an organization or when it is finally shipped to another organization.

This integration module is transparent to the EPCIS, and thus it does not affect its functionality. It filters the events to be sent to the DS which are the records that ought to be published to the DS. The IS-DS Module developed has been successfully

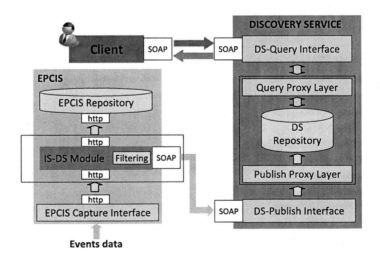

Fig. 6 Integration with EPCIS: IS-DS module

integrated with the EPCIS implementation developed by the Fosstrak open source project (Fosstrak 2010), formerly named Accada, and tested in RFID deployment trials conducted within the BRIDGE project (*http://www.bridge-project.eu*).

4 Security Issues

Mechanisms for implementing authentication, integrity, non repudiation and confidentiality of communications between different entities should be included in the core functionalities of the EPCglobal architecture. Clients using networked services, such as EPCIS and DS, will be required to provide authentication credentials. The amount of information exchanged will be subject to be filtered by access control policies based upon the authentication credentials they supply which will be related to the business relationship they have with each provider of information.

This section describes a security model for communications between different entities (Burbridge et al. 2007) and a security framework to handle and enforce access control policies (Soppera et al. 2009) in a manageable and scalable way.

4.1 Security Model for Communications in the EPCglobal Network

The security model designed to achieve secure communication between different entities in the EPCglobal Network (such as EPCIS and DS) is based on a Public Key Infrastructure (PKI) model. Using digital signatures, encryption and decryption technologies, a PKI provides:

- *Confidentiality*: Protects privacy by ensuring that electronic communications are not intercepted and read by unauthorized persons
- *Integrity*: Assures the integrity of electronic communications by ensuring that they are not altered during transmission
- *Authentication*: Verifies the identity of the parties involved in an electronic transmission
- *Non-repudiation*: Ensures that no party involved in an electronic transaction can deny their involvement in the transaction and it provides authenticated, confidential and non-reputable communications

In a first approach, the PKI model for the DS prototype is based on a single PKI architecture in which a unique Certificate Authority (CA) provides the PKI services to sign and issue the certificates for all the entities: EPCIS, DS and clients. Each entity has an X.509 certificate (ITU-T), with a private key and a public key, which are used to encrypt and decrypt different messages.

The technology used to integrate the security model in the DS prototype is WSS4J (Web Services Security for Java), using the SOAP protocol (W3C 2007). The WSS4J technology, that implements the OASIS Web Services Security (OASIS 2007), was applied to achieve the different security capabilities expected:

- *Confidentiality*, achieved with XML-Encryption (W3C 2002) encrypting the SOAP message.
- *Integrity* and *Non-Repudiation* is achieved with XML-Signature (W3C 2008) signing the SOAP message.
- *Authentication*, achieved with XML-Signature signing the SOAP message and the use of a PKI architecture.

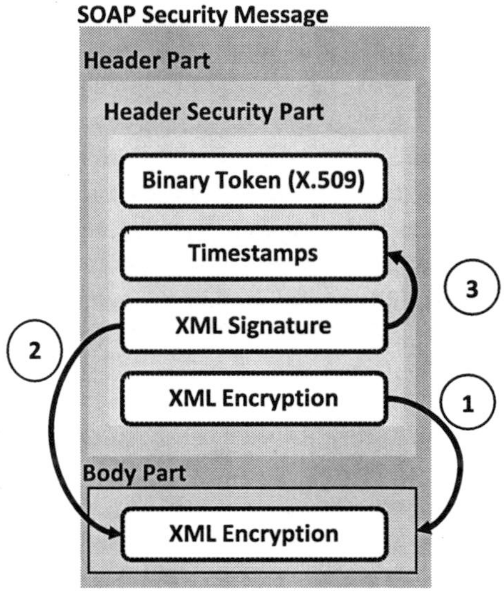

Fig. 7 SOAP message

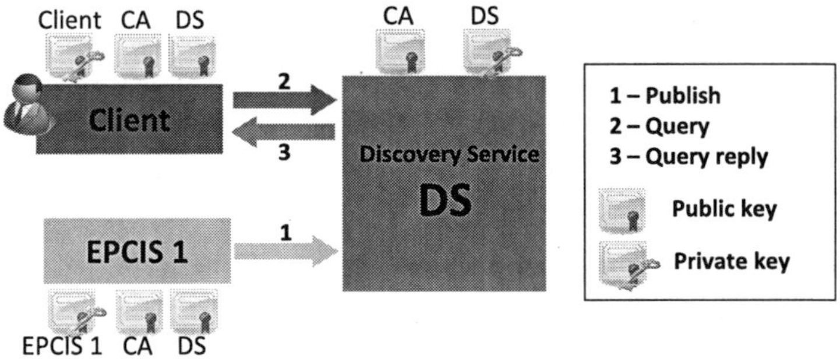

Fig. 8 Security model for communications

To prepare the message, the following general procedure is performed: (1) the SOAP message body is encrypted, (2) signed with the sender's private key and (3) a signed timestamp is added to the SOAP message header (thus creating a unique message). In addition, the ITU-T X.509 certificate (public key) of the sender is included in the message header.

Figure 8 illustrates the security scenario described above, showing the different certificates (public and private keys) with each entity. DS and CA public keys have been previously sent to EPCIS and to DS clients authorised to make queries.

Table 2 Communications between different entities

Code	Communication	Interface	Operation
1	EPCIS DS	DS-Publish	Publish
2	Client DS	DS-Query	Query request
3	DS Client	DS-Query	Query reply

Table 3 Security characteristics of different messages

Characteristic	(1) Publish	(2) Query request	(3) Query reply
Includes sender certificate	YES	YES	–
Body signed and encrypted	YES	YES	YES
Includes Timestamp	YES	YES	–

Secure communications in this scenario are described in Table 2. Table 3 describes the main security characteristics of different messages according to the security model defined.

4.2 Security Framework for Access Control

The aim of the security framework is to control access to data or computing resources within a security domain. The access control security framework is based on ITU-T X.812 "Access Control Framework" (ITU-T 1995) and the eXtensible Access Control Markup Language (XACML) (OASIS 2009) technology, defined by OASIS (Organization for the Advancement of Structured Information Standards).

Taking into account the XACML specification, the developed access control security framework is mainly composed of the following entities:

- *PDP (Policy Decision Point)*: it evaluates applicable policies and renders an authorisation decision (permit or deny). It is built using web services technologies and can be accessed through two separate interfaces implemented using XACML technology: Management interface (to add, delete and query policies) and Operational interface (to evaluate a concrete policy). A database, based on PostgreSQL, is used to store policies.
- *PEP (Policy Enforcement Point)*: it performs access control, by making decision request to PDP and enforcing authorization decisions. Petitions to access DS (Publish or Query) are intercepted by PEP, which generates an access request (implemented in XACML) to PDP, including relevant information extracted from the petition. Once decision is taken (permit or deny), PDP sends it to PEP, which resends the original petition to DS or stops the operation.

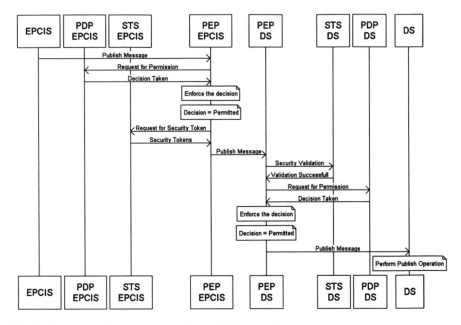

Fig. 9 Messages exchanged to authorise a publish operation

- *PAP (Policy Administration Point)*: it is in charge of managing the policies stored in a PDP using the PDP management interface (to create, remove or modify policies).
- *STS (Security Token Service)*: it is in charge of issuing and validating security tokens in order to provide secure communication between entities. The PEP requests the security tokens from the STS to secure DS communications.

Figure 9 illustrates an example of the behaviour of the security framework during the process needed to authorise an EPCIS to publish information into a DS.

4.2.1 Security Framework Validation Environment

To demonstrate the feasibility of this security framework, a validation environment was deployed and integrated with the DS prototype (see Fig. 10).

This validation environment is composed of two EPCIS, one DS, one PDP, two PEPs (one for publishing and the other for querying) and one PAP. Some applications have been developed to emulate the functionality of generating events (Publish Application) and query information (Query Application). The policy evaluation and the decisions are centralized in a unique PDP. Nevertheless, this validation environment does not take into account a STS entity due to its functionality being directly incorporated in each entity (EPCIS, PEP, DS and Query application) by means of ITU-T X.509 certificates included on the same entity.

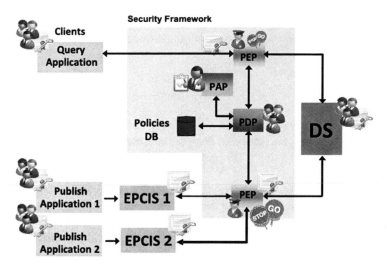

Fig. 10 Security framework validation environment

4.2.2 Access Control Policies

The definition of policies is the cornerstone of the security framework defined. XACML language was selected for policy definitions, providing syntax for a policy language as well as the semantics for processing policies. A policy is defined by means of: Subject (who is going to do the petition), Action (what the subject wants to do), Resource (where the action is made) and Rule (decision to take: permit or deny). The general structure of a XACML policy is shown below:

```
<Policy>
  <Target>...</Target>
  <Rule RuleId="Rule1" Effect="Permit">
    <Target>
    <Subjects>AnySubject</Subject>
    <Resources>AnyResource</Resources>
    <Actions>AnyAction</Actions>
    </Target>
    <Condition>
    </Condition>
  </Rule>
</Policy>
```

Defining policies to validate the security framework was a challenge. DS-Publish and DS-Query interfaces do not define a field to include an identifier for identifying the subject of the operation (an EPCIS trying to publish information or a Client making a query). Therefore it was unclear as to how to identify the subject. However, because entities involved in this scenario (DS, EPCIS and Clients), use ITU-T X.509

certificates for unequivocal authentication in their communications, a solution was to use the Subject information included in the certificate, because it is unique.

5 Conclusions

There are different models to design a Discovery Service (DS). The feasibility of the Directory of Resources model has been demonstrated by designing, implementing, deploying and validating a DS prototype integrated in the current EPCglobal Network

Communication security objectives (authentication, integrity, confidentiality and non-repudiation) have been achieved using a communications security model based on standardised technologies.

The security framework for access control presented provides publishers of information with mechanisms to define who can access what information in a scalable and secure way. This security framework is equally suitable for integration with other networked services, such as EPCIS or even services not related with the EPCglobal Network.

Policy definitions are a cornerstone of the deployment of access control security framework for networked services. XACML has demonstrated its flexibility to define fine-grained access control policies in order to fit challenging requirements needed in complex supply chain scenarios.

In addition, different areas where further research work could be done have been identified, such as:

- Design and prototyping of a DS-DS interface to create a federation of DS.
- Prototyping of a DS using DHT (Distributed Hash Tables) as information storage technology.
- Performing of DS trials in real scenarios with living supply chains, to obtain performance data and to evaluate benefits on business and logistic processes.
- Research on tools to help final users in the definition and management of complex fine-grained policies, based on XACML.

The DS prototype detailed design has been contributed to standardization organisations (such as EPCglobal and IETF) and the source code is being distributed by AT4 wireless under LGPL license to a number of companies and research initiatives interested in DS.

Acknowledgments This work has been developed within the BRIDGE project scope (http://www.,bridge-project.eu), partially funded by the European Commission 6th Framework Program (contract number IST-2005-033546). The authors would like to thank the BRIDGE project partners involved in the Work Packages "Serial-level lookup services" and "Security", and specially Mark Harrison (University of Cambridge), Trevor Burbridge (BT), Andrea Soppera (BT) and Oliver Kasten (SAP) for their collaboration in the design of the DS prototype and the access control Security Framework.

References

Burbridge T, Broekhuizen V, Farr J et al (2007) RFID Network confidentiality. http://www.bridge-project.eu/data/File/BRIDGE_WP04_RFID_Network_Confidentiality.pdf. Accessed 3 Mar 2010

Cantero JJ, Guijarro MA, Arrebola G et al (2008) Traceability applications based on Discovery Services. In: 13th IEEE international conference on emerging technologies and factory automation (ETFA), Hamburg

EPCglobal (2009) The EPCglobal architecture framework. http://www.epcglobalinc.org/standards/architecture/architecture_1_3-framework-20090319.pdf. Accessed 3 Mar 2010

EPCglobal (2008) Object naming service (ONS) version 1.0.1". http://www.epcglobalinc.org/standards/ons/ons_1_0_1-standard-20080529.pdf. Accessed 3 Mar 2010

EPCglobal (2007) EPC information service (EPCIS) version 1.0.1. http://www.epcglobalinc.org/standards/epcis/epcis_1_0_1-standard-20070921.pdf. Accessed 3 Mar 2010

Fosstrak (2010) http://www.fosstrak.org/epcis/index.html. Accessed 3 Mar 2010

Guijjarro MA, Arrebola G, Cantero JJ et al (2008) Working prototype of serial-level lookup service. http://www.bridge-project.eu/data/File/BRIDGE_WP02_Prototype_Serial_level_lookup_service.pdf. Accessed 3 Mar 2010

ITU-T (2005). Information technology – open systems interconnection – the directory: public-key and attribute certificate framework, recommendation ITU-T X.509. http://www.itu.int/rec/T-REC-X.509-200508-I/en. Accessed 3 Mar 2010

ITU-T (1995), Information technology – open systems interconnection – security frameworks for open systems: access control framework, recommendation ITU-T X.812. http://www.itu.int/rec/T-REC-X.812/en. Accessed 3 Mar 2010

Soppera A, Burbridge T, Bowman Paul et al (2009) Final report on network confidentiality. http://www.bridge-project.eu/data/File/BRIDGE_WP04_Final_Report_on_Network_Confidentiality.pdf. Accessed 3 Mar 2010

OASIS (2009) "eXtensible access control markup language (XACML) version 3.0". http://www.oasis-open.org/committees/document.php?document_id=32425. Accessed 3 Mar 2010

OASIS (2007) WS-Trust 1.3. http://docs.oasis-open.org/ws-sx/ws-trust/200512/ws-trust-1.3-os.pdf. Accessed 3 Mar 2010

University of Cambridge, AT4 wireless, BT Research et al (2007a) Requirements document of serial level lookup service for various industries. http://www.bridge-project.eu/data/File/BRIDGE WP02 Serial level lookup Requirements.pdf. Accessed 3 March 2010

University of Cambridge, AT4 wireless, BT Research et al (2007b) High level design for discovery services. http://www.bridge-project.eu/data/File/BRIDGE WP02 high level design discovery services.pdf. Accessed 3 Mar 2010

W3C (2008) XML Signature syntax and processing (Second Edition), W3C recommendation. http://www.w3.org/TR/2008/REC-xmldsig-core-20080610/. Accessed 3 Mar 2010

W3C (2007) SOAP Version 1.2 Part 1: Messaging framework (Second Edition) W3C Recommendation. http://www.w3.org/TR/2007/REC-soap12-part1-20070427/. Accessed 3 Mar 2010

W3C (2002) XML encryption syntax and processing. W3C recommendation. http://www.w3.org/TR/2002/REC-xmlenc-core-20021210/. Accessed 3 Mar 2010

Evaluating Discovery Services Architectures in the Context of the Internet of Things

Elias Polytarchos, Stelios Eliakis, Dimitris Bochtis, and Katerina Pramatari

Abstract As the "Internet of Things" is expected to grow rapidly in the following years, the need to develop and deploy efficient and scalable Discovery Services in this context is very important for its success. Thus, the ability to evaluate and compare the performance of different Discovery Services architectures is vital if we want to allege that a given design is better at meeting requirements of a specific application. The purpose of this chapter is to provide a paradigm for the evaluation of different Discovery Services for the Internet of Things in terms of efficiency, scalability and performance through the use of simulations. The methodology presented uses the application of Discovery Services to a supply chain with the Service Lookup Service Discovery Service using OMNeT++, an open source network simulation suite. Then, we delve into the simulation design and the details of our findings.

1 Introduction

The "Internet of Things" promises humans to live in a smart, highly networked world, which allows for a wide range of interactions with this environment (Quack et al. 2008). However, Internet of Things is still a new thematic area and there are many important challenges to be solved before it becomes a reality. EPCglobal (EPCglobal 2009b) is working on standards that facilitate the collection, storage, update, location and sharing of information regarding uniquely identified objects but there are still many open issues. EPCglobal's unique identification standards are based on RFID (Radio Frequency Identification) technology. RFID will enable computers to automatically recognize and identify everyday objects, and then track,

E. Polytarchos (✉)
Department of Management Science and Technology, ELTRUN Research Lab, Athens University of Economics & Business, Athens, Greece
e-mail: ipoli@intracom.gr

D.C. Ranasinghe et al. (eds.), *Unique Radio Innovation for the 21st Century*,
DOI 10.1007/978-3-642-03462-6_10, © Springer-Verlag Berlin Heidelberg 2010

trace, monitor, trigger events and perform actions on those objects (Meloan 2003) This technology will serve to extend the "Internet of Computers" to the "Internet of Things". Researchers study Internet of Things from different perspectives and address different aspects of the topic, including object recognition (Quack et al. 2008), security – privacy (Evdokimov et al. 2008; Grummt and Muller 2008; Kurschner et al. 2008), modular software development (Rellermeyer et al. 2008), cost and economical issues (Bottani and Rizzi 2008; Decker et al. 2008), discovery services (Beier et al. 2006; Kurschner et al. 2008; Polytarchos et al.) etc.

The "Internet of Things" will offer many opportunities to businesses and organizations, but also numerous problems may emerge, with the increasing amount of the available information. Aspects of information privacy, security and integrity will be critical for the adoption of the new services which will be provided through the "Internet of Things". These problems should be dealt with before the massive deployment of "Internet of Things" enabled systems. In order to address problems like privacy and information management, we have to provide a suitable, flexible underlying discovery service. Consequently, scalability and performance of discovery services are major factors which should be considered during the design of such services.

In this chapter, we look into alternative architectures and protocols that can be used to implement Discovery Services and compare them in terms of scalability and performance. The case of traceability in the supply chain is used as the context of using Discovery Services, as it is a challenging and unstable environment with many items (i.e. unique product instances) which are always on the move. In addition, there are many information and service providers which increase the complexity of the system. Simulation results allow drawing some conclusions, both on technical issues and business issues affecting or affected by the design of Discovery Services. By measuring the load on the Discovery Services infrastructure, the necessary equipment is estimated and the appropriate architecture is selected. Also by changing the structure of the supply chain, such as the amount of sales and the number of retailers and suppliers, we can estimate the impact of these changes on the overall architecture.

2 Discovery Services in the Context of the Internet of Things

The existence of highly efficient and scalable Discovery Services is fundamental to the existence and the ubiquity of the Internet of Things. Alternative terms used in the literature include "lookup service", "directory service" and "naming service" Discovery Services offer trading partners the ability to find all parties who have had possession of a given object and to share events about that product. They provide pointers to multiple providers of information across a supply chain (indicate the addresses of information services of all organizations that hold information about a given EPC (Electronic Product Code) (Bridge Project High Level Design for Discovery Services).

The EPC is an alphanumeric designation that uniquely identifies an object in the supply chain. Identification numbers are used to retrieve information associated with objects in the supply chain. This numbering scheme has been standardised by EPCglobal (EPCglobal 2008).

The main requirements for Discovery Services (Kurschner et al. 2008) can be viewed as the following:

- Organizations should be able to manage the data that they have stored in the Discovery Services infrastructure. Such data contain the addresses of the EPCIS nodes that belong to them, the data entries for objects that have been committed by them and the access rights for the same data.
- The confidentiality of the users of the infrastructure should be ensured by the infrastructure itself.
- The design of the Discovery Services should guarantee the availability and the reliability of the system.
- Changes in relations between entities that participate in the Internet of Things should not affect their interactions with the Discovery Services.
- The Discovery Services should incorporate an architecture that will encourage the participation in the EPCglobal network.
- The architecture of the Discovery Services should be scalable, in order to be able to cope with increasing numbers of objects and the new nodes (e.g. EPCIS servers) that participate in the EPCglobal network.
- The infrastructure replies to the requests should be complete, valid and consistent with the security rights of the stakeholders.

The amount of published research work about Directory Services is very limited and most proposed architectures (Cho et al. 2007; EPCglobal 2008: Version 1.0; Jin et al. 2007), follow a hierarchical structure that does not deviate largely from the general idea of a directory service. Most of these architectures are centralized and have not been designed in order to facilitate a distributed, dynamic and unstable environment. The amount of exchanged information within the Internet of Things is very high and demands more resources in contrast with a directory service approach such as UDDI (OASIS web site) and ONS (EPCglobal 2008: Version 1.0). In addition, the continuous changes of information and location generate new requirements for designing the discovery service. Also, many researchers try to alleviate the problem of security and privacy. There are no suggestions which focus on efficiency and scalability of information sharing atop such a heterogeneous collection of ubiquitous devices (Rellermeyer et al. 2008).

In the following sections we will concisely review some Directory Services designed for the Internet of Things.

2.1 Object Name Service

The ONS (Object Name Service) specification (EPCglobal 2009a: Version 1.0) describes a system based on the standard DNS (Domain Name System) infrastructure of the Internet, which provides a means to lookup EPCIS services for the EPC of the item in question.

The ONS points to a number of EPCIS, in a similar way to how the DNS points to an IP address. The main difference is that the ONS resolves an EPC to a list of URLs (Universal Resource Locators) as opposed to resolving a URL to an IP address, as done by a DNS. The most crucial problem is that ONS is designed as a centralised service that can only lookup information about a product type as opposed to information about an individual instance of an object. Using the ONS infrastructure to provide fine grain access to individual product level information does not yield a scalable solution as the number of object in the Internet of Things increase over time. Additionally, a security analysis (University of Cambridge, AT4 wireless, BT Research, SAP Research 2007) of the Object Name Service proposal shows that the Domain Name System is not a good skeleton for building Discovery Services. One other issue is that the movement of an item in the supply chain creates problems relating to the reliability and the accuracy of the data regarding the procedure to update the ONS records (Polytarchos et al.).

2.2 EPC Discovery Services

The EPC Discovery Services (Bridge Project High Level Design for Discovery Services) constitute an additional level to the network defined by EPCglobal. They have been designed to provide links to different sources of information on a specific object. These sources are entities (e.g. organizations, factories, retail stores) that collected data for an object at some point during the lifecycle of the object. This allows the client to discover information from multiple providers, thus yielding more complete and accurate information, using the standard EPCIS of the information providers.

The specification of Discovery Services by the EPCglobal has not been finalised yet, but there are two candidate architectures that have been described in (OpenSLP).

2.2.1 Directory Service

The first candidate is the RFID Directory Services. Figure 1 shows the operation of a directory-based Discovery Service proposed in (Bridge Project High Level Design for Discovery Services).

The RFID Directory Service also supports standing queries, where a client wishes to receive all future updates about new events received by an EPCIS repository, which match the client's specified query criteria. That requires a subscription request to be sent to the Directory Service from the client. Afterwards, whenever the EPCIS updates the information on the Directory Service, it will notify the appropriate clients regarding the change.

The Directory Service contains a list of pairs of EPCs and EPCIS servers that can provide services and data. The process of discovery is as follows:

1. The EPCIS (R) informs the Directory Service (I) for the EPCs it has information about,
2. The client (C) requests from the Directory Service the EPCIS that holds the data for an EPC and then
3. The client requests from that EPCIS a service.

Fig. 1 RFID Directory service

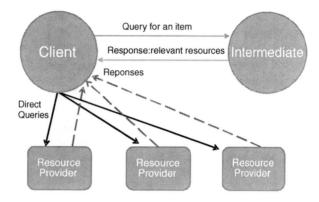

2.2.2 Query Relay

The second candidate is the Query Relay, which is depicted in Fig. 2. In the Query Relay Discovery Service scheme, the client's (C) queries are forwarded by the inter-mediary (I) to the appropriate EPCIS (R) and the EPCIS in turn connects with the client, in order to provide a service. The Query Relay architecture provides a more secure approach. The EPCIS may choose not to respond to a client's query. As a result the client is not capable of knowing whether there is an EPCIS available for a specific object.

As for the standing queries, the subscription is propagated to the EPCIS, who in turn notifies the required clients for the appropriate changes.

Given that information on the flow of products is commercially sensitive infor-mation, it is obvious that the companies will be reluctant to provide this information to anyone. Thus, in order for an EPCIS to appear on a result list of the EPC Discovery Service, the consent of the owner of the EPCIS will be necessary.

Although the specific requirements of the EPC Directory Service have been determined, the technical specification and a standard are still work in progress (EPCglobal 2009b).

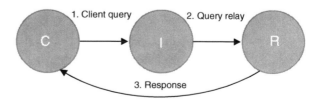

Fig. 2 Query relay

2.3 Service Lookup Service

SLS (Service Lookup Service) (Polytarchos et al.) is based on three-level hierarchical structure. It has been designed in order to resolve the deficiencies of the ONS and provide a more efficient and scalable counterpart.

In this chapter, we focus on the evaluation of the performance and scalability of the SLS architecture, as it shares a number of common characteristics with the first candidate architecture of the EPC Directory Service (the RFID Directory Service), since both essentially respond to queries regarding EPCs by providing them with a pointer (i.e. a URL) to the EPCIS (or the OSIS, as far as the SLS is concerned). The SLS though, should be more scalable, as it defines one additional abstraction layer, the OD, while the role of the ONS in the EPC network is not clearly defined yet.

In the sections below, the different layers of the SLS will be concisely defined. More information can be found in (Polytarchos et al.).

2.3.1 Object Directory

The highest level stores information about the location where someone could ask questions about a specific object.

The Object Directory (OD) is the DNS equivalent in the world of objects. But instead of providing naming resolution based on a URL, it provides another URL based on the object's EPC. The provided URL should point to this Object's Service Directory. The Object Directory should be built and operated as a DNS system, like the ONS. The difference with ONS is that records do not point to an EPC Information Server but to a specific Objects' Services Directory.

Every query to an OD server contains the EPC of an object and the OD should respond to it with the address of an Object Services' Directory Server (explained below).

2.3.2 Object Services' Directory

The second (hierarchically) level stores information regarding the information services that a specific object supports. Along with the information about the supported object services, this level holds information for the location of each object service information holder for each object that controls, i.e. the link to the third level of the infrastructure, the Object-Service Information Server.

Every Object Services' Directory (OSD) server contains information about a certain type of product (i.e. an object class, which is a fraction of the EPC code).

The third level is the actual object service information holder for a collection of object-service pair (Object-Service Information Server). The Object-Service Information Server is the level that holds the actual information.

2.3.3 Object-Service Information Server

The third level is the actual object service information holder for a collection of object-service pair (Object-Service Information Server). The Object-Service

Information Server (OSIS) is the level that holds the actual information. OSIS is the equivalent to the EPCIS in the EPCglobal architecture.

Optimally, every node of the supply chain (i.e. manufacturer, supplier, retailer) should have its own OSIS.

2.4 Other

Some additional proposals for Discovery Services in the Internet of Things are presented below.

The Service Location Protocol (SLP) (Afilias) was originally an Internet Engineering Task Force (IETF) standards track protocol that provides a framework to allow networking applications to discover the existence, location, and configuration of networked services in enterprise networks. However, SLP is not a global resolution system for the entire Internet; rather it is intended to serve enterprise networks with shared services, which makes it unsuitable for EPC Discovery Services (Kurschner et al. 2008).

In (Kurschner et al. 2008) C. Kurschner et al. propose the main requirements for the Discovery Service design and counter this issue by designing a "Query Relay" solution which focuses on privacy and security of information sharing. They suggest a "Query Relay Entity" which allows full access control at each organization.

In (Beier et al. 2006) S. Beier et al. provide the first ever implementation of Discovery Services and showcase their feasibility and usefulness. They try to address security, privacy and scalability, but it is a relatively simple implementation and not sufficiently mature in order to be adopted to support the whole "Internet of Things".

In (Evdokimov et al. 2008) S. Evdokimov et al. face the issue of ONS unipolarity which is a security – anonymity problem. They present MONS (Multipolarity for the ONS), which is an architecture designed in order to achieve multipolarity in the ONS and show how multipolarity in corresponding authentication extensions can be achieved. However, this approach inherits the aforementioned disadvantages of the ONS and cannot be used as a Discovery Service solution.

There is also a corporate solution by Afilias (Omnet++ web site). Afilias Discovery Services (ADS) leverages an open web services protocol, called the Extensible Supply-chain Discovery Service (ESDS) and provides a complementary Object Name Service (ONS) that is compliant with EPCglobal standards. The architecture of ADS depends on ONS and EPCglobal standards but does not address security and scalability issues.

3 Evaluating Discovery Services

The approach of each of the aforementioned architectures to the requirements of the EPC Directory Services, focuses on different issues, presents its own merits and has different priorities. Thus, there is a need to compare and evaluate the different architectures of Discovery Services, in order to be able to determine which one of them could be a better fit for the Internet of Things, or could better serve the need of an organization for an internal Discovery Service.

The evaluation of a Discovery Service should consider the following parameters, as they are drawn from the requirements in Sect. 2: Efficacy, Efficiency, Security, Scalability.

In this chapter we provide a paradigm for the evaluation of the **scalability** of Discovery Services through simulation and apply it on the SLS Discovery Services architecture.

In this simulation SLS represents the hierarchical approaches of the discovery services. We assume that SLS is equivalent to the ONS in terms of scalability at the worst case scenario. This assumption is valid because SLS has one more level than EPC architecture.

4 Simulation Design

In the following sections, the methodology of the simulation will be described. The detailed results of the simulation can be viewed in the Appendix A.

4.1 Simulation Environment

In order to evaluate the discovery services' architectures, a simulation of a supply chain was designed and used, in order to implement the SLS approach. The platform where the simulation was deployed is the OMNeT++. OMNeT++ is a suite that enables the simulation of discrete events. It has been widely used to simulate computer networks and other distributed systems. One of the biggest assets of the OMNeT++ is that it is open source, making it highly configurable. It has been designed in order to support large networks consisting of reusable elements. Monitoring the simulation models can be achieved through a powerful graphical user interface, which enables the user to monitor every component of the simulation, as well as the information flow between them. These properties of the OMNeT++ made it ideal for the creation of the simulation model.

Though, it is worthy to note that OMNnet++ failed to compute results when the number of retail stores got bigger than 5. This fact denotes that even in a relatively small supply chain, the load in the infrastructure of the Discovery Services is massive. Thus, a highly scalable architecture is more than desirable. The SLS copes with this load by allowing the addition of more Object Service Directory Servers.

4.2 Topology

The context of the simulation is a retail supply chain. Each retailer has a number of retail stores and a central warehouse. The stores and the warehouses are considered

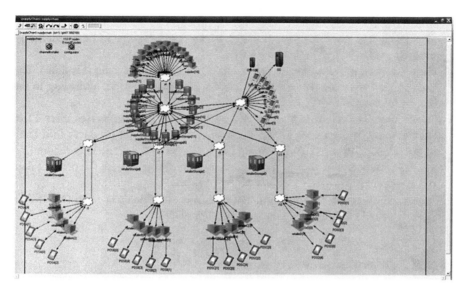

Fig. 3 Simulation environment

as nodes of the simulated network. The suppliers contribute to the network with two nodes each; the supplier's warehouse and the factory.

The simulation also contains the three levels of the SLS Discovery Service:

1. An Object Directory Server (OD – 1st level)
2. A number of servers that implement the Object Service Directory (OSD – 2nd level). In our experiment, product type is irrelevant and not capable of affecting the results. On the other hand the number of products is expected to be an important variable affecting the simulation; thus, if we accept that every OSD server manages products of a single type (provided by the object class), there will be a single OSD server.
3. The 3rd level of the SLS consists of the supply chain nodes, since they are responsible for providing the services and the information for the objects (Object-Service Information Servers). The structure of the network of the simulation is depicted in Fig. 3.

4.3 Main Simulation Scenario

In order to evaluate the SLS Discovery Service, the nodes of the model's supply chain, will adhere to the following rules:

- The simulation begins at time 0 with sales taking place in the retail stores for 12 h. The sales follow a Poisson(λ) distribution. Thus, the sale events are randomly distributed in the 12 h and their mean number is λ.

- After 12 h, the sales stop and ordering of products to the retailers' warehouses takes place. The number of the ordered products will be equal to the number of products sold.
- The products that were ordered arrive after 12 h (no sales during this time). The retail stores undertake the task to update the discovery service, according to the SLS architecture.
- The retailers' warehouses place orders to the suppliers warehouses after 12 h. The amount of products order equals to the total amount of products sold from all the retailers' stores. The distribution of the order to the retailers is random.
- After the retailers' warehouses receive the products, they update the product data in the SLS.
- Finally, whenever the factory produces a number of products in order to fulfill an order, it updates the SLS with the produced products.

4.4 SLS Infrastructure Simulation

The following rules have been established throughout the simulation of the SLS Discovery service:

- Every time a node receives an object:
  1. A query for this object's EPC is sent to the Object Directory Server, in order to locate the Object Service Directory Server that is responsible for this object.
  2. When the node receives the reply with the Object Service Directory Server, it subsequently informs it about the services for this object.
- Whenever a supplier produces an object (product), he undertakes the task to update the designated Object Service Directory Server (OSDS).
- Whenever an OSDS receives an update for a previously unregistered object, the Object Directory (OD) is appropriately informed for the existence of this new object.
- When the OD receives a query concerning a product, the address of the OSDS responsible for this object is returned.

4.5 Variables and Metrics of the Test Model

In order to extract useful conclusions from the architecture of Directory Services used, certain parameters of the experiment were modified and the implications of these modifications were documented.

The parameters that were modified are:

- The number of retail stores per retailer
- The mean number of sales per day
- The number of suppliers

- The method of updating the SLS infrastructure when receiving or producing new products. Specifically:

  - For every product received or produced, an update was transmitted.
  - For every 1000 products received, a cumulative update was transmitted.

The implications of the modifications were evaluated using the methods below:

- *Load* in the first level of the SLS hierarchy.
- *Load* in the second level of the SLS hierarchy.
- Maximum number of the retail stores that the infrastructure can support, in the context of the capabilities of the simulation platform.

The *Load* is defined as the number of messages or queries that an SLS node accepts during an hour.

5 Results

In Appendix A the results from the different simulation experiments are presented. A full statistical analysis of the results was not possible due to the limitations of OMNnet++ and the limited processing power of our servers, even though it would have been highly desirable and would have proved the statistical significance of the results. A full statistic analysis of a large scale simulation of a complicated supply chain with product movement that is based on scenarios that directly reflect a real world supply chain would be able to provide answers on the number of servers that would be required to achieve optimal performance.

Based on the results presented in the Appendix A, we can derive a number of useful conclusions and shed light on the issues that Discovery Services will have to deal with in the Internet of Things.

In the diagrams that depict the load of the infrastructure (see Figs. 5, 6, 8, 10, 12, 14, 16 and 18 in the Appendix), the higher peaks represent the transaction between the retailers and the SLS when the former receive new products. The lower peaks on the other hand represent the transactions between the suppliers and the SLS, where the suppliers inform the SLS of the new products that were produced. Based on these results, the fact that the SLS will deal with a very high load (Bottani and Rizzi 2008) for small periods of time can be deducted, as opposed to a uniform load throughout the day.

Two additional factors that lead to increasing the load on the SLS infrastructure is obviously the total number of sales and the number of the retail stores (see Table 1). Both, the number of sales and the number of retail stores, affect the average number of products being transferred through the supply chain and, thus, increase the load of the SLS.

Table 1 Average load while modifying retail store number

Retail store number modification	Object directory (average load)	Object service directory server (average load)
Retailers =3	165,477 hits/h	41,369 hits/h
Retailers =5	284,382 hits/h	71,095 hits/h

Increasing the suppliers did not affect the efficiency of the SLS infrastructure. The parameter that determines the load is the amount of the products in the supply chain, which depends from the amount of sales and the number of retail stores. More suppliers do not cause more sales and, thus, they do not increase the load in the Discovery Services infrastructure.

The mass update scheme of the SLS, i.e. updating the OSD server not for every single product instance event separately but a bundle of many events grouped together, has significantly improved the scalability of the system (Table 2). In the following table the impact of the mass update on the load of the servers is presented.

The required computational power for the process of 1000 EPCs is the same both in the per object and the mass update.

Beyond technological considerations, there are some managerial aspects that should also be considered.

An initial observation is that the load of messages, queries and replies that is caused by every node is directed in a shared infrastructure, the SLS. The payloads generated by each node can obviously be of different-varying sizes and, thus, issues of fairness in the distribution of the maintenance and administration costs could arise.

Additionally, it is worthy to note that the SLS Discovery Service treats all the products independently of their type-class-brand or any other distinguishing feature. Every object is just an EPC and is equal to every other object that bears an EPC. Though, another fact is that every object has a different production cost and generates a different profit. Thus, a node that produces cheap products with shallow profit margins generates the same payload to the SLS as a node that deals with expensive high-profile commodities with high profit. Assuming that these nodes-suppliers produce the same number of products, it is expected that both will deliver the same load to the SLS infrastructure. So, it would be reasonable that the cost of using the Discovery Services (or the Internet of Things) would not be the same in the two cases.

Table 2 Average Load while modifying the update scheme of the SLS

Modification of the update scheme of the SLS	Object directory (average load)	Object service directory server (average load)
Mass update	156021 hits/h	39 hits/h
Per object update	284,382 hits/h	71,095 hits/h

6 Conclusions and Further Research

Based on the simulation results we are able to conclude that the hierarchical approaches of the services will probably confront with the massive load of traffic and the highly volatile information. Solutions like "distributed hash tables" (Bridge Project High Level Design for Discovery Services) should be under consideration.

The simulation paradigm that has been presented can be used in order to determine and evaluate the implementation of an infrastructure for Discovery Services in the Internet of Things in a supply chain.

Beyond the effect of the Discovery Services in the supply chain, an interesting approach would be to determine the effects of the supply chain alterations and changes in the Discovery Services. Simulation experiments that could help determine the ability of the Discovery Services to adapt in supply chain modifications, such as supply chain merging or split, would further be very valuable.

Our goal was not to perform a full and detailed statistical analysis in order to provide a complete answer on the number and the exact topology of different OSD servers that would be needed to achieve the optimal performance based on statistically significant results. The primary purpose was to provide a paradigm on the performance and scalability analysis of the Discovery Services in the Internet of Things that can be adapted to different architectures (most notably the EPC Discovery Service). Even if the proposed architecture is more distributed than the existing architectures (EPC Discovery Service), it cannot be regarded as an appropriate approach because of scalability and performance issues. The amount of nodes and partners in the supply chain increases the complexity of the overall architecture and further research should be focused mainly on more scalable and distributed approaches.

Appendix A

Below, the results of the different simulation experiments are presented. On each diagram, the horizontal axis represents the time. The vertical axis can either be the number of objects, when the diagrams refer to sales, or number of hits, when the diagrams refer to load.

Scenario 1: 12 Retail Stores – Low Volume of Sales – EPC Addressing

Table 3 Scenario specifications

# of retail stores	3
SLS update method	Per product
# of suppliers	20
Average # of products sold	432/day

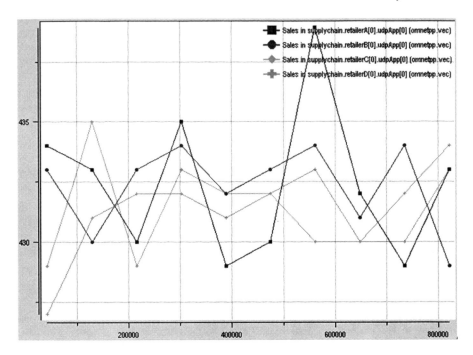

Fig. 4 Sales per retailer

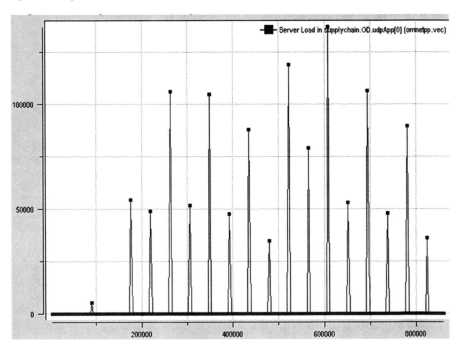

Fig. 5 Object directory load

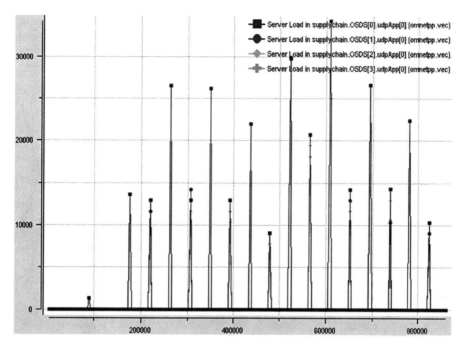

Fig. 6 Object service directory server load

Scenario 2: 12 Retail Stores – High Volume of Sales – EPC Addressing

Table 4 Scenario specifications

# of retail stores	3
SLS update method	Per product
# of suppliers	20
Average # of products sold	860/day

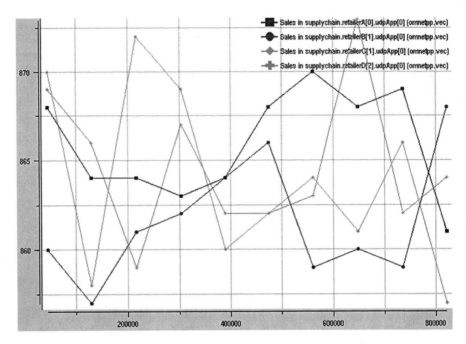

Fig. 7 Sales per retailer

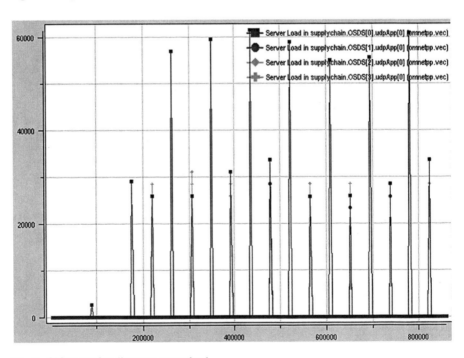

Fig. 8 Object service directory server load

Scenario 3: 12 Retail Stores – High Volume of Sales – EPC Addressing

Table 5 Scenario specifications

# of retail stores	3
SLS update method	Bundle of 1000 products
# of suppliers	20
Average # of products sold	860/day

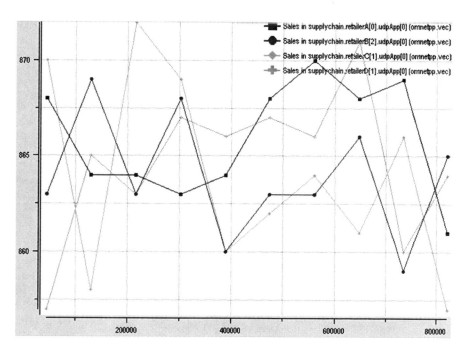

Fig. 9 Sales per retailer

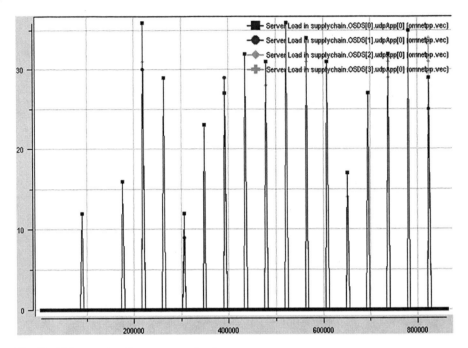

Fig. 10 Object service directory server load

Scenario 4: 20 Retail Stores – Low Volume of Sales – EPC Addressing

Table 6 Scenario specifications

# of retail stores	5
SLS update method	Per product
# of suppliers	20
Average # of products sold	432/day

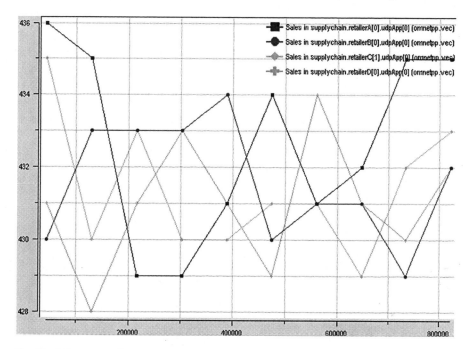

Fig. 11 Sales per retailer

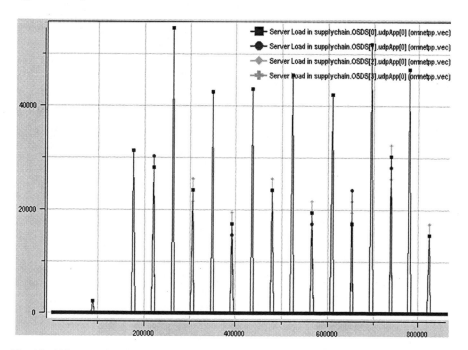

Fig. 12 Object service directory server load

Scenario 5: 20 Retail Stores – High Volume of Sales – EPC Addressing

Table 7 Scenario specifications

# of retail stores	5
SLS update method	Per product
# of suppliers	20
Average # of products sold	860/day

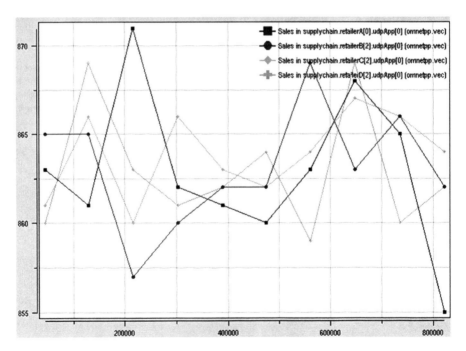

Fig. 13 Sales per retailer

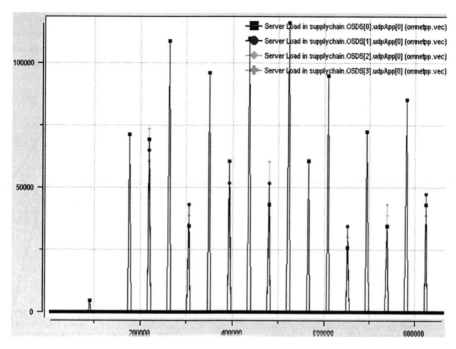

Fig. 14 Object service directory server load

Scenario 6: 20 Retail Stores – High Volume of Sales – EPC Addressing

Table 8 Scenario specifications

# of retail stores	5
SLS update method	Bundle of 1000 products
# of suppliers	20
Average # of products sold	860/day

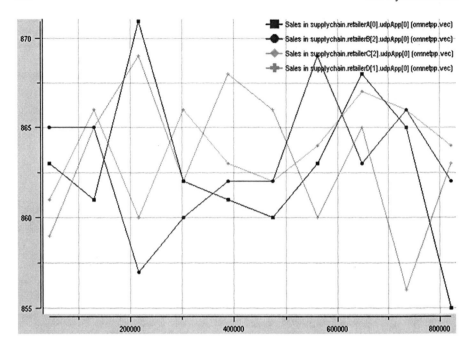

Fig. 15 Sales per retailer

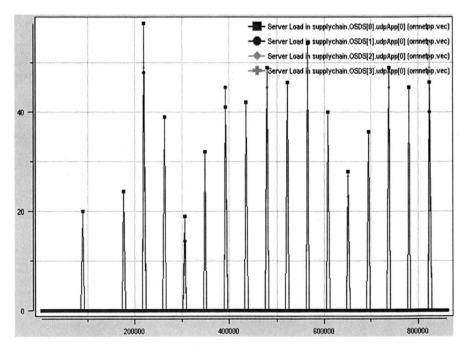

Fig. 16 Object service directory server load

Scenario 7: 12 Retail Stores – High Volume of Sales – EPC Addressing

Table 9 Scenario specifications

# of retail stores	5
SLS update method	Bundle of 1000 products
# of suppliers	30
Average # of products sold	860/day

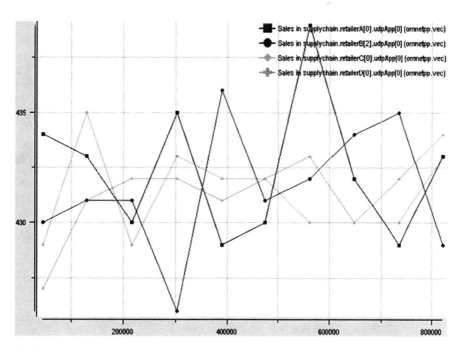

Fig. 17 Sales per retailer

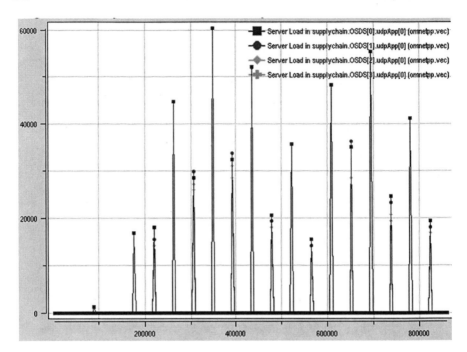

Fig. 18 Object service directory server load

References

Afilias (2008) How afilias discovery services works, information request on http://www.afilias.info/node/232 Accessed 16 Dec 2008

Beier S, Grandison T, Kailing K, Rantzau R (2006) Discovery services – enabling RFID traceability in PCglobal networks. In: 13th International Conference on Management of Data, Delhi, India, 14–16 December

Bottani E, Rizzi A (2008) Economical assessment of the impact of RFID technology and EPC system on the fast-moving consumer goods supply chain, Int J Prod. Econ 112:548–569

BRIDGE (2009) Bridge Project High Level Design for Discovery Services. IST-2005-033546. http://www.bridge-project.eu. Accessed 24 June 2009

Cho K, Pack S, Kwon† T Choi T (2007) SRMS: SIPbased RFID management System, IEEE International Conference on Pervasive Services, Istanbul, Turkey, 15–20 July, pp 11–18

Decker C, Berchtold M, Chaves LWF, Beigl M, Roehr D, Riedel T, Beuster M, Herzog T, Herzig D (2008) Cost-benefit model for smart items in the supply chain, IOT. Springer, Heidelberg

EPCglobal (2008) EPCglobal Tag Data Standard Version 1.4, June 2008. http://www.epcglobalinc.org/standards Accessed 28 July 2009

EPCglobal (2009a) Object Name Service (ONS), Version 1.0, 20051004. http://www.epcglobalinc.org/standards Accessed 28 July 2009

EPCglobal (2009b). http://www.epcglobalinc.org/standards Accessed 28 July 2009

Evdokimov S, Fabian B, Günther O (2008) Multipolarity for the object naming service, IOT. Springer, Heidelberg

Grummt E, Muller M (2008) Fine-Grained Access Control for EPC Information Services, IOT. Springer, Heidelberg

Jin B, Cong L, Zhang L, Zhang Y, Wen Y (2007) Towards an RFID-oriented service discovery system. UIC 2007, LNCS 4611, Springer, Heidelberg, pp 235–245

Kurschner C et al (2008) Discovery service design in the EPCglobal network towards full supply chain visibility, IOT. Springer, Heidelberg

Meloan S (2003) Toward a global "Internet of things", Sun Microsystem White Paper http://java.sun.com/developer/technicalArticles/Ecommerce/rfid/ Accessed 20 Sep 2009

OASIS (2009) web site. http://www.uddi.org Accessed 28 July 2009

Omnet++ (2008) web site. http://omnetpp.org/publications. Accessed 17 June 2008

OpenSLP (2009) An Introduction to the Service Location Protocol (SLP). http://www.openslp.org/doc/html/IntroductionToSLP/index.html. Accessed 24 June 2009

Polytarchos E, Leontiadis N, Eliakis S, Pramatari K (2008) A lookup service in an Inter-connected world of uniquely identified objects. 13th IEEE international conference on emerging technologies and factory automation, Hamburg, Germany, 15–18 September

Quack T, Bay H, Van Gool L (2008) Object recognition for the Internet of things, IOT. Springer, Heidelberg

Rellermeyer JS et al (2008) The software fabric for the internet of things, IOT. Springer, Heidelberg

University of Cambridge, AT4 wireless, BT Research, SAP Research (2007) High level design for discovery services. University of Cambridge, Cambridge

RFID-Enhanced Ubiquitous Knowledge Bases: Framework and Approach

Michele Ruta, Eugenio Di Sciascio, and Floriano Scioscia

Abstract In contrast to classical paradigms, in pervasive computing the user simultaneously interacts with several micro-devices such as RFID tags "integrated" into the environment, extracting data from them. Current approaches based on centralized control and information storage are utterly impractical. Here a ubiquitous Knowledge Base (u-KB) is defined as a distributed knowledge base whose individual facts are disseminated on RFID-tagged objects within the environment, without centralized coordination. The proposed framework includes: (i) components and operations specification of a u-KB, (ii) a distributed application-layer protocol for dissemination and discovery of knowledge embedded within RFIDs dipped in a MANET and (iii) a homomorphic algorithm for compressing on-tag data. The feasibility of the approach has been confirmed by large-scale simulation tests.

1 Introduction

In pervasive computing both information and computation are embedded into a given environment, that is into everyday objects and/or actions. The user simultaneously interacts with many micro-devices during ordinary activities, even not necessarily being aware of interaction details. Due to the volatility and unpredictability of such scenarios, centralized and fixed control is practically unfeasible.

Considering Radio Frequency IDentification (RFID) technology, nowadays tags with higher memory availability disclose the concrete possibility of building cooperative environments where autonomous objects can be discovered, interrogated and inventoried without any backend infrastructure. A proper dissemination protocol can allow the exact location of suitable descriptions directly on tags attached to objects thanks to a capillary diffusion of tag/reader basic information within the environment. Such a vision allows building an environment where tagged objects

M. Ruta (✉)
Politecnico di Bari, Via Re David 200, Bari BA I-70125, Italy
e-mail: m.ruta@poliba.it

D.C. Ranasinghe et al. (eds.), *Unique Radio Innovation for the 21st Century*,
DOI 10.1007/978-3-642-03462-6_11, © Springer-Verlag Berlin Heidelberg 2010

and readers make a self-contained retrieval architecture where external data links are not compulsory. The approach aims at the so-called Internet of Things and goes beyond common infrastructure-based RFID applications, because it allows users simply equipped with a personal device to perform an advanced object discovery.

We define a *ubiquitous Knowledge Base (u-KB)* as a distributed knowledge base whose individuals (assertional knowledge) are disseminated on the objects within a given context, lacking the need for centralized coordination. The framework presented here specify both elements and operations of a u-KB, as well as a distributed protocol for dissemination and discovery of knowledge embedded within RFID tags dipped in a Mobile Ad-hoc Network (MANET). RFID readers are considered as cluster-heads with respect to tags in radio visibility. They are also able to automatically build up multi-hop inter-reader wireless communication when placed in the same area (Ramanathan and Redi 2002) exploiting IEEE 802.11 (IEEE 802.11 1999), along with IP and UDP, for network infrastructures.

As formalism for resource annotation we use ontology languages based on Description Logics (DL) and originally conceived for the Semantic Web effort, particularly DIG (Description Logics Implementation Group) (Bechhofer et al. 2003), which is a more compact equivalent of OWL-DL Web Ontology Language (McGuinness and van Harmelen 2004). These languages were defined to provide a standard for semantically rich, formal and unambiguous description of Web resources through annotated metadata. According to the Internet of Things vision, we adopt them to describe objects and phenomena of the physical world, linking them to the digital world through semantic-enhanced RFID. From this standpoint, a relevant technical aspect is information compression, because XML-based formats adopted in the Semantic Web (such as RDF, OWL and DIG) are too verbose to allow efficient data storage and management in mobility. Compression techniques become essential in order to enable memorization and transmission of semantically annotated information on tiny mobile devices such as RFID tags or wireless sensors. Moreover, the benefits of compression apply to the whole ubiquitous computing environment, as decreasing data size means shorter communication delays, efficient usage of bandwidth and reduced energy consumption for handhelds.

The main contributions of the chapter are:

- to give an innovative KB (knowledge base) framework for pervasive, infrastructure-less environments preserving the distinction between factual and terminological knowledge and also adhering to the classical Tell/Ask paradigm;
- to propose some slight (backward compatible) modifications to current state-of-the-art RFID protocol to enable semantic enhancements;
- to integrate a compression algorithm into the framework and be able to store within a tag a semantically rich description of the object the RFID is attached to (using less than 2 kB of tag memory);
- to provide a case study clarifying and motivating the approach, possibly evidencing benefits it offers.

The remaining of the chapter is organized as follows: in the next section motivation and possible application scenarios are presented. In Sect. 2 a survey on related work is provided. In Sect. 3 significant theoretical aspects of the framework are described. The reference architecture and protocol are outlined in Sects. 4 and 5 respectively. The case study in Sect. 6 clarifies framework implementation in detail (including compression issues). Experimental methods and results are presented and commented upon in Sect. 7. Finally, conclusion closes the paper.

2 Motivation

In current applications (see Baader et al. 2002) §1.5 for a survey), Knowledge Representation Systems (KRSs) play a role which is similar to Database Management Systems. Both are used as central repositories where explicit domain knowledge is inserted with the aim of extracting the implicit one by means of inference procedures. Hence, in traditional KRSs, a KB is seen as a fixed entity immediately available, either in local storage or via a high-throughput network link. This approach is effective only as long as large computing resources and a dependable infrastructure are granted.

A different setting is needed to adapt KR tools and technologies to mobile and ubiquitous computing applications. They are characterized by:

- user and device *mobility*, which makes connections volatile and the (un)availability of resources unpredictable;
- dependency on *context*, which means that applications must adapt the way they support user tasks not only according to the availability of both nearby resources, but also to information characterizing environment and user;
- severe *resource limitations* of mobile devices affecting processing, storage, link bandwidth and power consumption.

Hence, knowledge-based systems designed for wired networks are not simply adaptable to wireless ones, due to intrinsic differences and performance issues.

The goal of a pervasive knowledge-based system is to embed semantically rich and easily accessible information into the physical world. This requires sharable vocabularies, otherwise the disagreement about the meaning of descriptions would impair system interactions (Vasudevan 2004). Logic-based languages for the Semantic Web endowed with formal semantics can provide such levels of interoperability. Advantages of semantic-aware approaches in RFID applications are widely acknowledged (Weinstein 2005). Basic services the technology offers could be significantly improved by introducing a semantically rich object description and discovery capabilities.

Let us consider the products lifecycle: manufacturing and quality control can exploit accurate descriptions of raw materials, components and processes; supply chain management could benefit from improved item tracking; verification of multi-factor service level agreements between commercial partners can be automated. Furthermore, sale depots could be easily inventoried also providing ubiquitous commerce (u-commerce) (Watson et al. 2002) capabilities – such as those introduced in (Venkataramani and Iyer 2007) and (Ruta et al. 2006) – without expensive investments in infrastructure. Finally, smart post-sale services can be provided to purchasers, by integrating knowledge discovery capabilities into home and office appliances (Ruta et al. 2007). In addition, asset management is greatly improved in those scenarios where retrieval should be based on relevant object properties and purposes, rather than mere identification codes.

In healthcare applications, equipment, drugs and patients can be thoroughly and formally described and tracked, not only to ensure that appropriate treatments are given, but also to provide decision support in therapy assignment. The pervasiveness of infrastructure helps to reduce costs and break barriers between patient management in the hospital and at home.

Likewise, in museums, libraries and archaeological sites, semantic-based content fruition can be granted either to local visitors or remote users connected through the Internet. In both cases the internal RFID infrastructure will be leveraged.

A knowledge-based approach can be integrated with other monitoring and sensing technologies beyond RFID. Wireless semantic sensor networks (Ni et al. 2005) are an emerging yet challenging technology. Semantic-based sensory data dissemination and query processing are needed to enable advanced solutions for e.g. environmental monitoring, precision agriculture, green supply chains and disaster recovery.

3 Related Work

Pervasive computing requires a decentralized and collaborative coordination between autonomous mobile hosts. Therefore, pervasive knowledge-based systems have to achieve high degrees of autonomic capability, providing transparent access to knowledge sources that may be present in a given area. This is achieved by exploiting a relatively large number of heterogeneous micro devices – for example RFID tags or wireless sensors – each conveying a small amount of useful information.

Middleware infrastructures for ubiquitous computing which can be found in the literature usually rely on centralized nodes for management and discovery of information (Chakraborty et al. 2006; Pallapa and Das 2007; Toninelli et al. 2008; Vazquez and Lopez-de-Ipina 2007). In particular, RFID technology currently links physical objects with their "virtual counterpart" (Römer et al. 2004) in the digital world. Tags trivially store an identification code, which is used to retrieve object properties from an information server, through a networked infrastructure.

The biggest obstacle toward decentralized approaches is the high cost of RFID tags with sufficient memory. The growing demand of RFID solutions will enable passive RFID tags with higher memory capacities at low cost in the next few years (M Ayoub et al. 2009).

In (De et al. 2004) a pervasive architecture is presented for tracking mobile objects in real-time for supply chain and B2B transaction management. A global and persistent IT infrastructure is needed in order to interconnect RFID systems within partner organizations through the Internet. These requirements make the approach unsuitable for mobile B2C and C2C applications.

More recent research works (Ruta et al. 2007; Venkataramani and Iyer 2007) proposed approaches for the integration of semantic-enhanced EPCglobal RFID into MANETs. Semantic annotations could be put into RFIDs attached to objects so that tagged goods stored a semantically rich description featuring the product the tag is clung to. In this way, objects equipped with RFID tags described themselves toward the "rest of the world" in a self-contained fashion. Nevertheless, a fixed central component was still needed for reasoning over a Knowledge Base (KB). This led to expensive information duplication within the environment: semantic annotations were simultaneously placed on tags and within the KB; moreover, the reasoning engine was a single point of failure. Such issues can be mitigated if a more distributed approach is followed, as the one proposed here.

For compression, lossless algorithms can be categorized into three families: static, semi-adaptive, and adaptive (Howard and Vitter 1991). *Static* compression uses either fixed statistics or no statistics. *Semi-adaptive* compression works in two steps: the input data is scanned once for statistics gathering and again for encoding. In *adaptive* compression, instead, statistics are calculated and dynamically updated along with compression.

gzip is one of the most popular universal compression tool, based on a variant of the *LZ77* adaptive algorithm (Ziv and Lempel 1977).

Huffman encoding (Huffman 1952) and *arithmetic encoding* (Witten et al. 1987) are two fundamental semi-adaptive compression techniques. The former uses a variable-length code table (the *huffman tree*), derived from evaluating the frequency of each symbol. In the latter case, the whole message is represented by a real value between 0 and 1. In this class of algorithms and tools, the *PAQ* family (Mahoney 2005) currently has the best compression rates, but its computing and memory requirements are far beyond capabilities of mobile devices.

Higher compression rates can be achieved by algorithms specifically designed for XML encoding. *XMill* (Liefke and Suciu 2000) is an efficient schema-aware XML compressor. It splits XML content into different *containers*, which are separately compressed and sequentially stored in the output file (hence XMill is non-homomorphic). XMill works better than generic compressors for medium and large XML documents, while for small files (up to 20 kB approximately) it is penalized by the small size of containers. In *DPDT* (Harrusi et al. 2006) the XML document is encoded using *Partial Prediction Matching* (PPM), an adaptive technique. Reported compression rates are higher than XMill, though PPM-based compressors are generally slower.

XPRESS (Min et al. 2007) adopts *reverse arithmetic encoding*, a semi-adaptive algorithm. Furthermore, a *type inference* module detects data types of XML attribute values and a specialized encoding method is applied for each type (numbers, dates, and so on). Experimental results showed XPRESS achieves high query performance for compressed XML data and updates can be directly performed on compressed XML. Compression rates, though, are lower than other methods.

4 Theoretical Framework

Figure 1 depicts the proposed framework for knowledge dissemination and discovery in a pervasive context. The u-KB layer provides common access to information embedded into semantic-enhanced EPCglobal (Traub et al. 2005) RFID tags populating a smart environment. UDP is used for data exchange in IEEE 802.11 mobile ad-hoc networks. Further identification and sensing systems can be interfaced to the general framework, through a semantic support micro-layer added to standard protocols. Applications can use the knowledge provided by a u-KB, by means of DL-based reasoning services exploiting the semantic discovery protocol featuring the u-KB layer. They can be performed either by hosts in the local MANET or by a remote entity through a gateway exposing a high-level interface (e.g. Web Services of RPC[1] or REST[2] type) and translating remote methods into operations on the u-KB.

This chapter focuses on definition and experimentation of the u-KB layer and the RFID semantic layer in Fig. 1. Parallel research effort is being spent into the adaptation of reasoning procedures to resource-constrained mobile devices.

4.1 KB Components and Operations

Given a reference problem in a specified domain, a DL *Knowledge Base* models it in two components (Baader et al. 2002):

- a *TBox* (terminological box), containing intensional knowledge, i.e. general knowledge about the domain. It has the form of an ontology – also called terminology – describing concepts and their relationships;

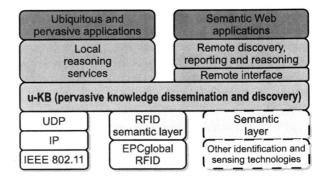

Fig. 1 Architecture of the proposed approach

---

[1] Remote Procedure Call

[2] REpresentational State Transfer

- an *ABox* (assertion box), containing *extensional* knowledge, which is specific to the particular problem. It consists of an assertions set concerning the domain individuals.

It is useful to recall that intensional knowledge is usually thought not to change whereas extensional knowledge is usually contingent, or dependent on a set of circumstances, and therefore subject to change (Baader et al. 2002, p. 17).

Current KRSs are characterized in terms of the *functions* they provide to applications, instead of system data structures and allowed low-level operations (Levesque 1984). As a result, the design separates functionality from implementation, leading to both better predictability of system behaviour and greater ease of use and integration of a KRS within larger application solutions. Two basic functions were identified for KB management:

- *Tell*: build the TBox and the ABox by means of explicit terminological axioms and assertions about individuals;
- *Ask*: extract (implicit) knowledge. Queries are answered by the system through inference procedures, determining if the query meaning is implied by the information that has been told.

This paradigm has led to detailed and formal interface specifications for Knowledge Representation Systems – such as *KRSS* (Patel-Schneider and Swartout 1993) and, more recently, *DIG* (Bechhofer et al. 2003) – implemented by most KRSs.

The capability to selectively remove information from a KB is also desirable. Current systems allow to *Un-Tell* (i.e. retract) only information that has been previously told (see Brachman et al. 1991, for a discussion). Experience has shown that, for the vast majority of applications, the TBox seldom or never changes after an initial knowledge acquisition phase (see Baader et al. 2002, chap. 8, for a review).

4.2 u-KB Components and Operations

In our approach we preserve the differences between TBox and ABox.

The *TBox* is expressed by means of an ontology document, which can be owned by one or more mobile hosts. Based on the previous discussions, it can be reasonably assumed that ontologies defined before object annotation and u-KB deployment seldom change during normal system activity. Nevertheless, in order to improve the system flexibility, the TBox management could be performed also following paradigms and approaches devised in peer-to-peer protocols such as BitTorrent (Cohen 2008). That is the ontology document file can be entirely available on a single host or it can be fragmented in one or more chunks scattered within the MANET. It is worth noting that since several object classes, described with respect to different ontologies, can co-exist within the same physical space, multiple u-KBs can actually populate a given environment sharing the system infrastructure.

Ontology Universally Unique Identifiers (OUUIDs) are adopted to unambiguously mark ontologies and to associate each individual to its reference ontology (Ruta et al. 2006).

The *ABox* is deployed within a smart context, as KB individuals are physically tied to micro devices disseminated in the field. In RFID-based scenarios, each individual consists in semantically annotated metadata describing object/product features, stored within the RFID transponder. Each annotation refers to an ontology providing the intensional knowledge. In detail, each individual is characterized by:

- a globally unique item identifier (the EPC – Electronic Product Code – in the case of RFID tags);
- the OUUID;
- a set of contextual attributes which allow the extension of logic-based reasoning services with application-specific information (e.g. price is a typical parameter in mobile commerce, whereas content duration is more useful in mobile learning);
- semantic annotation, stored as a compressed document fragment in the DL-based language DIG.

In our u-KB approach, we adhere to the fundamental *Tell/Ask paradigm*. These operations, however, are implemented in a novel way, coping with the peculiarities of pervasive computing scenarios. Next section explains in detail the data structures and protocol devised to build a u-KB system.

Tell/Un-Tell operations are *hidden* from users, i.e. no explicit knowledge declaration/retraction is needed. Intuitively, a u-KB is built with knowledge fragments carried by individual micro devices that populate the environment in a given instant and by the ontology they refer to. The system allows autonomic and adaptive knowledge base maintenance, by means of a data alignment protocol between cache memories of multiple mobile devices. Each node advertises the individuals detected in its proximity (e.g. via RFID), other hosts store advertisements in their cache and forward them to hosts at one-hop distance. The protocol tackles the issues of moving tagged objects entering/leaving the environment by keeping track of the freshness of advertised individuals through sequence numbers, so that updates will be automatically propagated. Furthermore, each individual also has a limited time-to-live, so that it will be automatically removed (un-told) from the u-KB if not renewed. Similar mechanisms are employed to disseminate ontology chunks within the network, aiming at a trade-off between fault tolerance and generated overhead. The system aims for a steady state where every host is aware of both ontologies and individuals in the environment. To reduce storage requirements and network load, semantic annotations are not included in advertisements; when needed they are provided *on-demand* instead.

Ask operations require a preliminary resource discovery step. The requester specifies the ontology identifier and a range for each attribute it is interested in. The system then returns IP addresses of hosts owning the ontology chunks and all the individuals that meet the specified criteria. *On-demand* provisioning of KB individuals reduces transmissions for data alignment and prevents propagation issues in the case of description update or individual removal. Additionally, filtering avoids unnecessary data transfers. Then the requester can fetch ontology chunks and filtered individuals by their respective providers, so that it can reconstruct a local subset of the whole KB, containing only the TBox and annotations which are actually needed. Finally, it is able to submit any Ask-type request to a local or remote reasoning engine.

5 Architecture Details

The details of the proposed architecture are provided next. Particularly, an exhaustive description of the infrastructure underling the framework is presented along with dissemination and discovery protocols specification.

5.1 Layered Infrastructure

Nodes participating in an ad-hoc network are moving and their inclusion within the MANET does not require configuration procedures. Each node could act as requester/provider or play the role of a router forwarding the traffic coming from nearby hosts.

The proposed framework presents a two-level infrastructure as shown in Fig. 2. RFID is exploited at the *field layer* (interconnecting tags dipped in the environment and readers in range), whereas the *discovery layer* is related to inter-reader ad-hoc communication. Tags in the field and readers are interconnected via the semantic-enhanced EPCglobal RFID protocol data exchange (Ruta et al. 2007), whereas the data propagation among readers is performed following a data dissemination paradigm in 802.11 (IEEE 802.11, 1999). Resources are autonomously exposed via the enhanced RFID and, in the same way readers should be able to perform a discovery thanks to the preliminary propagation of data each cluster-head has seen in its range. Note that really pervasive environments do not necessarily provide stable dependable communication links. On the contrary more probably and realistically these pervasive environments are infrastructure-less. Hence, an autonomous and decentralized approach to object discovery should ensure the availability of required information also when a connection to a central remote server is missing.

Thus, the resource discovery is based on three stages:

1. the extraction of the object parameters (for carrying object features from field layer to discovery one);
2. the resource data dissemination (to make the overall nodes fully aware about the "network content");

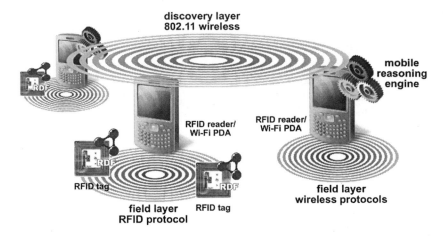

Fig. 2 Field and discovery layers in the proposed framework architecture

3. the extraction of resource annotations (for carrying semantic-based descriptions from field level to the discovery one) for the further matchmaking. This phase is performed *on-demand* when a request is addressed in unicast to a selected node.

Each reader has a central role in the whole architecture as it advertises contextual parameters referred to tags in its radio range (at the field layer) and during further phases, it will receive requests from nearby nodes (via 802.11 at the discovery layer). It will extract semantic annotations from tags in range (using RFID air interface protocol such as EPCglobal's Class 1 Generation II protocol) and replies to the requester. A reader maintains a cache containing the advertisements which will be matched against requests.

The proposed approach is fully decentralized: address and main characteristics of each resource/tag are autonomously advertised by related readers using small-sized messages throughout the network composed by the other readers.

The use of a broadcasting mechanism to advertise object features could be inefficient in terms of bandwidth and power consumption (both precious resources in pervasive environments). So in the proposed approach, only resource parameters are advertised in broadcast mode in order to unambiguously identify both the location and the category of a resource/tag. If a node is explicitly interested in a resource, it will download in unicast mode the semantic annotation it is associated to. In this way, advertisement flooding (due to the broadcasting mechanism) is reduced.

5.2 Semantic-Based Dissemination and Discovery Protocol

Each resource in the MANET is labelled by means of the triple *{SOURCE_ADDRESS, OUUID, EPC}*, that is the IP address of the RFID reader

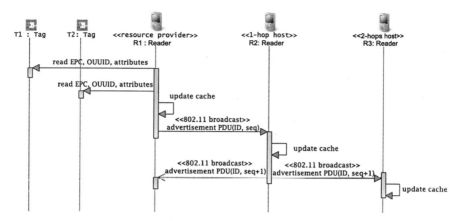

Fig. 3 Tag data dissemination phase

which has "seen" the resource, the ontology the resource is associated with and the Electronic Product Code, respectively.

To perform the data dissemination, resource providers periodically send *advertisement PDUs* (Protocol Data Units) traveling a given number of hops (MAX_ADV_DIAMETER) at most. As shown in Fig. 3, for each managed resource, a reader advertises the reference OUUID and some context-aware parameter (i.e. the resource Time-To-Live (TTL)). An early selection chooses only semantically compatible annotations (OUUID matching) with suitable values for contextual attributes.

During their travel, advertisements are forwarded using MAC broadcasts and they are stored in the cache memories of nodes they go through. Figure 4 shows a cache entry. Each field consists of:

- *source address*: address of resource provider;
- *size*: annotation size (in byte);
- *OUUID*: ontology identifier;
- *lifetime*: remaining time to live of a resource/tag.;
- *timestamp*: last reference to the entry (read/write). That is, when a new resource is stored or when an existing one is invoked, the field is updated;
- *travelled hops*: distance (hops number) between provider and cache holder;
- *sequence number*: reference to the last resource provider;
- *EPC*: Electronic Product Code of a resource/tag;
- *resource description*: semantic annotation of a resource. It has a variable length, but in some case it could be a pointer to a compressed DIG file.

An entry is added to the cache table whenever the node receives either an advertisement or a cache content frame. The corresponding entry is updated if the PDU byte

1	2	3	4	5	6	7	8	9	1 0	1 1	1 2	1 3	1 4	1 5	1 6	1 7	1 8	1 9	2 0	2 1	2 2	2 3	2 4	2 5	2 6	2 7	2 8	2 9	3 0	3 1	3 2	3 3	3 4	3 5	3 6	3 7	3 8	3 9	4 0	4 1	...

| SOURCE ADDRESS | SIZE | OUUID | LIFETIME | TIMESTAMP | SEQUENCE NUMBER | EPC | RESOURCE DESCRIPTION ... |

TRAVELED HOPS

Fig. 4 The structure of a cache record

carries a sequence number higher than the stored one or if the route the packet suggests is shorter than the previously stored one.

Before starting whatever reasoning task, the requester has to properly recompose the TBox. To achieve this goal techniques of hybrid peer-to-peer sharing are exploited. The goal is to avoid collecting the whole TBox before reasoning, but only the actually needed portions. For brevity, further details about reasoning are not provided here, the interested reader can refer to (Di Noia et al. 2007).

Figure 5 shows the resource discovery phase. When starting a matchmaking, a node attempts to cover the request by using resource descriptions stored within its own cache. If a requester has no stored descriptions or if managed resources do not satisfy the request, the node sends a *solicit PDU* with a maximum travel diameter (MAX_REQ_DIAMETER) to get new resource locators.

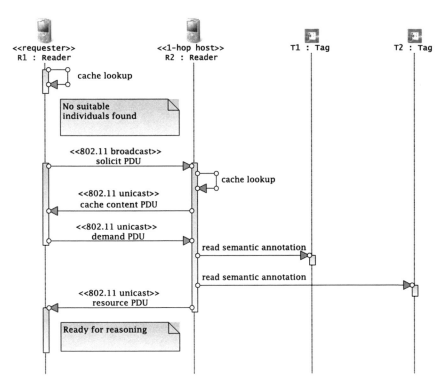

Fig. 5 Retrieval of advertised semantic annotations from tags

A node receiving a solicit replies (in unicast) with a *cache content PDU,* providing cache entries whose parameters match the ones contained within the solicit frame. On the contrary, it will reply with a "no matches" message. Note that, during their travel, replies to a request and solicitation PDUs are used to update the cache memory of forwarding nodes.

Finally, if some semantic annotation is missing, it can be retrieved by sending a unicast *request PDU* to the node "owning" the resource(s). The latter will reply with a *resource PDU,* containing the annotation(s).

After receiving required information from hosts in its search range, the requester performs the matchmaking again. If the result is still unsatisfactory, the node could try to forward a new solicit by increasing the search diameter and previous steps can be repeated progressively expanding the diameter, until a maximum extension is reached.

6 Case Study

The proposed framework has been specifically studied in pervasive computing environments where a wide range of objects/products are endowed with RFID transponders complying with the EPCglobal standard for class I – second generation UHF tags (EPCglobal Inc. 2005a). Mobile RFID readers equipped with IEEE 802.11 wireless connectivity create and manage the u-KB. Tagged objects are KB individuals in our system.

Previous works (Ruta et al. 2007) proposed enhancements to the EPCglobal standard which recur in our definition of u-KB individuals, as explained above. In the following subsections, the main features of semantic-enhanced RFID protocol are recalled, and interactions between RFID technology and the proposed framework are outlined, respectively.

6.1 RFID-Based u-KB

Tag memory is organized in four logical banks (EPCglobal 2008):

1. *Reserved,* storing optional kill and access passwords;
2. *Electronic Product Code (EPC);*
3. *Tag IDentification (TID),* storing tag manufacturer and model identification codes;
4. *User,* optionally present for custom application data.

Contents of TID memory up to $1F_h$ bit are invariable. For tags having class identifier value $E2_h$ stored in the first byte of the TID bank, optional information could be stored in additional TID memory from 20_h address. We store the following at this memory address:

- a 16 bit word for *optional protocol features*, stored starting from 20_h address most significant bit first: currently only the most significant bit is used to indicate whether the tag is semantic-enabled or not; other bits are reserved for future uses;
- the OUUID of the ontology referred by the semantic annotation in the tag.

Finally, contextual attributes are stored within the User memory bank, along with the semantically annotated object description expressed in DIG. Due to verbosity of DIG language, the annotation is compressed using the encoding algorithm proposed in (Scioscia and Ruta 2009), since it allows direct queries over the compressed format.

EPCglobal UHF RFID air interface protocol consists in three steps:

1. *Pre-selection* by a reader of a subset of the tag population currently in range, by means of the *Select* command;
2. *Inventory* loop, where the reader detects one tag in range for each iteration and reads its EPC code;
3. *Access* to the tag memory content.

In our approach, original commands are exploited in novel ways, while keeping full backward compatibility. The *Select* command is used to preselect semantic-enabled tags. The inventory step is performed in the standard way. *Read* and *Write* commands in the access step can be used to extract and update the OUUID, the contextual attributes and the semantic annotation of the tagged object.

6.2 Interaction with EPCglobal's RFID Framework

6.2.1 Dissemination

After each advertisement period DEFAULT_RTIME, a reader scans RFID tags in its range. Only semantic-enabled ones are preselected, by means of a *Select* command whose parameters are shown in Table 1.

Values for the (MemBank, Pointer, Length) triple identify the bit at 20_h address in the EPC memory bank, which is compared with Mask value 1_2.

Table 1 *Select* is command able to detect only semantic-enabled tags

Parameter	Value	Description
Target	100_2	SL flag
Action	000_2	If bits match then set target flag, else reset
MemBank	10_2	TID memory bank
Pointer	00100000_2	Start address
Length	00000001_2	Number of bits to compare
Mask	1_2	Bit mask

Table 2 *Read* command able to extract OUUID from the TID memory bank

Parameter	Value	Description
MemBank	10_2	TID memory bank
WordPtr	00000011_2	Starting address (4th memory word)
WordCount	00000010_2	Read 2 words (32 bits)

Table 3 *Read* command to extract contextual attributes from user memory bank

Parameter	Value	Description
MemBank	11_2	User memory bank
WordPtr	00000000_2	Starting address (1st memory word)
WordCount	00001000_2	Read 8 words (16 bytes)

Since semantic-enabled tags are identified by having the bit set, Target and Action parameters have the effect to set the SL tag status flag only for them and to clear it for the other ones.

The subsequent inventory step skips tags having SL flag cleared. EPC codes of semantic-based tags are then individually scanned and TTL of corresponding cache table entries are refreshed.

If the reader detects a new EPC code is not in the cache memory, it will read its OUUID and contextual attributes, exploiting two *Read* commands, as shown in Tables 2 and 3, respectively.[3]

Data extracted from the RFID tag will be stored in a new cache entry with a fresh sequence number. Conversely, if the EPC code is not detected for an existing local entry in the cache table, the reader will wait for the TTL to expire before removing the entry. This prevents the well-known issue of "RFID event flickering"[4] (Römer et al. 2004) from causing incorrect removal/addition of u-KB individuals. At the end of the loop, the cache table is fully updated and the reader can issue an *advertisement* PDU to notify individuals to neighbouring hosts.

6.2.2 Discovery

When a reader receives a *request* PDU, it starts an RFID scan of semantic-enabled tags only, as seen above. During inventory, for each detected EPC among those listed in the PDU payload, it reads the compressed semantic annotation stored in the

---

[3]The *Read* command allows reading one or more 16-bit memory words from any of the four tag memory banks. MemBank parameter identifies the memory bank (as in *Select* command). WordPtr and WordCount are the starting address and the number of memory words to be read, respectively; if WordCount is 0, all the memory words will be read up to the end of the selected bank.

[4]Due to collisions, a tag might not be detected in every consecutive scan. This phenomenon can trigger spurious leave-enter event pairs.

Table 4 *Read* command to extract the compressed semantic annotation from user

Parameter	Value	Description
MemBank	11_2	User memory bank
WordPtr	00000000_2	Starting address
WordCount	00000000_2	Read up the end

User memory bank of the tag, with a *Read* command as shown in Table 4. Finally, it replies to the requester.

In addition to cache tables described in the previous section, the reader may have an optional cache for the most recently used semantic annotations in order to reduce RFID accesses. This may improve response latency and battery life.

6.2.3 Implication of RFID Data Locality

In common scenarios, the content of a particular RFID tag can be relevant to a user only if he/she is in the same environment as the tagged object is. For instance, in a large shopping mall, buyers are implicitly interested in products of the department they are currently in. The proposed data dissemination protocol exploits this locality property and delivers advertisements only to readers within a maximum hop distance from the advertiser (which is supposed to be physically close to the RFID transponder). The value of this operational parameter can be adjusted to balance network load and physical extension of a u-KB, according to specific application requirements. In the above shopping example, a different ubiquitous knowledge base will be built in each store department by IEEE 802.11-enabled RFID readers placed on shelves. Two departments far from each other should not and will not share information. Nevertheless, each department may have a gateway node allowing interactions between the local u-KB system and a back-end warehouse information system.

6.3 Encoding of Semantic Annotations

In spite of the fact that RFID tags integrate memories with an increasing capacity (nowadays RFIDs with at least 4 kB memory already have wide diffusion: M Ayoub et al. 2009), the compression of semantic annotations is a relevant technical issue for our proposal. XML-based languages adopted in the Semantic Web are too verbose to allow efficient data management in mobile environments. Encoding techniques become essential in order to enable proper storage and transmission of annotations on tiny devices such as RFID tags. Moreover, compression reduces communication delays, implies an efficient usage of bandwidth and improves battery performances.

When evaluating approaches to compression, besides compression ratio and required processing/memory resources, *efficiency of queries* on compressed data must be taken into account. Recent research has been increasingly focused on compression schemes for XML documents allowing to directly query encoded annotations (Min et al. 2007), which would eliminate the need for decompression before query evaluation. To achieve this objective, we designed an approach for compression of XML-based semantic annotations able to support queries directly on compressed data (Scioscia and Ruta 2009). Reverse Arithmetic Encoding (RAE), a semi-adaptive and homomorphic technique (Min et al. 2007) was adapted to that purpose. *Semi-adaptive* compression works in two steps: the input data is scanned once for statistics collection and again for encoding. *Homomorphic* compression (Tolani and Haritsa 2002) preserves the structure of the original XML data. On the contrary, with non-homomorphic algorithms the document structure is not recognizable after compression. Homomorphism is the feature that avoids decompression before detecting document pieces which satisfy given query conditions. The approach was implemented into a tool called *COX* (Compressor for Ontological XML-based languages) and another one, *DECOX* (DECompressor) which can be used for documents in RDF, DIG and OWL formats.

COX distinguishes data structures and data (particularly, for data structures -XML tags and attributes- RAE is used). It adopts two different compression solutions. Both techniques require a two-stage process. In the first step the XML document is parsed and statistics are gathered. In the second step compression is performed and output is written.

6.3.1 First Step

DIG and OWL languages only contain tags, attributes and attribute values (excluding other syntax elements such as comments and processing instructions). COX deals with tag and attribute names in the same way, only distinguishing the latter by means of a "@" prefix added to the name. Therefore, from now on the term "tag" will refer either to tags or attributes.

Let us consider a very short DIG document instance, describing a refrigerated cow milk delivery with respect to a reference ontology in the food products domain (not reported here):

```
<tells xmlns="http://dl.kr.org/dig/2003/02/lang">
<defindividual name="milk_delivery_M2"/>
  <instanceof>
     <individual name="milk_delivery_M2"/>
     <and>
        <catom name="cow_milk"/>
        <all>
          <ratom name="processed_with"/>
          <catom name="refrigeration"/>
        </all>
     </and>
  </instanceof>
</tells>
```

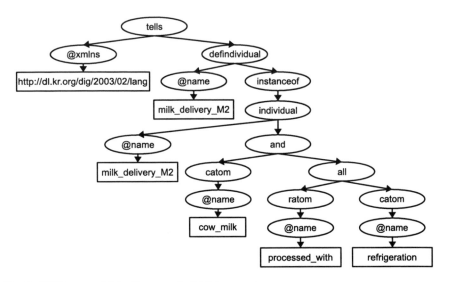

Fig. 6 COX tree model for the example DIG document

In the first step the document is parsed and the syntactic tree in Fig. 6 is built in memory.

After parsing, according to the arithmetic encoding scheme (Witten et al. 1987), an adjusted frequency of each tag name is calculated as ratio between the number of occurrences of a tag and the total document tags. Then the interval $[d, D) = [1.0 + 2^{-7}, 2.0 - 2^{-15})$ is split into disjoint sub-intervals (each associated to a single tag) using the sub-interval calculation function:

$$a_i = kf_i + \frac{1 - k}{n}$$

where a_i is the length of the ith sub-interval, f_i is the adjusted frequency of the ith tag, and n is the number of different tags. The weight k is defined as:

$$k = 1 - \sqrt{\sigma}$$

where σ is the standard deviation of absolute frequencies of all document tags. The formula is a linear combination of two components: the first one is proportional to tag frequency, whereas the second one is fixed for all tags. When the frequency distribution of tags in a document is regular ($\sigma \to 0$ and $k \to 1$), the proportional term dominates. Conversely, for irregular frequency distributions (the most common case) σ increases and k decreases, so that the second component of the formula increases and even very rare tags receive a not-too-small sub-interval. This prevents

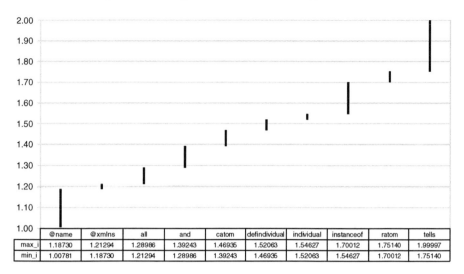

	@name	@xmlns	all	and	catom	defindividual	individual	instanceof	ratom	tells
max_i	1.18730	1.21294	1.28986	1.39243	1.46935	1.52063	1.54627	1.70012	1.75140	1.99997
min_i	1.00781	1.18730	1.21294	1.28986	1.39243	1.46935	1.52063	1.54627	1.70012	1.75140

Fig. 7 Computed intervals associated to tags in the example document

errors in decompression for tags with a very low frequency (in the order of 10^{-4} or lower).

The minimum value of the sub-interval assigned to each tag is computed with the formula:

$$min_{i+1} = min_i + a_i(D - d), \qquad i = 1, \ldots, n - 1$$

where n is total number of tags and $min_1 = d$. Hence, values referred to opening tags fall in the interval $[d, D)$.

Values of sub-interval limits for the above example document are reported in Fig. 7.

The interval $[1.0, d)$ is reserved to encode closing tags. It can be verified that every possible value strictly falls between 1.0 and 2.0. By exploiting 32-bit floating point representation, the first byte will always be 01111111_2, so it can be truncated without loss of information (Min et al. 2007).

Concluding the first step, a header is written at the beginning of the output file. It contains a sequence of records composed of: 1 byte for both tag name length and tag name itself; 3 bytes (after truncation) for the encoding of tag sub-interval minimum value.

Attribute values are interpreted as ASCII strings; most recurrent words are identified and encoded with 16-bit sequences. The statistical collection of their frequencies is performed concurrently with the document structure analysis. Since a simple code substitution is applied, homomorphism is obeyed also in this case.

Note that although a significant size saving can be obtained in ontologies with recurrent elements, on the other hand the use of a header could lower compression performance for short documents. As a consequence, a heuristic rule was adopted: only attributes above both a length threshold and a frequency threshold are encoded. A header is thus built, containing correspondences between each attribute string and the related code.

6.3.2 Second Step

The body of the output file is produced. Opening and closing tags, attributes and attribute values are encoded in the order as they appear in the input document. An *opening tag* T (or the beginning of an attribute) is encoded with RAE. We follow the same approach of XPRESS (Min et al. 2007) to encode values referred to tags or sequences of tags in a path. The applied algorithm is given below:

> **input** T tag to be encoded, *subint* associative array of sub-intervals assigned to tags
> **output** $n \in [d, D)$ real number that encodes the tag path from T up to the root tag
> 1. $min, max) := subint[T]$
> 2. **while** has_parent_tag(T)
> 3. $T := $ get_parent_tag(T)\$
> 4. $(begin, end) := subint[T]$\$
> 5. $min := min + (max - min) * begin$
> 6. $max := min + (max - min) * end$
> 7. **end while**
> 8. $n := min$

Basically, the interval associated to the tag path from T to the root (hence the *reverse* adjective), is computed by progressively narrowing the starting interval according to the size of sub-intervals assigned in the first step. In this way, every value in the final numeric interval unambiguously identifies the sequence of tags from T up to the root. Given a 32-bit floating point representation of the output value, to encode a tag our actual implementation only takes the two central bytes. In addition to the first one (which is fixed), the last byte (which is the least significant mantissa byte) is truncated to save space. We verified if this intentional loss of precision does not prevent a correct tag reconstruction in decompression phase.

With reference to the previous DIG document excerpt, the following Fig. 8 clarifies the compression algorithm behaviour, showing how the compression procedure progressively calculates the intervals associated to each document portion. Figure 8 illustrates the processing of the name attribute under the defindividual tag, at line 2 of the document. The path from the leaf to the root element is built as @name – defindividual – tells. Then the algorithm loop starts: firstly, in iteration (a) the sub-interval corresponding to @name is taken (as from Fig. 7). In

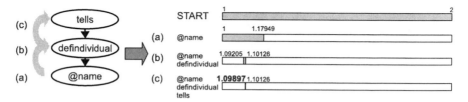

Fig. 8 Encoding procedure example

the next iteration (b), `defindividual` is considered and the interval computed in (a) is reduced in a proportional way to the sub-interval of `defindividual` (from Fig. 7) with respect to the base interval $[d, D)$. Similarly, in step (c) tells is considered and the interval computed in (b) is reduced proportionally to the sub-interval of `tells` (again from Fig. 7) with respect to the base interval. Once the root element of the document has been reached, the loop terminates: the numerical interval that identifies unambiguously the above path is $[1.09897, 1.10126)$. The lower bound is finally taken to encode the path.

A *closing tag* or the end of an attribute is encoded by the two central bytes of the 32-bit floating point representation of 1.0070 and 1.00444, respectively. These values were chosen because truncating does not cause loss of precision.

Finally, an *attribute value* is processed as follows: if it was encoded in the first step, it is replaced by its 1-byte code followed by the delimiter byte fe_h, otherwise the ASCII string is copied to output, followed by the delimiter ff_h.

7 Experiments

Preliminary tests about read and write time from/to semantic-enhanced tags have been performed in (Di Noia et al. 2008) to provide an early evaluation of the impact that EPCglobal protocol evolution may have on RFID system performance. They are used as initial evidence that adoption of compressed semantic resource annotations on RFID tags does not impair semantic-based RFID applications with respect to the most common ones.

Furthermore, the u-KB overall framework proposed here has been verified and tested using *ns-2* network simulator (Network Simulator 1995). Performance evaluation includes the following metrics (Ruta et al. 2009).

1. *Network load*, assessed by means of the total number of packets generated at discovery layer. Results show that traffic has higher correlation with the number of resource providers rather than requesters (clients). This happens because advertisement frames are regularly sent in a proactive way by providers, even if there are no requests. On the other hand, solicit, cache content and request packets are produced *on-demand* by client hosts and their neighbours.

2. *Hit ratio*, i.e. the percentage of successful resource retrieval in a variety of scenarios. Obtained hit ratios are very high, with values above 90% in all tests and
   above 95% in more than half the tests. As a general consideration, this clearly
   indicates the effectiveness of the proposed approach.
3. *Duration* of service discovery. For a given number of resource providers, the service time decreases as requesters increase. This happens because, when a solicit
   PDU is answered by cache content PDUs, intermediate nodes cache related
   resource records. This reduces latency in replying to later requests. Overall values are still slightly high with respect to typical needs of pervasive scenarios.
   This result may be influenced by the fact that current implementation is not
   specifically optimized for execution speed.

A second experimental campaign was performed to assess benefits and draw-
backs of different compression approaches for semantic annotations. The proposed
COX tool was compared to: (i) *gzip*,[5] a widely-adopted general-purpose compressor; (ii) *XMill* (Liefke and Suciu 2000), an XML-specific compression tool and (iii)
our own previous tool *DIGCompressor* (Ruta et al. 2007), which was optimized for
high compression rates and execution speed.

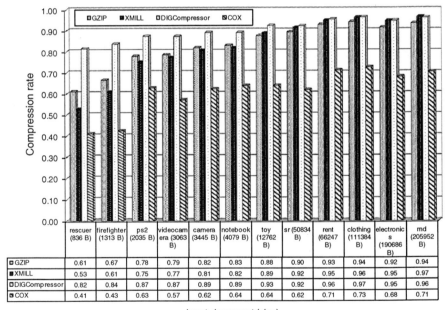

	rescuer (836 B)	firefighter (1313 B)	ps2 (2035 B)	videocamera (3063 B)	camera (3445 B)	notebook (4079 B)	toy (12762 B)	sr (50834 B)	rent (66247 B)	clothing (111384 B)	electronics (190686 B)	md (205952 B)
GZIP	0.61	0.67	0.78	0.79	0.82	0.83	0.88	0.90	0.93	0.94	0.92	0.94
XMILL	0.53	0.61	0.75	0.77	0.81	0.82	0.89	0.92	0.95	0.96	0.95	0.97
DIGCompressor	0.82	0.84	0.87	0.87	0.89	0.89	0.93	0.92	0.96	0.97	0.95	0.96
COX	0.41	0.43	0.63	0.57	0.62	0.64	0.64	0.62	0.71	0.73	0.68	0.71

Input document (size)

Fig. 9 Performance comparison – compression rate

---

[5]GZIP compression utility: http://www.gzip.org/.

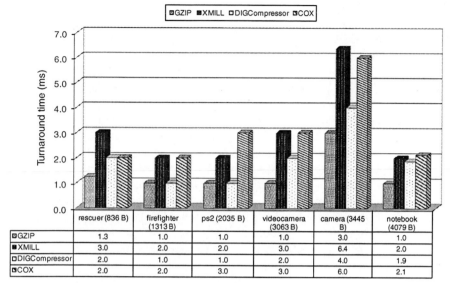

	rescuer (836 B)	firefighter (1313 B)	ps2 (2035 B)	videocamera (3063 B)	camera (3445 B)	notebook (4079 B)
GZIP	1.3	1.0	1.0	1.0	3.0	1.0
XMILL	3.0	2.0	2.0	3.0	6.4	2.0
DIGCompressor	2.0	1.0	1.0	2.0	4.0	1.9
COX	2.0	2.0	3.0	3.0	6.0	2.1

(a) Input document (size)

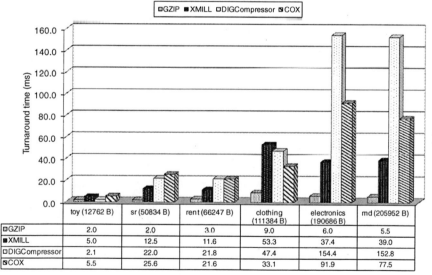

	toy (12762 B)	sr (50834 B)	rent (66247 B)	clothing (111384 B)	electronics (190686 B)	md (205952 B)
GZIP	2.0	2.0	3.0	9.0	6.0	5.5
XMILL	5.0	12.5	11.6	53.3	37.4	39.0
DIGCompressor	2.1	22.0	21.8	47.4	154.4	152.8
COX	5.5	25.6	21.6	33.1	91.9	77.5

(b) Input document (size)

Fig. 10 Performance comparison – turnaround time. **(a)** Instance annotations. **(b)** Ontologies

Performance comparison has been carried out estimating three basic parameters: (1) *compression rate*, defined as $(1 - size_{compressed}/size_{uncompressed})$; (2) *turnaround time*, that is the overall duration of the compression process; (3) *memory usage* by the compressor process.

Tests were performed using a laptop PC equipped with an Intel P8400 Core 2 Duo CPU (2.27 GHz clock frequency), 3 GB RAM at 800 MHz and Ubuntu 9.04 GNU/Linux operating system with 2.6.27 kernel version and *Valgrind* (Nethercote and Seward 2007) 3.4.1 profiling toolkit. To obtain a complete picture of performance, 12 DIG documents of different size were used, 6 semantic annotations of individual objects/products and 6 domain ontologies, which are significantly longer.

Figure 9 shows the resulting compression rate. For each DIG file, the original size in byte is reported. COX and DIGCompressor were respectively the worst and the best performer. Compression rate of COX is somewhat sacrificed in order to allow general-purpose compression with support for direct queries over the encoded format.

For turnaround time, each test was run 10 times consecutively, and the average of the last 8 runs was taken. Results for annotations are reported in Fig. 10a. gzip is the most optimized tool. XMill and COX suffer from the intrinsic complexity of their algorithms, but absolute values are still acceptable. For ontology documents, (see Fig. 10b), the scenario changed. COX provides acceptable performance, whereas DIGCompressor has significantly higher turnaround times for input longer than 100 kB.

In conclusion, memory usage analysis was performed using *Massif* tool of Valgrind toolkit. For our comparison, only the memory peak was considered.

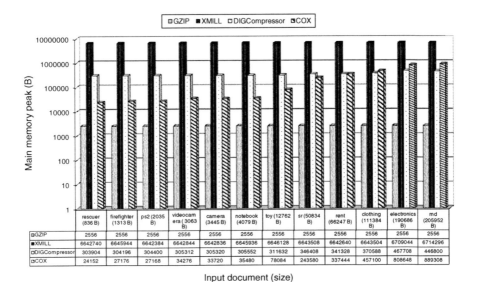

Fig. 11 Performance comparison – main memory peak

Results are shown in Fig. 11. XMill and gzip, due to their specific structure, use an almost constant memory amounts regardless of the input size. DIGCompressor uses about 300 kB for small annotations, growing up to 460 kB for bigger ontologies. COX is much more efficient (about 35 kB) for small documents, but its memory peak exceeds DIGCompressor for the largest inputs.

8 Conclusions

This chapter presented an approach to carry out an advanced matchmaking using semantic metadata stored in RFID tags without requiring unique and fixed knowledge bases. An advanced resource discovery framework is supported by a knowledge dissemination protocol, so allowing an *on-demand* retrieval of suitable descriptions directly from tags located on the objects. An analytical model has been developed for a preliminary assessment of resource requirements. An experimental validation of the proposed approach – including dissemination, discovery and RFID data compression evaluations – has been performed in order to test its feasibility.

References

Baader F, Calvanese D, Mc Guinness D, Nardi D, Patel-Schneider P (2002) The description logic handbook. Cambridge University Press, Cambridge

Bechhofer S, Möller R, Crowther P (2003) The DIG description logic interface. In: Proceedings of the 16th international workshop on description logics (DL'03). CEUR Workshop Proceedings, vol 81. Rome, Italy

Brachman R, McGuinness D, Patel-Schneider P, Resnick L, Borgida A (1991) Living with CLASSIC: when and how to use a KL-ONE-like language. In: John S (ed.) Principles of Semantic Networks. Morgan Kaufmann, San Fransisco, CA, pp 401–456

Chakraborty D, Joshi A, Yesha Y, Finin T (2006) Toward distributed service discovery in pervasive computing environments. IEEE Trans Mobile Comput 5(2):97–112

Cohen B (2008) The BitTorrent protocol specification, version 11031. http://www.bittorrent.org/beps/bep_0003.html. Accessed 20 November 2009

De P, Basu K, Das S (2004) An ubiquitous architectural framework and protocol for object tracking using RFID tags. In: The 1st annual international conference on mobile and ubiquitous systems, Networking and Services (MOBIQUITOUS 2004). Cambridge, MA, pp 174–182

Di Noia T, Di Sciascio E, Donini FM (2007) Semantic matchmaking as non-monotonic reasoning: a description logic approach. J Artif Intell Res (JAIR) 29:269–307

Di Noia T, Di Sciascio E, Donini F M, Ruta M, Scioscia F, Tinelli E (2008) Semantic-based bluetooth-RFID interaction for advanced resource discovery in pervasive contexts. Int J Semantic Web Inf Syst 4(1):50–74

EPCglobal Inc (2005) Object naming service (ONS) – version 1.0. EPCglobal Ratified specification

EPCglobal Inc (2008) EPC Radio-frequency identity protocols class-1 generation-2 UHF RFID protocol for communications at 860 MHz-960 MHz Version 1.2.0 EPCglobal Ratified specification

Harrusi S, Averbuch A, Yehudai A (2006) XML syntax conscious compression. In: Data compression conference (DCC 2006). Snowbird, UT, pp 10–19

Howard P, Vitter J (1991) Analysis of arithmetic coding for data compression. In: Data compression conference (DCC'91). Snowbird, UT, pp 3–12

Huffman D (1952) A method for the construction of minimum redundancy codes. In: Proceedings of the Institute of Radio Engineers (IRE) 40(9):1098–1101

IEEE 802.11 (1999) Information technology telecommunications and information exchange between systems local and metropolitan area networks specific requirements part 11: wireless LAN medium access control (MAC) and physical layer (PHY) specifications. ANSI/IEEE Std. 802.11, ISO/IEC 8802-11. First edn

Levesque H (1984) Foundations of a functional approach to knowledge representation. Artif Intell 23:155–212

Liefke H, Suciu D (2000) Xmill: an efficient compressor for xml data. SIGMOD Rec 29(2):153–164

M Ayoub K, Manoj S, Brahmanandha PR (2009) A survey of RFID tags. Int J Recent Trends Eng 1(4):68–71

Mahoney M (2005) Adaptive weighing of context models for lossless data compression. Technical report, Florida Tech University, Technical Report CS-2005-16

McGuinness DL, van Harmelen F (2004) OWL web ontology language, W3C recommendation. http://www.w3.org/TR/owl-features/. Accessed 20 November 2009

Min J, Park M, Chung C (2007) A compressor for effective archiving, retrieval, and updating of XML documents. ACM Trans Internet Technol (TOIT) 6(3):223–258

Nethercote N, Seward J (2007) Valgrind: a framework for heavyweight dynamic binary instrumentation. In: Proceedings of the 2007 ACM SIGPLAN conference on programming language design and implementation – PLDI 07. ACM Press, New York, NY, USA, pp 89–100

Network Simulator (1995) The network simulator – ns-2. http://www.isi.edu/nsnam/ns/. Accessed 20 November 2009

Ni LM, Zhu Y, Ma J, Li M, Luo Q, Liu Y et al (2005) Semantic sensor net: an extensible framework. Lecture Notes in Computer Science, vol 3619. Springer, Heidelberg, pp 1144–1153

Pallapa G, Das S (2007) Resource discovery in ubiquitous health care. In: 21st international conference on advanced information networking and applications workshops, AINAW'07, vol 2. Niagara Falls, ON

Patel-Schneider P, Swartout B (1993) Description-logic knowledge representation system specification. KRSS group of the ARPA knowledge sharing effort

Ramanathan R, Redi J (2002) A brief overview of ad hoc networks: challenges and directions. IEEE Commun Mag 40:20–22

Römer K, Schoch T, Mattern F, Dübendorfer T (2004) Smart identification frameworks for ubiquitous computing applications. Wireless Netw 10:689–700

Ruta M, Di Noia T, Di Sciascio E, Donini FM (2006) Semantic-enhanced bluetooth discovery protocol for M-commerce applications. Int J Web Grid Serv 2:424–452

Ruta M, Di Noia T, Di Sciascio E, Scioscia F, Piscitelli G (2007) If objects could talk: a novel resource discovery approach for pervasive environments. Int J Internet Protocol Technol (IJIPT), Special Issue on RFID: Technol Appl Trends 2:199–217

Ruta M, Scioscia F, Di Noia T, Di Sciascio E, Piscitelli G (2009) Ubiquitous knowledge-based framework for RFID semantic discovery in smart u-commerce environments. In: Proceedings of the 11th international conference on electronic commerce. ACM, San Diego, CA, pp 9–18

Scioscia F, Ruta M (2009) Building a semantic web of things: issues and perspectives in information compression. Semantic web information management (SWIM'09). In: Proceedings of the 3rd IEEE international conference on semantic computing (ICSC 2009). Berkeley, CA, pp 589–594

Tolani P, Haritsa J (2002) XGRIND: a query-friendly XML compressor. In: Proceedings of the 18th international conference on data engineering (ICDE.02). IEEE, San Jose, CA, pp 225–234

Toninelli A, Corradi A, Montanari R (2008) Semantic-based discovery to support mobile context-aware service access. Computer Communications. Elsevier, Amsterdam

Traub K, Allgair G, Barthel H, Bustein L, Garrett J, et al (2005) EPCglobal architecture framework. EPCglobal Ratified specification

Vasudevan V (2004) Ensembleware: contextual service provisioning in an ubiquitous services world. In: Symposium on Applications and the Internet, Tokyo, Japan, 26–30 January

Vazquez JI, Lopez-de-Ipina D (2007) mRDP: an HTTP-based lightweight semantic discovery protocol. Comput Netw 51(16):4529–4542, Elsevier

Venkataramani G, Iyer P (2007) Semantics-aware RFID middleware for personalized web services to retail customers. In: International workshop on service-oriented engineering and optimization, Goa (India), December 2007

Watson R, Pitt L, Berthon P, Zinkhan G (2002) U-Commerce: expanding the Universe of marketing. J Acad Mark Sci 30:333–347

Weinstein R (2005) RFID: a technical overview and its application to the enterprise. IT Prof 7:27–33

Witten I, Neal R, Cleary J (1987) Arithmetic coding for data compression. Communications of the ACM 30(6):520–540

Ziv J, Lempel A (1977) A universal algorithm for sequential data compression. IEEE Transactions on Information Theory 23(3):337–343

RFID Middleware Systems: A Comparative Analysis

Nova Ahmed and Umakishore Ramachandran

Abstract In recent years, there is a widespread adoption of RFID technology in several key application domains. Deployment of RFID technology in large-scale applications is attractive due to its low cost; however RFID technology, especially the one that uses passive tags, is error prone in nature. Large-scale deployment of RFID technology requires efficient middleware solution that is able to handle the large amount of data, process the data in a timely fashion and deliver the required information to the applications. We discuss the state of the art in RFID middleware, and present the opportunities and challenges of the RFID technology.

1 Middleware Solution for RFID Applications

RFID technology is rapidly becoming part of today's life in many applications such as smart homes, airports, and supply chain management systems. With the ubiquity of RFID deployments, there exists a growing amount of data being generated from multiple sources that need to be processed in order to support user queries. The different applications require an abstraction to manage the large stream of data (as the RFID readers are often error prone Cherniack et al. 2003; Franklin et al. 2005) in a timely fashion.

A middleware solution is considered to be a layer that interacts with the underlying system transparent to applications. There have been many interesting research works pertaining to generic middleware solutions. Middleware designed for large-scale RFID based systems have some unique properties: (a) it has to be aware of the nature of the underlying RFID hardware, (b) it has to be aware of the error-prone nature of RFID generated data, and (c) it has to meet application requirements in a timely manner.

N. Ahmed (✉)
College of Computing, Georgia Institute of Technology, Atlanta, GA 30332, USA
e-mail: nova@cc.gatech.edu

D.C. Ranasinghe et al. (eds.), *Unique Radio Innovation for the 21st Century*,
DOI 10.1007/978-3-642-03462-6_12, © Springer-Verlag Berlin Heidelberg 2010

1.1 Applications Requiring RFID Middleware Systems

Many different applications are considering RFID technology as a viable solution for identification, tracking, monitoring etc. Large-scale deployment of this technology that requires a middleware framework can be divided among two major application categories: *item tracking* and *item location* as shown in Fig. 1.

- *Item Tracking*: In a typical supply chain scenario, as an example of an object tracking application (Gribble et al. 2001), readers are placed along conveyor belts for item detection. Airport baggage claim system is another example of object tracking applications (Jeffery et al. 2006a).
- *Item Location*: In the case of item location applications, tagged items are placed statically and the set of readers are moving along a path locating items in their wake. Robots equipped with RFID readers looking for lost objects in an environment is an example of object location. Another scenario is using rescue personnel in a disaster situation: tags deployed in the environment may guide the person in recovery operation.

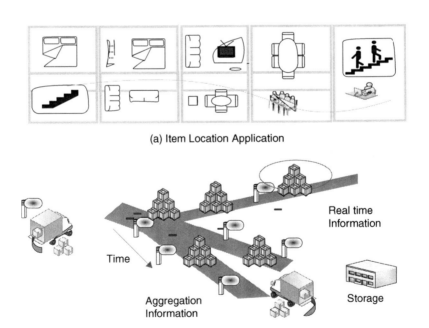

(a) Item Location Application

(b) Item Tracking Application

Fig. 1 Large-scale RFID deployment in, **a** item location applications, **b** item tracking applications

1.2 Middleware Requirements

A middleware for RFID deployment requires careful design considerations taking into account the inherent vulnerabilities of these devices along with application requirement. We illustrate various application requirements that drive the classification of RFID middleware systems discussed in the following section. A typical middleware system is shown in Fig. 2.

1.2.1 Basic Middleware Requirements

A middleware designed for a large-scale deployment of RFID devices must take into account a set of basic requirements that are addressed by all the research based middleware systems as well as commercial middleware systems discussed in the literature.

- *Data Organization*: For large amounts of data there is a need to properly sort and organize data for efficient query responses.
- *Scalability*: System should be capable of handling the addition of new readers or to dynamically ignore malfunctioning readers without significantly impacting performance.
- *Load Balancing*: System load should be balanced across the computational elements when handling a large amount of data.
- *High Throughput*: System is expected to provide extremely high update rates from the readers and therefore requires high throughput to handle the sheer numbers of items.

1.2.2 Advanced Requirements

RFID middleware systems may be required to provide a richer functionality depending on various application requirements triggering interest in different classes of middleware advancement.

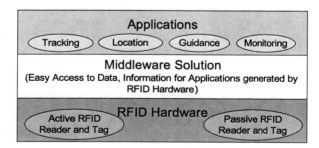

Fig. 2 RFID enabled system

- *Fine Grain Data Management*: Depending on application requirements, RFID middleware systems may need to have a greater control over the large volume of data generated in a meaningful way.
- *Dynamic Resource Management*: In many different applications, RFID data may overload the system requiring an ability to dynamically manage the system resources.
- *Human Centered Deployment*: While a majority of large-scale RFID deployments are concerned with item tracking applications, there are emerging opportunities for use of RFID in application domains such as indoor guidance and monitoring. Such human-centric deployment of RFID technology requires handling of data in a privacy-conscious manner.

Our discussion on the middleware systems for RFID enabled systems focuses on various middleware approaches driven by application requirements. We elaborate on different aspects of the current middleware solutions.

1.3 Summary of Classification

RFID middleware systems provide a layer of abstraction to hide the underlying RFID related hardware complexity. These middleware systems can be classified as *general* middleware solutions, *event based* systems, *dynamic resource management* systems, *special purpose* systems and *commercial* systems as shown in Fig. 3.

Classification Name	Properties	Example
General Middleware Solutions	Scalability, Data management, Item Tracking applications	Fosstrak [28], WinRFID [37], Adaptive RFID Middleware [18], RF²ID [12]
Event Based Systems	Treats RFID data as Complex events	Real time event handling system [4], RMS [5], ESN [6], Contextual Events framework [7,8]
Dynamic Resource Management Systems	Uses mechanism to Handle unexpected Large amount of data	Agent based load balancing System [10], Connection pool based system [11], RF²ID [13]
Special Purpose Systems	Considers item location applications	GuardianAngel [9]
Commercial Systems	Developed by various Commercial vendors	IBM WebSpehere [20], OATs System [33], SAP AutoID [39], BizTalk [38]

Fig. 3 Classification of RFID middleware systems

- *Generic RFID Middleware*: General purpose middleware systems are dedicated for scalable solution for handling large amount of RFID generated data. These solutions provide the basic requirements of being able to: organize data, provide scalable solution, manage large-scale data and have high throughput as discussed in the basic requirements.
- *Event Based RFID Middleware*: Event based systems are specialized middleware solutions handling RFID data as events. This class of solution is able to have a greater control of RFID data in a meaningful way. A greater control of fine grain data management enables a wide range of query operations that may be required for certain classes of applications.
- *Dynamic Resource Management RFID Middleware*: Dynamic resource management systems aim to provide an ability to handle unexpected system load on top of existing middleware solutions. The asynchronous nature of RFID generated data may overload a system in a particular period of time and it may be required to have specialized algorithms to manage the load across the system transparent to the applications.
- *Special Purpose RFID Middleware*: There are special purpose systems that consider application specific requirements such as human centeredness in design and architecture. The emerging use of RFID technology in applications involving indoor facilities consider privacy as a major concern and must address solutions that take into account concerns of the human users.
- *Commercial RFID Middleware*: Commercial middleware solutions are offered by various corporations and are challenging to evaluate due to the limited amount of available information about the underlying system.

We briefly cover these various categories of systems. The format of the discussion follows a summary of a particular category followed by a concise description of individual systems. An analysis of various systems is presented at the end of the chapter.

2 Generic RFID Middleware Solutions

Generic middleware solutions are mostly designed for large-scale RFID deployment for item tracking applications across a large geographic space. The main focus of these systems is to provide a way to efficiently process RFID generated data in a scalable manner to meet application specific requirements. Systems like Fosstrak, WinRFID (Prabhu et al. 2006), Adaptive middleware (Jeffery et al. 2008) follows a hierarchy of components as shown in Fig. 4a. RF^2ID (Ahmed et al. 2007), on the other hand, considers a data flow oriented approach as illustrated in Fig. 4b. Interestingly, the adaptive middleware (Jeffery et al. 2008) functionally follows a pipelined scheme which is close to the data flow oriented approach.

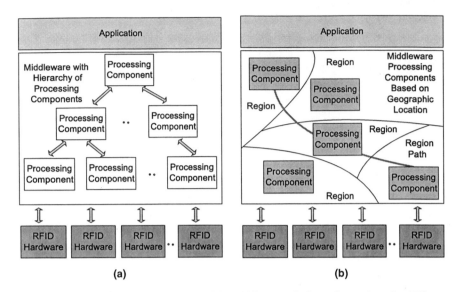

Fig. 4 a Hierarchy of components in RFID middleware. **b** Data flow oriented middleware connecting through path

2.1 Fosstrak Formerly Known as Accada

Fosstrak (previously known as Accada Floerkemeier et al. 2007), is an open source implementation based on the RFID standardization body EPCGlobal (Official website for EPCglobal) that leads development of Electronic Product Code. It has three major modules: *reader module, filtering and collection middleware* and *data for application context* (Application Level Events 1.1.1 Standard 2009; EPCglobal 2007, 2008, 2009; EPC Information Services Standard 2007; EPC Tag Class Definitions 2007; EPC Tag Data Standard 2003; EPC Tag Data Translation Standard 2009; Fosstrak).

Reader module is in charge of management of data that takes care of data dissemination, filtering and aggregation at the reader level. Filtering and Collection middleware decouples the readers and applications. This provides a way to incorporate additional functionality for aggregation and filtering of the RFID generated data. Data for application context layer of the framework allows for application specific processing to be applied to the RFID generated data.

Fosstrak provides a general solution to deal with the large amount of data using a modular approach to handle different activities.

2.2 WinRFID

WinRFID (Prabhu et al. 2006) provides a multi layered middleware system where each layer has a separate functionality and level of interaction. The architecture consists of *RFID hardware layer, protocol layer* and *data management layer.*

RFID hardware layer is the lowest layer dealing with the actual hardware such as RFID readers, tags and additional sensor elements. It provides a common interface for supporting various hardware devices. *Protocol layer* provides an abstraction to manage various available RFID protocols. It also allows flexibility in adding additional protocols and capabilities as new ones are introduced. *Data management layer* deals with error prone RFID generated data by processing the data (e.g., removing duplicate reads, verifying the tags, etc.). It has several special components that are responsible for various data management functionalities such as managing XML data, presenting the data to the applications and support for creating various services through *XML framework*, *data presentation* and *services* components respectively.

WinRFID system provides the infrastructural and data management support for large-scale RFID deployment.

2.3 The Adaptive RFID Middleware System

Adaptive RFID middleware (Jeffery et al. 2008) considers a layer named metaphysical data independence (MDI) to abstract the RFID generated data from the applications. The basic philosophy is to hide the actual physical device and present appropriate interfaces to the application. *MDI SMURF* is the middleware and is organized as a pipeline of processing stages. It has several components for achieving various functionalities.

RFID data flows to the temporal SMURF that uses statistical framework to correct RFID data stream for its error prone nature. A data model consists of *objects*, *attributes* and *uncertainty estimates*. An object is a unique entity of the physical world. Every object has a set of attributes associated with it. Primary attributes are defined as time and space. Attributes carry meanings of objects in the physical world. Uncertainty estimate is required for the nature of the sensing devices. It is used at multiple levels based on the attribute of the object.

Adaptive RFID middleware provides a way to analyze the RFID generated data at the same time providing a powerful distributed structure for communication.

2.4 RF^2ID

RF^2ID is a middleware framework that refers to *Reliable Framework for Radio Frequency Identification (RF^2ID)* (Ahmed et al. 2007). It uses the nature of the data flow at system level to improve the reliability of the system. It provides two major system abstractions named *virtual readers* and *virtual paths*.

Virtual Readers (VR) is an abstraction to capture the static and potentially errorprone nature of the physical readers and antennas in a scalable manner. Virtual reader is the distributed computing element of the system. VRs are distributed across the geographic area and each VR is in charge of the RFID readers deployed in that particular area. Virtual Paths (Vpath) is used to capture the logical flow of information among the virtual readers as RFID-tagged objects move throughout the environment. Using a notion of path at system level gives several advantages. First, system load can be distributed among multiple virtual readers that constitute a specific virtual path. Second, different QoS attributes can be defined for a path, such as accuracy and priority levels that the virtual readers use to operate on data flowing through the corresponding path. Finally, as there is an internal representation of data based on path attributes, it becomes trivial to support path-related operations on the data, e.g., searching for query results, or making a future projection of data behavior based on history.

RF^2ID is able to provide a scalable way to communicate across a large geographic region in a scalable manner taking the data flow specific approach to handle the RFID generated data.

2.5 Other Systems

There are other general purpose middleware systems which mainly fall into the hierarchical system architecture category based on their internal architecture. *RFIDStack* (Floerkemeier and Lampe 2005) tries to address the inherent challenges faced due to the nature of RFID technology itself (e.g., unreliability, low bandwidth for communication, etc.). It presents the requirements and corresponding desired features to handle such needs. The system description stops short of describing actual mechanisms to achieve the desired features (e.g., content-based routing of RFID data).

Wang and Liu (2005), discuss how RFID data need to be managed in a different way based on various application needs. They mention a method of allowing temporal information in the data itself that would enhance various query mechanisms in the system. Such mechanisms are very useful for handling RFID data but the system is not a complete middleware solution as it considers only data management aspects and not communication among components of the system.

REFiLL (Anagnostopoulos 2009) considers a programming mechanism that uses filtering based methods to support various applications. It employs a hierarchy of components as well as it allows system events to be determined similar to an event based systems.

3 Event Based RFID Middleware

There are set of middleware systems that focus on the complex events to organize the middleware framework. We describe some of these systems in the following sections. A basic structure of event based system is shown in Fig. 5 consisting of middleware with event processing components to support applications handling RFID data. Complex event processing (CEP) is used in the discussion of the event based systems.

Fig. 5 RFID middleware
handling data as events

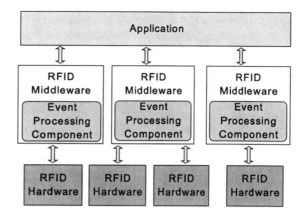

3.1 Real Time Event Handling System

Dutta et al. (2007) consider an RFID middleware that treats RFID activities as
events, and performs real-time operations on such events. It considers item tracking applications for RFID based activities. It consists of two major components:
RFID event handler (EH) and *Events of Interest database (EIDB)*.

Basic events are generated by the RFID hardware such as the RFID readers and
non RFID hardware such as the clocks. These basic events are sent to the event
handler. *EH* is the most complex part of the system which can be centralized or distributed. Current systems report a centralized EH that takes basic incoming events
and matches events of interests (EI) in them. *EIDB* contains EI along with the corresponding details. EIs are broken into primitive EIs and then they are stored in a
multi-dimensional R-tree structure. Semantics of composite EI is captured by a state
transition diagram.

The system is able to demonstrate a scalable and efficient way to handle complex
events in the system.

3.2 RFID Middleware based on Complex Event Processing

RFID Middleware System (RMS) described by Dong et al. (2006) illustrates complex
event processing (CEP) as a mechanism to handle large-scale RFID data. The CEP
based methodology enables the system to work across multiple streams of data for
discovering event correlation and event relationships.

RMS components are detailed here. *Reader adapter* manages components from
various RFID readers and the driver can be installed and uninstalled through this
adapter. It also receives history information from the active database for analysis.
Cache strategy is used to improve the system performance by allowing to process
data streams from the reader adapter. *Event Processor Manager* is the key part of the
middleware – it is in charge of handling the data such as filtering, aggregating and

grouping of data and discovering CEP events. *Subscriber Manager* uses a publish-subscribe strategy to send out reports based on the event processor. *Active Database* is different from the passive database as it is able to generate data that focuses on the past, present and future events. The event processor is implemented using the event processing language (EPL).

RMS is able to provide a framework to manage complex RFID data generated from multiple data streams.

3.3 Complex Event Processing in EPC Sensor Network Middleware ESN

EPC sensor network is used as the base system which incorporates complex event processing mechanism in ESN middleware (Wang et al. 2008). The argument is to use CEP method to enable the handling of the events easier at the same time being able to extend the system for wide area network. It has an event driven architecture and it uses complex event processing to meet the advanced application requirements.

ESN has several basic components: Events from the RFID readers are sent to the *event handler*. It uses a queue to manage the processing of events. *Event Database* is the repository to store data for operations such as aggregation that require data storage over a period of time. *Event Processing Engine* is the kernel of the middleware which filters, formats and constructs complex events of data in real time. *Event Action Unit* connects to the upper layer enterprise applications. It is also able to send out reports using the publish/subscribe actions. The system also provides a way to handle the data streams that considers the specific system architecture supported by the EPCglobal body.

3.4 Contextual Events Framework

Moon et al. (2006), Kim et al. (2006) consider the contextual events framework to deal with RFID based data. The goal is to enable a system to be able to process the RFID generated data in real time and thus being able to define the context.

The system architecture is discussed here which has four major components. *Contextual Event Notification* layer delivers contextual events to the RFID applications. *Contextual Event Management* collects the RFID events and transforms the events using reference data. *RFID Event Collection* uses the middleware interface to get RFID event information. The *Reference Data Collection* component receives reference data from external data servers. *CESpec Registry/CE Repository* is the repository to store contextual event specifications and generated contextual events. This system uses a notion of contextual event generated by the RFID data to process the events in real time.

4 Dynamic Resource Management of RFID Data

RFID generated data, in many applications such as supply chain management, are considered to be asynchronous in nature. This asynchronous incoming data can cause the system to handle a large amount of data over a particular interval of time while the system is under-loaded in a different time window. An RFID middleware must be able to dynamically manage unexpected data load to make sure the applications are not affected. There have been research works to manage unexpected RFID generated load using load migration-based techniques as presented in agent system (Feng et al. 2007) and connection pool based system (Park et al. 2007) and load shedding based technique as presented in load shedding based resource management for RF²ID (Ahmed et al. 2007, 2009). The basic technique employed in resource migration based dynamic load management is to make a balanced assignment of load from heavily loaded components to lightly loaded components. A basic resource management system is presented in Fig. 6. Agent based (Feng et al. 2007) system and connection pool based system (Park et al. 2007) fall into this category. These systems consider residing among a set of general purpose RFID middleware systems and a set of RFID hardware components (RFID readers and tags). We discuss these methods in the following sections.

4.1 Agent-based Load Balancing for RFID Middleware

Agent based load balancing scheme (Feng et al. 2007) considers the presence of mobile agents to make resource management decisions in the system. It uses several policies to manage the system load.

Policies are described in brief here. *Information Gathering Policy* gathers workload information from the middleware hosts and broadcasts this information to make proper decisions. *Initiation Policy* determines load balancing initiator which can be

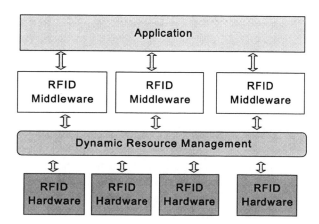

Fig. 6 Dynamic resource management component in RFID middleware

by source node, destination node or by a symmetric process. *Job Transfer Policy* figures out reallocation of the job after initialization is over. *Selection Policy* determines all jobs to be reallocated. *Location Policy* determines specific middleware to transfer a particular job. Load balancing middleware considers an agent container within each host that is above the agent platform. In this way, RFID readers are able to dynamically determine the middleware to contact based on the current workload. Two load management agents named load gathering agent (LGA) and reader load balance agent (RLBA) are designed. LGA is a mobile agent that traverses the system and RLBA is the stationary agent residing in the middleware host. The proposed system presents a framework for designing and implementing a set of middleware systems that are able to dynamically adjust the resources using load balancing mechanism. The proposed system stops short of describing an actual implementation and evaluation; thus it is not obvious what the overhead would be for such reallocation in real-time.

4.2 Load Balancing Using Connection Pool

Connection pool based load balancing (Park et al. 2007) readjusts load among different nodes from heavily loaded one with a lighter load. In this method, tag data read by the RFID readers enter a connection pool. The connection pool is connected to several middleware that are managed by connection pool manager. It has several major components: connection pool, connection pool manager, connection pool table and middleware manager table and the interrupt check component.

Connection Pool operates like a switch where lots of data are transmitted from RFID readers to the middleware through this pool. *Connection Pool Manager* is in charge of allocating tag data to the middleware based on the resource status. *Connection Pool Table* keeps the current information about the tag data in the connection pool. *Middleware Manager Table* keeps information on tag data to middleware assignment information. *Interrupt Check Component* plays a role to monitor the middleware and notify such information to the middleware manager table. The system also presents a way to manage the load among multiple middleware systems. Connection pool based system does not present any implementation of such system. It gathers all the data and disseminates it accordingly which can have significant performance bottleneck if not managed properly.

4.3 Load Shedding based Resource Management

RF^{2}ID presents a resource management technique that uses load shedding (Ahmed et al. 2009) instead of load balancing as a RFID middleware resource management technique. It uses two different methods that leverage the distributed computing elements virtual readers (VR) and virtual path (Vpath) to make intelligent load

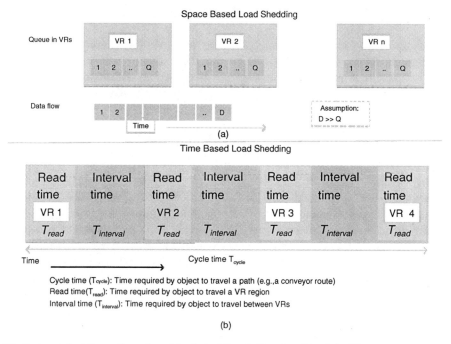

Fig. 7 Load shedding. **a** Space based load shedding. **b** Time based load shedding

shedding decisions. It uses a *space based load shedding mechanism* and *time based load shedding mechanism* as shown in Fig. 7.

In space based load shedding technique, VRs cooperatively make a decision to process a certain amount of load in a way that is able to provide a predefined system performance. This technique is able to provide reasonable performance in a large distributed system setup with unexpected heavy load. The time based load shedding technique is a time equivalent of space based technique where the VRs make a decision to process items for a specific period of time.

This technique works within the environment where components are able to synchronize based on operational time. Load shedding based techniques have been designed and implemented and the system demonstrates acceptable performance in the presence of heavy load.

4.4 Other Systems

A heuristic based solution for near optimal reader to tag assignment to deal with the load in the system is presented by Dong et al. (2008). However, the solution only addresses statically placed readers and tags. Producing a system that accounts for tags in motion adds another layer of complexity.

5 RFID Infrastructure for Item Location Applications

Research efforts on item location applications using RFID technology, are gaining momentum. Most of the existing systems provide a solution that deploys the RFID enabled system without a middleware approach. The GuardianAngel middleware system provides a solution strategy that targets indoor guidance and monitoring applications using virtualization techniques.

5.1 GuardianAngel

Guidance and monitoring applications for deployment in environments such as an assisted living center require a way to share data that is unobtrusive for the user from a privacy and comfort point of view. Conventional solutions of asking the users to wear a badge or a video-based system are not able to provide the level of comfort users ask for. Incorporation of the human actors to the system brings in unique considerations that are not addressed in general system solutions. The challenge lies in providing fine grain guidance information to the user at the same time providing coarse grain information of the user for monitoring purposes. GuardianAngel (Ahmed 2009, Ahmed et al. 2010) middleware is designed to incorporate guidance and monitoring capability in an RFID tagged environment for users with mobile RFID devices.

GuardianAngel works in two layers as can be seen in Fig. 8. The lower layer named as the *Guidance Layer* provides the locality information to the user to make guidance decisions. And an upper layer known as the *Monitoring Layer* has the global knowledge of the environment. Environment is equipped with low cost RFID tags. Guidance layer is supported by a handheld device equipped with RFID reader. It is thus able to provide information regarding the resident's current location and immediate objects by sensing the environment. Monitoring layer has the information about entire environment. Guidance layer periodically updates coarse grain information defined as the *virtual location* about the resident to the monitoring layer. Guidance layer with its limited capability only keeps partial map information that is acquired on demand from the monitoring layer. GuardianAngel is able to provide a different view to the user and towards the system that enables an unobtrusive way for guidance and monitoring services.

5.2 Other Systems

There is a rich variety of work in the field of context aware computing which focuses on a generalized solution for a diverse set of devices, interfaces, and operations. A supportive environment for individuals requiring cognitive assistance is presented in Chang et al. (2008). This system considers using a tag enabled environment, similar to the GuardianAngel approach, but focuses more on creating a unified interface

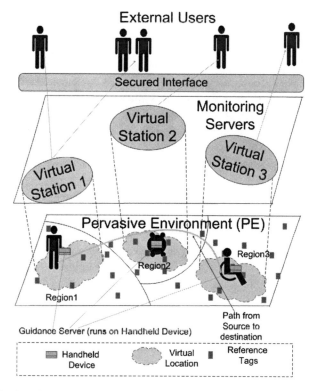

Fig. 8 GuardianAngel

that improves the user experience as opposed to system specific details of a middleware. The system described in Abascal et al. (2009) presents an environment for people with cognitive disability which provides context-aware information and intelligent decision making. Also worth noting are the programming abstractions described in Acharya et al. (2008), which provide for a high level virtualization architecture that enables a layer to aggregate queries. This is a more generalized solution than an RFID technology specific solution. Other research has been done with RFID-based middleware systems. The work presented by Thiem et al. (2008) discusses a framework that combines RFID infrastructure with a wireless communication interface. This allows for the expansion of existing infrastructure, increasing environment coverage at a significantly lower cost than other options.

6 Commercial RFID Middleware Solutions

Commercial development of RFID middleware shows the demand for this solution approach in recent years. There have been RFID middleware solutions provided by many of the major solution providers. IBM WebSphere Premises Server provides a service-oriented architecture. It presents a solution that offers real time tracking, development of multisensing system, ability to incorporate distributed environment and ability to process data that matches the business requirements. A large-scale supply chain systems like Airbus has considered using IBM WebSphere infrastructure along with business applications of OATs systems. Similarly, Microsoft provides an RFID solution named BizTalk that relies on the event based technology and promises to deliver a scalable, robust and secure system. SAP AutoID Infrastructure (2009) considers features like fast, connected, scalable for its customers as an RFID middleware solution. It considers programming language based support and usage of savant technology.

7 Analysis

Growing research and development in RFID middleware indicate how important this topic is in terms of today's applications. We have taken a look at the various architectural components of the RFID middleware systems as the major differences lie in how the system is organized. In terms of performance, there is no standard metric that can be used to study individual systems as many of the systems are geared for specific application domains. The unavailability of real world data for large-scale RFID deployments is another limitation that impairs a comparative study of research systems in the light of real world application scenarios. We have taken a qualitative approach to discuss the current middleware solutions to understand their uniqueness and differences from one another.

7.1 Analytical View of General Middleware Solutions

Fosstrak which has been known as Accada or Savant in previous literature (Application Level Events 1.1.1 Standard 2009; EPCglobal 2007, 2008, 2009; EPC Information Services Standard 2007; EPC Tag Class Definitions 2007; EPC Tag Data Standard 2003; EPC Tag Data Translation Standard 2009; Fosstrak) is a pioneer in the RFID middleware based work. It provides a detailed view of the middleware system that follows a hierarchy of various components. This architecture has been incorporated as a baseline system for building different enhancements on top of (such as event based systems Wang et al. 2008). It provides a way to define a generic middleware solution for large-scale RFID deployment. It has played a major role in standardizing the various specification of such a system deployment when this technology was in its infancy. It is the most well addressed middleware solution among the research based middleware in terms of specification and description of components and interactions.

Metaphysical data independence (MDI) (Jeffery et al. 2008) is aimed to decrease the complexities of hardware from the applications. Implementation of MDI is called MDI-SMURF. There is a large gap in the way data is handled in digital systems from what is needed in the application layer. The MDI-SMURF provides abstractions around the physical data required by applications. It works as a pipeline (Jeffery et al. 2006b) of stages starting from *temporal-SMURF* where RFID data enters, and then it forwards data to *spatial-SMURF* that takes care of proximity information in data. *Virtualize* module manages the data based on temporal and spatial information. The advantage of the approach mentioned here takes care of the data organization and management which is very challenging in large-scale deployment. The system is able to support queries to address data specific concern such as the amount of uncertainty to expect in a set of data. This close coupling to data operation is required for any large-scale RFID deployment.

RF^2ID (Ahmed et al. 2007, 2009), on the other hand considers the deployment of large-scale RFID applications from the systems perspective. It takes a look at the scalability, communication and data organization that are complementary to the previous approaches. RF^2ID takes the unique approach of organizing the computational components that mimic the physical flow of data. The data flow oriented organization provides the system with interesting capabilities such as real time response on item's location (e.g., missing or misplaced item detection), dynamic resource management in case of unexpected system load and the ability to provide temporal and spatial application support. The communication that follows the physical route of items can be accommodated in any baseline system for increased level of data access operations. RF^2ID would be very useful in applications that require a real time view of the system status such as real time inventory.

All the general purpose middleware solutions provide a way to meet the core application requirements in terms of data management, communication and deployment of a large-scale system. A middleware solution will not be complete without addressing these key functionalities. As can be seen, there has been enhancement on these generic solutions based on various factors which are discussed in the following sections.

7.2 Analytical View of Event Based Systems

Event based systems are able to label events occurring in the system at various levels. The ability to define events from the stream of data provides a granular way to process data as well as design operations suitable for various events.

An event based system can be a standalone middleware system (Dong et al. 2006; Jeffery et al. 2006a; Kim et al. 2006; Moon et al. 2006) or it can be integrated on existing general purpose RFID middleware systems as shown in ESN (Wang et al. 2008) that is based on EPC network defined by EPCglobal (Official website for EPCglobal). Event based systems provide a way for fine grain data analysis in real time. It also enables a way to generate complex events that are composed of collection of simple events. These systems are suitable for real time data processing of a large-scale system as well as designing and analyzing a set of complex events. An addition of fine grain computation capability using the event based framework to an existing general purpose middleware solution is expected to provide a flexible way to handle large-scale data.

7.3 Analytical View of Dynamic Resource Management Systems

Dynamic resource management based methods consider the systems capability of stabilizing in the presence of data overload. Systems that use various resource management methods consider existing middleware solutions to handle generic task of managing data, connection and communication of the components. Dynamic resource management systems can be considered as a specialized component added to the general purpose middleware solutions.

Performance of dynamic resource management based systems is dependent on the application scenarios and level of load of the system. Resource migration based methods (Feng et al. 2007; Park et al. 2007) are best suited when the middleware system has some time to make dynamic assignment of RFID data to middleware systems. Applications (e.g., aggregation information of inventories) that allow a window of time to process RFID generated data are well suited for such techniques. The heuristic based load balancing method considers static RFID reader and tag placement which may be suitable for applications that have static assignments (e.g., monitoring of a stable environment). Load shedding based resource management technique is suitable for applications that follow a constraint environment for real time data delivery. This technique is suitable to applications that need to provide real time status of the system (e.g., real time inventory of a system).

7.4 Analytical View of Special Purpose Systems

Special purpose systems are triggered by requirements of different ongoing effort to incorporate RFID technology in different application domains on top of the generic tracking based solutions. Privacy concern is a major concern in any indoor support systems. As the RFID technology matures, there will be a wider range of special purpose applications as well as middleware systems to support them.

7.5 Analytical View of Commercial Middleware Solutions

Commercial RFID middleware solutions are used in various applications. It is challenging to compare and contrast the commercial middleware solutions to research based solutions, or compare different commercial solutions among each other based on the limited available information on the working mechanism or performance parameters of these systems. Commercial middleware solutions are expected to have a larger code base compared to research prototypes or home grown middle-ware solutions which make the likelihood of malicious code insertion a possibility for larger source codes. But the smaller solutions are less likely to be tested thoroughly and may be vulnerable to attacks (Rieback et al. 2006). The growing interest in RFID technology is causing increasing malicious attacks on such systems. Growth and development of commercial solutions provide us with the emerging requirement and interest in middleware for RFID based technology. Sensor server solution developed by Oracle provides an extensible, fast and scalable solution for large-scale RFID deployment. The discussion on commercial solutions provides a great insight towards the application requirements that drives the research efforts.

8 Open Problems

RFID based applications are growing and middleware systems are emerging in recent years to meet various application requirements. However, there are remaining challenges to be addressed in near future that are of great importance. We consider *security*, *energy awareness* and *generality* to be of greater interest.

Security is a major concern in RFID systems as it is for any other system. A middleware system must be cautious to deal with large amount of data as some of it may be malicious. Large amount of information available in RFID middleware systems make it very attractive for malicious users for accessing sensitive information. Current middleware systems discussed from research perspective do not incorporate security as a feature in the system. The commercial solutions do not mention security as a major service that they offer as well. Security can be incorporated on top of a middleware solution. A middleware layer is very suitable for incorporating security such that malicious devices, data or queries can be eliminated at the very low level.

Energy conservation is becoming a major concern in the large-scale systems of recent years. Large-scale RFID based systems are being designed and deployed – these systems must consider pathways to minimize the energy footprint. A middleware solution is best suited for such measures as it couples closely with hardware devices. It is believed that there will be new design considerations for energy aware RFID middleware or existing middleware systems enhanced for energy conservation features in the near future.

A system that may have a diverse set of applications is hard to generalize. There are many existing solutions offered for RFID based systems. However, there is still a requirement for a generic middleware system that offers solutions for a wide range of applications such as tracking, monitoring, guidance etc. In the future we expect to see more generic solutions that can be tuned to specific needs.

Open problems indicate the growth of interest in improvement in the area of middleware for RFID technology.

9 Concluding Remarks

The various middleware solutions focus on different aspects of the problem while major goals that these systems shoot for are quite similar. The unreliable nature of RFID data imposes challenges on top of the large amount of data to handle. A middleware solution must deliver data in a timely manner at the same time being scalable. There are many interesting applications being developed in recent years which are driving the search for an adequate middleware solution for RFID devices. An interesting question remains in the applicability of the middleware solution in the light of real application deployment. Availability of industry specific requirements would strengthen the study further based on feasibility and acceptability.

References

Abascal J et al (2009) Towards an intelligent and supportive environment for people with physical or cognitive restrictions. In: Proceedings of the 2nd international conference on pervasive technologies related to assistive environments, Athens, Greece, 16–19 July

Acharya A et al (2008) Presence virtualization middleware for next-generation converged applications. In: Proceedings of the ACM/IFIP/USENIX middleware/08 conference companion, Leuven, Belgium, 1–5 December

Ahmed N et al (2007) RF^2 ID: a reliable middleware framework for RFID deployment. In: IEEE International parallel and Distributed processing Symposium, Long Beach, CA, USA, 26–30 March

Ahmed N (2009) Defining virtualization based system abstractions for an indoor assistive living for elderly care. In: Proceedings of the 11th international ACM SIGACCESS conference on computers and accessibility, Pittsburgh, Pennsylvania, USA, 25–28 October

Ahmed N et al (2009) Load shedding based resource management techniques for RFID data. In: Proceedings of the IEEE international conference on RFID Orlando, FL, USA, 27–28 April

Ahmed N et al (2010) GuardianAngel: an RFID based indoor guidance and monitoring system, PWN, Percom, Mannheim, Germany, 29 Mar–2 Apr 2010

Anagnostopoulos AP (2009) REFiLL: a lightweight programmable middleware platform for cost effective RFID application development. Pervasive Mob Comput 5(1):49–63

Application Level Events Standard v.1.1.1 – Part 1 (2009) http://www.epcglobalinc.org/standards/. Accessed 1 Mar 2010

Chang Y et al (2008) A context aware handheld wayfinding system for individuals with cognitive impairments. In: Proceedings of the 10th international ACM SIGACCESS conference on computers and accessibility, Halifax, Nova Scotia, Canada, 13–15 October

Cherniack M et al (2003) Scalable distributed stream processing. In: Proceedings of the CIDR conference, Asilomar, CA, USA, 5–8 January

Dong L et al (2006) Design of RFID middleware based on complex event processing. In: Proceedings of cybernetics and intelligent systems, Bangkok, 7–9 June

Dong Q et al (2008) Load balancing in large-scale RFID systems. Comput Netw 52(9):1782–1796, Anchorage, AK

Dutta et al (2007) Real-time event handling in an RFID middleware system. Lect Notes Comput Sci Databases Networked Info Syst 4777:232–251, doi: 10.1007/978-3-540-75512-8_17

EPC Information Services Standard v. 1.0.1 (2007) http://www.epcglobalinc.org/standards/. Accessed 1 Mar 2010

EPC Tag Class Definitions (2007) http://www.epcglobalinc.org/standards/. Accessed 1 Mar 2010

EPC Tag Data Standard v. 1.4 (2003) http://www.epcglobalinc.org/standards/. Accessed 1 Mar 2010

EPC Tag Data Translation Standard v. 1.4 (2009) http://www.epcglobalinc.org/standards/. Accessed 1 Mar 2010

EPCglobal Architecture Framework v. 1.3 (2009) http://www.epcglobalinc.org/standards/architecture/. Accessed 1 Mar 2010

EPCglobal, Low Level Reader Protocol Standard v. 1.0.1 (2007) http://www.epcglobalinc.org/standards/. Accessed 1 Mar 2010

EPCglobal, Object Naming Service Standard v. 1.0.1 (2008) www.epcglobalinc.org/standards/ons/. Accessed 1 Mar 2010

Franklin MJ et al (2005) Design considerations for high fan-in systems: the HiFi approach. In: Proceedings of the CIDR conference, Asilomar, CA, USA, 4–7 January, 2005, pp 290–304

Feng CJ et al (2007) Agent-based design of load balancing system for RFID Middlewares. In: Future trends of distributed computing systems, Sedona, AZ, USA, 21–30 March

Floerkemeier C, Lampe M (2005) RFID middleware design: addressing application requirements and RFID constraints. In: Proceedings of the 2005 joint conference on smart objects and ambient intelligence: innovative context-aware services: usages and technologies, Grenoble, France, 12–14 October, 2005, pp 219–224

Floerkemeier C, Roduner, C, Lampe M (2007) RFID application development with the Accada middleware platform. IEEE Syst J 1(2):82–94

Fosstrak, An open source RFID platform (2010) http://www.fosstrak.org/. Accessed 1 Mar 2010

Gribble SD et al (2001) The Ninja architecture for robust internet-scale systems and services. Comput Netw 35(4):473–497

IBM press release (2008) Airbus selects IBM and OATSystems for world's largest RFID-enabled manufacturing initiative, http://www-03.ibm.com/press/us/en/pressrelease/23875.wss. Accessed 1 March 2010

IBM Webspere premises server (2010) http://www-01.ibm.com/software/webservers/appserv/was/. Accessed 10 Aug 2010

Jeffery SR et al (2006a) Adaptive cleaning for RFID data streams. In: Proceedings of the 32nd international conference on very large data bases (VLDB'06), Seoul, Korea, 12–15 September 2006

Jeffery SR et al (2006b) A pipelined framework for online cleaning of sensor data streams. In: Proceedings of the 22nd international conference on data engineering, Atlanta, GA, USA, 3–7 April 2006

Jeffery SR et al (2008) An adaptive RFID middleware for supporting metaphysical data independence. VLDB J 17(2):265–289

Kim Y et al (2006) A framework for rapid development of RFID applications. Lect Notes Comput Sci, Computational Science and its Applications, 3983:226–235

Moon M et al (2006) Contextual events framework in RFID system. In Proceedings of third international conference on information technology, Las Vegas, NV, pp 586–587

Microsoft BizTalk (2009) http://www.microsoft.com/biztalk/en/us/rfid.aspx. Accessed 1 Mar 2010

OAT systems (2009) http://www.oatsystems.com/. Accessed 1 Mar 2009

EPCglobal (2009) http://www.epcglobalinc.org. Accessed on 1 Mar 2009

Oracle Sensor Edge server (2009) http://www.oracle.com/technology/products/sensor_edge_server/index.html. Accessed 10 Aug 2009

Park S et al (2007) Load balancing method using connection pool in RFID middleware. In: Proceedings of the 5th ACIS international conference on software engineering research, management and applications, Busan, South Korea, 20–22 August

Prabhu BS et al (2006) WinRFID: a middleware for the enablement of radiofrequency identification (RFID)-based applications. In: Wireless internet for the mobile enterprise consortium (WINMEC), Forum UCLA, Los Angeles, CA, USA, 9 May

Rieback MR et al (2006) Is your cat infected with a computer virus? In: Proceedings of the 4th annual IEEE international conference on pervasive computing and communications, Pisa, Italy, 13–17 March, 2006, pp 169–179

SAP AutoID Infrastructure (2009) http://www.sap.com/platform/netweaver/autoidinfrastructure.
    epx. Accessed on 1 Mar 2009
Thiem L et al (2008) RFID-based localization in heterogeneous mesh networks. In: Proceedings
    of the 6th ACM conference on embedded network sensor systems, New York, NY, USA, 5–7
    November
Wang F, Liu P (2005) Temporal management of RFID data. In: Proceedings of the 31st interna-
    tional conference on very large data bases, Trondheim, Norway, 30 Aug-2 Sept, pp.1128–1139
Wang W et al (2008) Complex event processing in epc sensor network middleware for both RFID
    and WSN. In: Proceedings of the 11th IEEE symposium on object oriented real-time distributed
    computing (ISORC), Orlando, FL, USA, 14 May

Part IV
Innovative Applications

RFID in the Apparel Retail Industry: A Case Study from Galeria Kaufhof

Jasser Al-Kassab, Philipp Blome, Gerd Wolfram, Frédéric Thiesse, and Elgar Fleisch

Abstract This contribution deals with the business value of radio frequency identification (RFID) technology in the apparel retail industry. We present a case study of an RFID project at Galeria Kaufhof, a subsidiary of Metro Group and one of the largest department store chains in Europe. The trial shows that operational efficiency gains through the automation of logistical in-store processes, such as inventory counting or goods receipt, are possible. Moreover, RFID enables new customer applications on the sales floor, which allow for a redesign of the customer interface, and thus an improvement of the service processes and the service quality. In addition, the analyses of the gathered data on the sales floor help to close the "data void" between the goods receipt and the point of sales of the department store, thus offering the opportunity to directly observe and analyze physical in-store processes. The RFID data analyses allow for deriving valuable information for the department store management in the areas of inventory management, category management, store layout management, and department store processes.

1 Introduction

In today's retail operations, automatic identification technologies have become omnipresent (Manthou and Vlachopoulo 2001; Wyld 2006). It is mainly the introduction of the barcode and the UCC/EAN numbering scheme for consumer products that – despite the initial cost concerns of grocery retailers, suppliers, and distributors in the 1970s – fundamentally changed the industry and laid the foundation for several novel supply-chain concepts, such as direct store delivery, continuous replenishment and vendor-managed inventories 35 years ago (Nelson 2001). The barcode's industry-wide acceptance, however, was only achieved by setting the focus on a strong business case based on "hard" and quantified benefits, which were

J. Al-Kassab (✉)
Institute for Technology Management (ITEM-HSG), Auto-ID Labs St. Gallen/Zurich, and SAP Research CEC St. Gallen, University of St. Gallen, St. Gallen, Switzerland
e-mail: jasser.al-kassab@unisg.ch

D.C. Ranasinghe et al. (eds.), *Unique Radio Innovation for the 21st Century*,
DOI 10.1007/978-3-642-03462-6_13, © Springer-Verlag Berlin Heidelberg 2010

likely to quickly result in a positive return on investment (Brown 1997; Haberman 2001; Hicks 1975; Pommer et al. 1980).

Recently, Radio Frequency Identification (RFID) technology has attracted an enormous interest as a successor for the barcode. Despite broad standardization activities, numerous research projects, high-publicity pilot projects, and mandates from major retailers (i.e., obligations towards the supplier to use RFID technology), the intense debate regarding the benefits to be expected still continues. While savings through operational efficiency gains on the level of individual processes through automation could already be proven, the limitation of the application potential of RFID on automating effects does not reflect the business value potentials of this technology. Integrated RFID systems enable the collection of fine-granular, real-time information about physical processes in the supply chain, which cannot be observed and gathered with conventional approaches. Moreover, RFID enables the redesign of the customer interface, e.g., by recommender systems on the sales floor. However, neither in academia, nor in practice, do adequate experiences or models exist with which the benefits of RFID can be made concrete.

Against this background, the Metro Group started an RFID trial together with Galeria Kaufhof Warenhaus AG (referred to as Galeria Kaufhof in the following sections) in one of its distribution centers (DC) and one of its department stores in September 2007, in order to assess the potentials of a fully integrated RFID system on the item-level. For that purpose, 250,000 items were tagged in the distribution center with RFID transponders and followed on their way through the supply chain, from the DC to the POS. Identification points were installed at the goods commissioning, the goods receipt, the transition between back store and front store, in the fitting rooms, at all elevators, escalators, exists, checkouts and several merchandise fixtures. All collected RFID data during the run-time of the period from September 2007 until December 2008 were collected and analyzed in a central repository. The findings of this trial will be presented in the following.

2 Fundamentals

2.1 RFID Technology

In the past 10 years, the diffusion of RFID technology was mainly fostered by the activities of the "Auto-ID" center research project and the industrial organization "EPCglobal". The goal of the Auto-ID center was to promote RFID in practice by developing interface standards. The basis of this family of standards, also referred to as "EPC Network", is the Electronic Product Code (EPC). In contrast to the conventional 13-digit barcode, the EPC is a sequence of digits that is stored on the microchip of the RFID transponders, and which can identify physical objects on the item-level (Auto-ID Labs 2003). Since the RFID transponders usually do not store any information besides the EPC, the code moreover represents a reference to further data sets, which are available in the network.

2.2 Research on the Business Value of RFID

Several authors have explored the business value of RFID in various industries. This chapter provides an overview of the research conducted in RFID's most important industries and application areas besides retail. Thereby, the chapter focuses on the identification of quantifiable benefits (business value) of RFID technology in these areas.

2.2.1 Customer Service

Heim et al. (2009) examine the effects of RFID on customer value in service environments. They demonstrate that the benefits are not limited to cost savings and inventory availability, but that individuals will recognize value from RFID service applications, including personalization, convenience, enjoyment, safety, information access, VIP treatment, service availability, ease of use, or product availability. Gigalis et al. (2002a, b), for example, explore the use of RFID for mobile service applications. Loebbecke (2004), Nagy et al. (2006), and Wang et al. (2006) conduct research on the RFID-enabled service environment of the future. Despite the promising outcomes of this research, only a few businesses invest in RFID to improve customer service. One reason might be that RFID has not been listed as a critical service research topic (Menor et al. 2002; Roth and Menor 2003) or that service management researchers are still in the need of theoretical foundations and frameworks for subsequent empirical work (Curtin et al. 2007; Ngai et al. 2008; Tajima 2007). As a consequence, the benefits of RFID in improving customer services are still unclear.

2.2.2 Libraries

Curran and Porter (2007), Falk (2005), Wyld (2008), and Yu (2007) conduct research on the usage of RFID in libraries. They find that RFID technology can help streamline major library processes such as stocktaking, book search (including the location of misplaced and lost books), improvement of circulation and maintenance, guiding and personal service, simplification of checkouts, keeping collections in order, and using statistics. They come to the conclusion that the functionalities and benefits offered by RFID systems match the needs and areas of improvement for libraries. However, Falk (2005) admits that the investments into the technology "cannot be justified by the monetary benefits these chips bring". Besides the above mentioned qualitative benefits, the only quantifiable value he lists is the reduction of compensation claims caused by stress injuries during the manual handling of books by library employees, since self-checkouts are expected to reduce these injuries sub-stantially (the San Francisco Public Library, for example, could save 265,000 USD over a 3-year period).

2.2.3 Aircraft Maintenance

Wheatley (2008), for example, describes Airbus' use of RFID technology in the aircraft industry to locate tagged spare parts, to conduct post-incident repairs, and to conduct periodic checks. Moreover, Airbus uses RFID technology to tag tools to carry information including the date a tool was made, or the time it was last calibrated (Ngai et al. 2007). However, the authors do not provide quantified or monetized business values.

2.2.4 Healthcare

RFID benefits in the healthcare industry can be found along the whole patient lifecy-cle in the hospital, including patient care, recovery, and discharge (Cangialosi et al. 2007; Tzeng et al. 2008). Once patients receive an RFID wristband, they can be tracked and monitored in every room in the hospital, their medication can be auto-matically tracked, and the inventory counting and billing in the hospital is facilitated through a reduced number of media breaks. Moreover, RFID can help to locate crit-ical medical devices in a hospital instantly, wherever they are located throughout the facility (Wyld 2008). Amini et al. (2007) conduct a simulation study in order to analyze how RFID data might enable process reengineering in health care services. However, besides listing qualitative benefits, no publication quantifies the business value of RFID in the healthcare industry.

2.2.5 Anti-Counterfeiting and Pharma

Several authors have researched the potential of RFID technology in securing the licit supply chain from counterfeit products, including drugs (Al-Kassab et al. 2008; Cole and Ranasinghe 2008; Lehtonen et al. 2007) or cigarettes (Wyld 2008), for example. Obviously, counterfeit drugs or cigarettes threaten the life of cus-tomers, thus making quantified business case calculations almost obsolete. However, Lehtonen et al. (2008) find that the business case for case-level tagging is positive for anti-counterfeiting applications.

2.2.6 Return Logistics

Based on a field case study of a logistics company, Langer et al. (2007) investi-gate the potential of RFID in return center logistics. They find that the number of customer claims fell substantially. RFID technology helps to reduce warehouse employee errors at the loading dock, to reduce the required time to research claims, and to provide a disincentive for salvage dealers to place fraudulent claims. As a result, the claims incidence decreased by 54.29% and the dollar value of claims decreased by 29.7%. Karaer and Lee (2007) investigate the potential of RFID as an enabler technology for the reverse channel of a manufacturer, using a mathematical model. They find that increased visibility can be to the advantage of the manufacturer's return process. They state, however, that the true return on RFID can only be analyzed through a comprehensive analysis in the whole supply chain.

2.2.7 Supply Chain Management

Numerous publications describe the potential benefits of RFID in supply chains, such as improvements in the efficiency, accuracy, and security of material and information flows (Angeles 2005; Asif and Mandviwalla 2005; Delen et al. 2007; Gaukler and Seifert 2007; Heese 2007; Karaer and Lee 2007; Kärkkäinen 2003; Kelly and Erickson 2005; Lee and Özer 2007; Michael and McCathie 2005; Prater et al. 2005; Reyes and Frazier 2007). However, only a small number of publications list quantified benefits to assess the business value of RFID in this application. Hou and Huang (2006), for example, conduct a quantitative analysis of costs and benefits for RFID applications in different logistics activities in the printing industry supply chain. Their derived benefits based on savings of operation time, labor and costs, since they consider RFID as a mere barcode replacement technology. They conclude that retailers with revenues of annually more than 200,000 Euros can save 17–92 working days per year with RFID. Venables (2005) describes the example of a Scottish brewery which uses RFID to track its costly and prone to loss aluminum kegs in the extended supply chain from brewery to distribution center and on to pubs. Through the use of RFID, the brewery has achieved annual savings of up to 2.3 million Euros on the costs of replacement kegs.

Other application areas include defense (e.g., Barratt and Choi 2007), quality management (e.g., Wang 2007), farming operations (e.g., Wyld 2008), restaurants (e.g., Ngai et al. 2008), automation, manufacturing and facility coordination (Gaukler and Hausman 2008; Hozak and Collier 2008; Thiesse and Fleisch 2008), sports, lifestyle and leisure (e.g., Srivastava 2007), and construction (Song et al. 2005), for example.

The research on the business value of RFID in different industrial sectors and application areas show that the majority of publications fail to quantify the benefits of RFID technology for the corresponding use cases. This is mainly due to the fact that data from real-world applications and installations does not exist. Moreover, according to Peppard et al. (2007), the difficulty of a quantitative assessment of such effects in practice lies in the fact that, depending on the technology and application area, to reconstruct the chain of causation in individual cases can be very complex due to the large number of other factors and the distance between IT use and profitability. The complexity thereby depends on the technology and field of application considered. Moreover, numerous authors have also pointed out that it cannot be assumed that an automatic link between an "IT Impact" and the occurrence of a desired business benefit effect exist, but that these effects are only made possible through additional complementary factors in the organization and contextual factors in their environment. Based on the resource-based view of the firm (Barney 1991), several studies investigated the existence of a specific so-called "IS / IT capability", whose presence allows individual firms to benefit more from their IT investments than others. These include, for example, works of Bharadwaj (2000), Bhatt and Grover (2005), Mata et al. (1995), Santhanam and Hartono (2003), and Zhu (2004). The attempt of a merger of the process-based approach and the resource-based view was made based on a comprehensive literature analysis of Melville et al. (2004). More research in this area is expected in the coming years.

Similarly, quantifiable benefits of RFID technology in the retail industry are limited to automational effects on the operational level. The next chapter will describe the background of the underlying real-world RFID trial and illustrate which additional benefits can be expected from RFID in this industry.

3 Case Background

3.1 Challenges in the Retail Industry

3.1.1 Challenges in the Apparel Retail Industry

The apparel retail industry is characterized by high seasonality, high volatility, high-impulse purchasing, and complicated distribution and logistics operations (Christopher et al. 2004). Moreover, the challenges encompass (i) short product life-cycles of fashion products (typically only about three months), (ii) a high variety of apparel product characteristics, (iii) product counterfeiting in the supply chain, and (iv) lack of knowledge about the behavior of fashion shoppers, yet considered to be "mysterious and unpredictable" (Moon and Ngai 2008). Multi-optional fashion shoppers are more demanding and their behavior has become increasingly difficult to assess (Gagnon and Chu 2005). Several aspects complicate the understanding of consumer behavior. First of all, there is an increased diversity of customer groups. This is due to the fact that long-standing life-patterns have become less predictable. For example, people tend to marry later, divorces are increasing, they start new careers, and move to new locations more often than in the past (Gagnon and Chu 2005). Furthermore, the aging of populations in Western countries forces retailers to consider the special needs of elderly people. At the same time, ethnic diversity in Western countries has increased during the last decades. Those demographic changes have contributed to shifts in decision patterns of consumers (Samli 2004).

3.1.2 Challenges in the Apparel Warehouse Retail

Regarding department stores, the competition between the retail formats (such as mass discounters, specialty stores, boutiques, or mom-and-pop stores) is of particular interest. As Metro points out in its annual report (2007) "specialist retailers and some textile department stores, as well as the rising number of shopping centers are intensifying the competition in the sector and putting increasing pressure on the broadly assorted department store operators. The classic department stores lost market share again last year, compared with retail in general". In their apparel business, the department stores are competing with vertically integrated "fast fashion" retailers. These are able to react faster to consumer trends by sharing data from the POS all over their value chain and to produce at lower costs (McAfee et al. 2004). Their total lead time from inspiration (prior to design) to the sales floor is about only 12–15 days. Traditional retailers on the other side require a lead time of about 60–90 days from the production at the manufacturer to their sales floor (KPMG 2005).

As a consequence, small and medium sized retailers observe high losses of market share and vertically integrated fast fashion retailers can integrate themselves into the market. Moreover, non-traditional outlets such as discounters are moving into the apparel market to also successfully sell clothing items.

Department stores and other traditional retail formats try to reduce their lead times and to improve their market orientation with new distribution and retailer-manufacturer cooperation concepts like shop-in-shops or efficient consumer response (ECR), but have not yet succeeded in coping with the "fast fashion benchmark". In conclusion, the issues raised impose the necessity for continuing efforts to improve retail operations. In this regard, supply chain and inventory management are the key drivers to increase efficiencies, enhance customer service and to reduce costs in this industry (Ellram et al. 1999; Samli 2004).

3.1.3 Perception of RFID in the Retail Industry

Due to the above mentioned challenges in the apparel industry, retailers are looking for new ways to increase their efficiency and to offer new shopping experiences to their customers. The benefits of RFID technology seem to be particularly apparent in fashion retailing (Moon and Ngai 2008). Besides cost savings, retailers expect new services for customers through RFID-based applications and new insights into customer behavior via the analyses of RFID data gathered on the sales floor. The results of these analyses are supposed to improve the retailers' department store processes, inventory management, layout management, and category management, as well as to put in place customer-specific marketing programs (Li and Visich 2006). Industry associations expect that RFID technology may also help fashion retailers solve a number of problems that are unique to their industry (IDTechEx 2005). Attracted by the promising benefits of the Radio Frequency Identification (RFID) technology, which is for example supposed to raise efficiency and to enable innovative customer service offerings, apparel retailers have started to investigate this technology resulting in several trials and pilots world-wide. Figure 1 provides an overview of this development. The numbers are based on counting the press releases (e.g., in RFID Journal) on RFID apparel retail trials over the last decade.

The majority of the RFID trials either focus on the back store, i.e., the extended supply chain and in-store logistics, such as inventory management and department store processes, or on the front store, i.e., RFID customer applications, for example. Only a minority of the trials covers both back and front store. The Galeria Kaufhof trial, which will be presented in the following, does so by covering the supply chain from the distribution center to the transition gate (from back store to front store in the department store) as well as the front store.

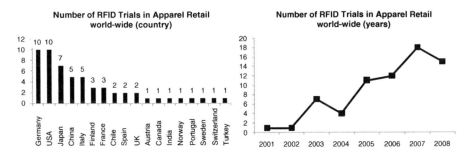

Fig. 1 Overview of number of trials and tendency in the apparel retail industry

3.2 RFID at Galeria Kaufhof

Kaufhof Warenhaus AG operates 126 department stores in Germany, 113 of them under the brand name "Galeria Kaufhof", and 15 in Belgium. Galeria Kaufhof is a subsidiary of Metro Group, the fifth largest retailer worldwide (Deloitte 2008). It is visited – as one of Europe's leading department store chains – by more than two million customers each day and has a total sales floor of 1.5 million square meters. Its assortment primarily consists of international brands of middle to upper price level. The company employs about 25,000 people and generated €3.6 billion in sales in 2007 (Kaufhof 2009).

Kaufhof's first RFID activities – a collaboration with the fashion merchandise manufacturer Gerry Weber (Loebbecke 2007; Loebbecke and Palmer 2006a) – date back to 2003. The main objective of this first project was to examine to what extent RFID can contribute to the acceleration and simplification of supply chain operations under real-world conditions (Metro 2005), including the assessment of achievable read rates and time-savings due to automatic object identification, and to investigate the technology's suitability for replacing conventional Electronic Article Surveillance (EAS) systems, a topic that has traditionally been of particular interest in fashion retailing (Bamfield 2004).

The project team also considered the idea of utilizing RFID to realize novel customer applications and to change in-store processes. However, these scenarios remained in the abstract and were not part of the trial itself. The quantitative business case analysis concentrated on time and labour cost savings due to process automation at various points in the supply chain. The overall evaluation of the RFID business potential, including the many qualitative results gathered by the project team, was very positive.

This experience motivated Kaufhof to join the consortium of "Building Radio Frequency Identification Solutions for the Global Environment" (BRIDGE), an R&D project funded by the European Commission in July 2006 for a duration of 36 months. BRIDGE is a 13 million Euro RFID research project running over 3 years and partly funded by the European Union. The objective of the BRIDGE project is to address ways to resolve barriers to the implementation of the EPCglobal Network in Europe by conducting research and by developing and implementing tools that enable the deployment of EPCglobal applications in Europe (BRIDGE 2009). The project comprises several chapters composed of working packages, in which thirty interdisciplinary partners from 12 countries (Europe and Asia) work together on hardware development, serial look-up service, security, anti-counterfeiting, supply chain management, manufacturing process, reusable asset management, item-level tagging for non-food items as well as policy recommendations (BRIDGE 2007), for example. The company's main interest in BRIDGE was to implement RFID in its own processes, to speed up RFID adoption in the textile sector, and to share its experiences with industry organizations and standardization bodies. As part of the project, Galeria Kaufhof implemented an RFID infrastructure in the menswear department on the third floor of its store in Essen, Germany, with a total of 2000 square meters in September 2007.

3.3 Project Objectives

Various executives of Metro and Kaufhof were present during the opening of the trial and sketched out their ideas regarding the project's objectives. The participants agreed that by using RFID in the sales brand, customers should be given a totally new kind of shopping experience and a considerably improved customer service. Accordingly, the CEO of Galeria Kaufhof was interested in evaluating to what extent the staff could be freed from non-value-adding activities and, thus, gain additional time for direct customer service.

The pilot was expected to demonstrate how a department store fully equipped with RFID on the item level can work. Hence, he was interested in the customers' reaction regarding RFID technology on the sales floor and to learn whether the expected advantages of item-level RFID were also feasible in practice.

Further, it was expected that an end-to-end RFID infrastructure at the item level can fill the "data void" (i.e., missing transparency) that exists between products being received and products being sold. RFID is expected to help seeing products in those steps in the process chain that so far have not been illuminated by the inventory management system. Hence, RFID is supposed to offer operational benefits related to inventory management and in-store logistical processes.

Despite these predefined goals, the realisation of the project had the character of an "innovation-based implementation" (Peppard and Ward 2005). As such, its objective was not primarily to solve a single, clearly identified business problem but, rather, to focus on the capabilities of the technology first and then to consider ways to exploit the resulting opportunities. Hence, a workshop with members from different organizational units was organized. The initial objectives that all participants agreed on at the beginning were increased stock availability, the identification of new revenue sources, the optimization of existing processes and the creation of new ones, and the design of additional management reports. Ideas regarding how the value of RFID in addressing these issues could be investigated were collected and roughly specified by the workshop participants and grouped into three categories (continuous analyses of RFID event data, field experiments, and surveys of customers and employees).

3.4 Galeria Kaufhof RFID Infrastructure

From a technological point of view, the trial is the first one worldwide (Wessel 2007) that makes use of so-called "*near-field UHF*" transponders, which combine high read rates for tags in the HF frequency band with the low cost of standardized UHF tags (Nikitin et al. 2007). Approximately 30,000 individual RFID-equipped apparel items are constantly available on the sales floor. In order to identify them and to link them in the merchandise information system with the presented merchandise, about 500 merchandise fixtures are likewise RFID-tagged.

RFID-tagged merchandise items are seamlessly tracked at the item level on their way from the DC to the POS. Those intended for immediate sale are sent via the freight elevators to the menswear department, where they leave the back store and enter the front store via a transition gate. The menswear department is equipped with a total of 55 readers and more than 200 antennae. Readers are installed at the goods issue of the distribution center (2 readers), at the goods receipt of the department store (2 readers), at the labeling points (3 readers in the distribution center, and 3 at the department store), at the gateways between the sales floor and the backroom (2 readers), at escalators (3 readers) and elevators (2 readers), on "smart shelves" (10 readers), in all 20 fitting rooms (20 readers), at the storage (1 reader), and at the points of sales (7 readers). Some RFID readers are equipped with photoelectric barriers and motion detectors in order to determine the direction of the flow of goods, e.g., upon transit from the back store. The installed RFID readers recorded the movements of thousands of individually RFID-tagged clothing items. A schematic overview of the installation is given in Fig. 2.

All products to be sold in the Essen menswear department are shipped via the regional DC, where passive RFID tags are attached on the clothing. With every movement of products captured by the readers, RFID events are generated and stored in the company's "EPC Information Service (EPCIS)" (EPCglobal 2007),

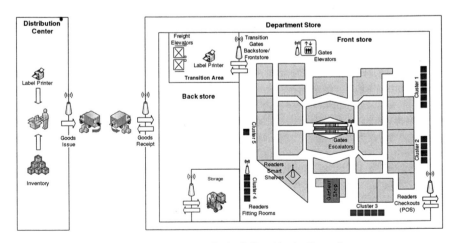

Fig. 2 Galeria Kaufhof RFID infrastructure in the DC and in the Essen department store

a repository for the collection and sharing of RFID data. The system is installed in parallel to the existing merchandise information system and supports store management and sales floor processes. Every item arriving at the Essen store is identified at the dock door, compared with the corresponding Electronic Data Interchange (EDI) message on the delivery, and added to the virtual store inventory. The distinction between back room and sales floor inventory is made feasible by readers at the entries, exits, and checkout counters of the menswear department. An extension of the merchandise information system (MIS) was developed that detects products that are not available on the shelf even though they are in stock. This information, in principle, enables the fully automated triggering of replenishments.

4 RFID Applications

4.1 Customer Applications

For an improved customer experience and for the improvement of customer service, Galeria Kaufhof developed based on the described infrastructure various RFID customer applications and installed them on the sales floor. The goal of these applications was to practically demonstrate and evaluate the customer value of RFID.

4.1.1 Smart Monitor

RFID readers in the fitting rooms recognize the product codes of clothing that is brought in. A screen inside each fitting room at the Gardeur shop displays additional information on the product and other available sizes and colors. Depending on the set of detected items, the system offers recommendations regarding alternative or complementary products (Fig. 3a).

4.1.2 Smart Shelf

Several shelves in the Gardeur shop are equipped with RFID readers and screens. A screen of each shelf shows the current availability of items on the particular shelf. Thus, customers looking for a specific item do not need to check the entire shelf to know which sizes and colors are available (see Fig. 3b).

4.1.3 Smart Mirror

A large mirror with a built-in screen on the sales floor offers product information. Once a customer approaches the so-called "Magic Mirror" with a tagged item receive information about materials, care instructions, available sizes, and colors (see Fig. 3c).

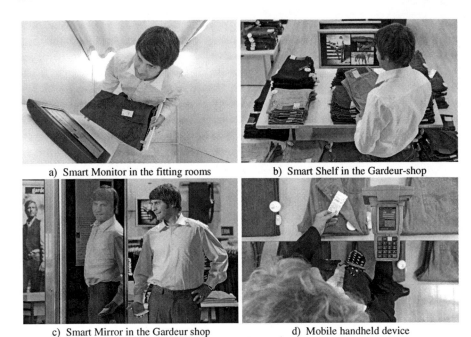

| a) Smart Monitor in the fitting rooms | b) Smart Shelf in the Gardeur-shop |
| c) Smart Mirror in the Gardeur shop | d) Mobile handheld device |

Fig. 3 RFID Customer applications on the sales floor of the Galeria Kaufhof Essen department store (© METRO AG / Photographer: John M. John). **a** Smart monitor in the fitting rooms, **b** Smart shelf in the Gardeur-shop, **c** Smart mirror in the Gardeur shop, **d** Mobile handheld device

4.1.4 Mobile Handheld Devices

Using mobile handheld devices ("MDE"), store employees can scan the item's RFID tag in order to send a query to the RFID system regarding the availability (e.g., different size and color) and location of other items in the same category on the sales floor or in the backroom. Furthermore, MDE devices are also used for goods receipt and inventory counting to accelerate the process beyond what is possible when manually counting each item using a mobile barcode reader (see Fig. 3d).

4.2 Customer and Employee Surveys

For the evaluation of customer attitudes towards the RFID-based applications, Galeria Kaufhof conducted a survey among 250 customers (50% women, 50% younger than 40 years) in May 2008. The RFID-based applications were presented to the customers on the front store. The customers were asked to answer questions regarding the applications using a fully structured interview guideline with primarily open questions. The three RFID customer applications (i.e., smart shelves, magic mirror, and smart displays) were to a large extent evaluated positively by customers, with smart shelves having the best ratings. 56% of all respondents regarded these items as constituting a great improvement in their shopping experience.

Smart displays were equally successful, with 49% recording very positive answers. The customers appreciated the detailed information on product availability in particular. Customers were also asked if they had perceived any negative aspects of either application. This question was answered in the negative by 83% and 79% of the respondents, respectively. In contrast, the magic mirror was seen as an improvement by only 33%. As some of the comments indicated, this can be attributed to the limited usefulness of the information displayed and the position of the display in the mirror area. Privacy issues were not considered in the survey, but Kaufhof informed their customers about the presence of RFID readers and prepared their employees to answer specific questions. Perhaps as a result, only three privacy-related customer complaints were observed during the whole trial period.

Kaufhof also conducted interviews with their sales employees. They were asked to report on their experiences with RFID. The respondents appreciated most of all the use of the mobile devices for inventory counting and the search functionality, which allows them to easily locate items on the sales floor. The interviews clearly showed that the employees' attitude towards RFID was dependent on its benefits in the completion of everyday processes. Moreover, a number of suggestions for improvement were gathered – e.g., regarding the available search criteria.

4.3 RFID Data Analyses

The continuous data collection by the described RFID infrastructure allowed Galeria Kaufhof to conduct several analyses of in-store processes that go beyond what is already being done with the help of barcode-based sales and inventory data. While the analyses of RFID raw data stored in the EPCIS repository alone did not provide much business value, it helped the company to monitor the performance of their data collection infrastructure and the quality of the data generated. For the technology to be usable for process-related analyses, the raw data have to undergo a sequence of processing steps.

On the one hand, filtering mechanisms are required in order to cleanse the data from stray readings, unreliable reads, and data redundancy. For example, for the analysis of customer try-ons in fitting rooms, all RFID read events from the 20 fitting rooms on the sales floor were considered. Fitting rooms were individually identifiable by the read event's attribute "ReadPoint" (e.g., fitting room number 2). When analyzing the product life-cycle of an item-level tagged pair of jeans we discovered that these jeans had been read within several seconds in different fitting rooms of one fitting room cluster. The reason proved to be the broad read range of the fitting room readers, resulting in stray readings and redundant read events. In total, four different kinds of read errors resulted from the fitting room readers. These included:

1. reads by adjacent cabins,
2. reads by customers or employees passing by,
3. reads from nearby merchandise fixtures, and
4. reads by left back items.

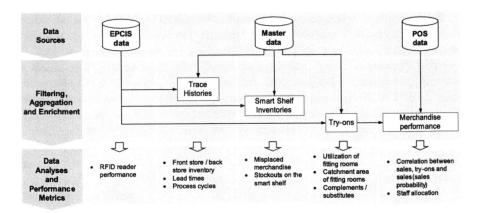

Fig. 4 Overview of RFID data analyses and performance metrics

These fitting room reading errors (noise) had to be filtered for the try-on-based data analyses to be conducted. Therefore, filtering algorithms that iterated through the fitting room read events and filtered the errors accordingly were designed.

Thus, in a second step, the necessary data filtering and aggregation techniques and mechanisms for the treatment of RFID raw data were defined in order to provide the necessary data basis for the business analyses. The following three sections are structured based on Fig. 4.

Accordingly, we briefly present the available data sources (EPCIS Raw Data, Master Data and POS Data), then we present the data filtering, the conducted data analyses and performance metrics.

4.3.1 Data Sources

EPCIS Raw Data

EPCIS events are the data constructs stored in the EPCIS repository. Four generic event types are specified in the EPCIS data model. An "ObjectEvent" corresponds to the detection of EPC-equipped items, e.g., on a dock door upon arrival of a shipment. Since the granularity of individual EPC identification is not always needed, a "QuantityEvent" can alternatively be used that includes only information on the product type and the number of objects, which equals the level of detail obtained from barcode scanning. In contrast to that, an "AggregationEvent" is not associated to a particular tag read but rather denotes a (dis-)aggregation of a group of items, e.g., cases that are put on a pallet for shipping. In a similar way, a "TransactionEvent" links EPCs to a specific business transaction, e.g., a purchase order. Business documents can be linked to any recorded event, which are represented by instances of the "BizTransaction" type. All event types contain mandatory fields on date and time of an event, which objects or entities were subjected to the event, where the event occurred, and in which business context the event happened. A complete description of EPCIS event types is given by EPCglobal (2007). These events are enriched by master and POS data, which provide the corresponding business context.

Master Data

Master data are usually provided by the supplier or the retailer's procurement department. They describe a product on an article level, i.e., each data set refers to one specific EAN (European Article Number) code. This is in contrast to EPCIS event data, which refers to individual items. While the structure of master data records is the same for all product types the range of values for single attributes (e.g., size and color codes) is supplier-specific, which complicates comparisons between products and categories.

POS Data

POS data are captured using barcodes, which encode traditional EAN numbers. Other data attributes are entered manually by sales staff. Despite the existence of master data, size information is collected redundantly since some products carry the same EAN code for all sizes.

4.3.2 Filtering, Aggregation and Enrichment

The following data sets are generated as follows (see Fig. 4):

- *Trace histories*. A product's lifecycle over time, starting from the labeling point in the DC and ending at checkout on the sales floor, is described by RFID trace histories. In order to generate trace histories, read events associated with a tag ID are first aggregated by reader location, sorted by their corresponding time stamps, and then transformed into a linked list. Trace histories are necessary for analyses in the area of process execution and inventory management.
- *Smart shelf inventories*. Besides tracking inventories and inventory changes over time (e.g., for the detection of "out of shelf"), data generated from smart shelf reads also allow for the identification of misplaced merchandise.
- *Try-ons*. For the generation of try-on data, the fitting room reads are first aggregated over time and then filtered in order to remove items carried by passing customers or employees, as well as items from nearby merchandise fixtures. Try-ons allow for the analysis of complements and substitutes that are tried on together, and of fitting room utilization.
- *Merchandise performance*. In order to investigate the correlation between sales, try-ons, and inventory levels on the sales floor, try-on events are combined with point of sales data and the inventory level at the time of the try-on. The resulting data allow for various analyses of the attractiveness and performance of individual items or entire categories.

4.3.3 Data Analyses and Performance Metrics

An exemplary analysis using these aggregated data is depicted in Fig. 5. During the data filtering and enrichment process, sales information, try-on information, and inventory-level information for each article and read event are merged into one data table. As such, these data can be analyzed using logistical regression models in order to identify a level of inventory on the front store that would be optimal in terms of number of try-ons and sales, and eventually customer satisfaction. With these results, Galeria Kaufhof can systematically approach the optimal level of articles on the front store, thus increasing sales and customer satisfaction, while at the same time decreasing the capital costs of items on the sales floor. For the optimization of the category management, the number of try-ons and sales for each article group are correlated (see Fig. 5), resulting in the sales probability (a number of sales/number of try-ons ratio). With this information, Kaufhof can compare articles within one article group by brand, supplier, color, and size, for example, thus optimizing his product range by removing articles from the assortment that were often tried on but rarely bought.

Using trace histories, the front store and back store inventory of items can be calculated in real-time for a given time and date. For example, a list can be generated for items on the level of article groups, i.e. shirts or pants, for each day of the year and plotted as a graph. Figure 6 provides an overview of such a graph. The analysis helped Galeria Kaufhof to identify:

- inefficient promotion execution (i.e. articles that were put too early on the front store),
- stockout and overstock (when articles are out of store or exceed a certain amount), and

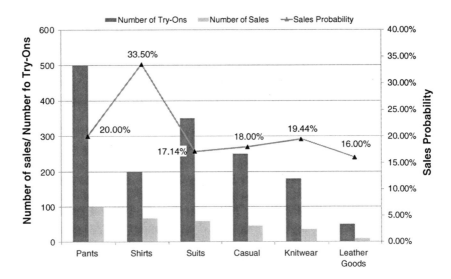

Fig. 5 Correlation between try-ons and sales

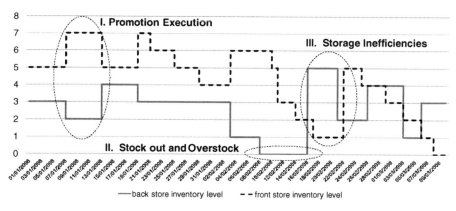

Fig. 6 Back store and front store inventory over time

- storage inefficiencies (i.e., when the number of articles in the back store exceeds the number of articles on the front store – replenishment problems).

By providing more visibility regarding the distribution of the department's inventory stock between the front store and the back store, this analysis supports Galeria Kaufhof in reducing the number of out-of-shelf but in stock situations, which is responsible for lost sales and reduced customer satisfaction Moreover, this analysis can help to detect promotion execution errors (i.e., promotional articles, which are put on the sales floor too early) and storage inefficiencies (i.e., low inventory levels on the front store, but high inventory levels in the back store).

In addition to employing process analyses, Kaufhof made use of RFID to collect data from various experiments on the sales floor. For example, inventory levels and product presentation were periodically changed over several weeks according to a predefined scheme to investigate the impact that these factors have on try-ons and sales. In these cases, RFID was not a facilitating technology but was rather a sophisticated measurement instrument.

4.4 Evaluation and Experiences

Despite their extensive experience with the technology gathered in recent years, implementing and operating an RFID system like the one in Essen posed a new challenge to Metro and Kaufhof. In the first weeks, some configuration efforts were necessary to optimize the quality of the collected data. However, after these issues were properly addressed, Kaufhof's evaluation of reliability and performance throughout the entire 15 months of the trial was very positive. RFID read rates at different reader locations were virtually 100%. This was partly due to the physical characteristics of textiles, but also due to the use of novel near-field UHF technology and other state-of-the-art equipment. The only case of insufficient read rates was observed at the checkout counter, which could be attributed to the fact that

RFID was installed in parallel to conventional barcode readers with staff at the POS having to activate the RFID reader device manually using a foot pedal.

Against the background of these results and low transponder prices, the idea of using RFID as a substitute for existing electronic article surveillance (EAS) systems was regarded as technologically feasible. EAS was even regarded as a major driver of any positive ROI calculation. However, it was also evident that before any large-scale roll-outs could realistically be planned, supply chain partners would have to agree on appropriate standards. There are standardization activities on an industry level to recommend the use of standard near-field UHF transponders without additional modifications for EAS. With such an RFID-based EAS solution, about half of the business case would be done as well. This, however, presumes an adoption of the technology by all suppliers.

The high level of detection accuracy also allowed testing the applicability of RFID for inventory management on the item level for the entire department. On the one hand, readers between the backroom area and the sales floor were used to monitor products moving between the two areas in order to calculate estimates of sales floor inventory, which was not possible with the existing inventory management system. These estimates were then validated using manual inspections and POS data. The results indicate that – if precise information on item locations is not required – backroom monitoring poses a cost-efficient alternative to inventory monitoring using smart shelves. Smart shelf readers were also used to analyze the frequency and duration of product misplacements on the shelves. The results of both tests helped the retailer to develop a better understanding of the performance of in-store processes. They also showed that RFID provides the necessary data quality for process transformation, such that necessary tasks on the sales floor are automatically triggered by real-time information.

In what concerns the use of mobile devices, this trial was also the first time that the performance of RFID in inventory counting for an entire assortment was tested. For this purpose, all items in the menswear department were recorded twice by RFID and barcode, respectively. The comparison of the measurements indicated time-savings of more than 87% on average by the use RFID. The least improvements were observed for slackly hanging textiles. However, the advantages of RFID became evident the more items were hanging on a particular merchandise fixture. Also, items that had to be taken from the fixture and items that required employees to additionally enter size information clearly showed the benefits of RFID. An additional advantage of RFID was seen in the fact that item-level identification prevents multiple reads of the same EAN code, which poses a potential source of inventory inaccuracies.

With the testing of bulk identification at goods receipt and inventory counting, it was possible to show that RFID allows for significant time-savings here as compared to the barcode. In contrast, the benefits of RFID at the POS were verifiable but nevertheless evaluated as negligible due to the small number of items per purchase.

The trial results that attracted the most interest, besides the automation of operational processes, came from the examination of large amounts of RFID data and the managerial implications thereof. The RFID data analyses described above provided a number of new insights into store logistics and customer processes, e.g., the differences between try-ons and sales across categories, the flow of goods through the store, the utilization of fitting rooms, the frequency of execution errors, and others. The information has a high value for supply chain coordination and marketing. In the case of applications in the textile domain, the vertical integration of the value chain is crucial. It is expected that the more store brands a retailer distributes, the more quickly the advantages of RFID can be realized. Another decisive factor will be the fact that modern UHF transponders can be used for both logistics and EAS solutions. The increase in precise information – be it about inventory levels or product properties – offers many novel possibilities, making it indispensable for the marketing, once such an infrastructure is in place.

On the other hand, it also became clear that RFID data would not always lead to unambiguous findings and that the phase of interpreting and drawing conclusions from RFID data would require more time. A major issue with the processing of collected data was the quality of master data to which RFID codes refer. For example, for some products, only category-level information was available; for others, suppliers were using only one common EAN code for all available sizes. Another issue arose from the discussion regarding the value of management-related information. The need to continuously analyze and aggregate data collected by the RFID infrastructure as well as field experiments and surveys into various reports was intuitively affirmed by members of the Galeria Kaufhof project team. However, a general answer regarding how to structure and quantify the impacts of such information on performance could not be provided. This was partly due to the lack of appropriate performance metrics. The performance of category managers, for instance, is not reflected by any specific process metric but is rather measured by the turnover generated within the respective product category. Also, members of the Galeria Kaufhof team mentioned that the quantification of the value of such marketing-related information is difficult. From the previous introduction of IT systems, however, they had already made the experience that MIS projects are significantly harder to justify quantitatively than operational systems.

4.5 Lessons Learned From the Trial

The Galeria Kaufhof RFID case provides a number of lessons relevant to management practice. These include, on the one hand, familiar factors that have often been cited in traditional IS development, such as top management support, training, etc., which also apply to RFID implementation. On the other hand, some lessons are distinctive and deserve more detailed consideration. In order to highlight the novelty of these implications with regard to prior studies, we first contrast our findings with those from the earlier RFID trial conducted by Kaufhof in 2003. Loebbecke (2007), following a concept first suggested in Loebbecke and Palmer (2006b), discusses the results of the 2003 trial in terms of the so-called "5 P's of RFID": physics, price, processes, performance, and privacy. In this section, we highlight to what extent the Galeria Kaufhof trial changed or extended the company's view of RFID using the five categories. Second, we propose to complement these categories using what we call the "4 I's of integrated RFID": integration, interpretation, interoperability, and involvement. These four additional categories allow us to discuss further practical implications that can be drawn specifically from our study of a fully integrated system (Thiesse et al. 2009).

Lesson 1: Physics

As Galeria Kaufhof noticed in the present trial, the problem of insufficient read rates still requires some configuration efforts. Through recent technological advances, however, read rates of virtually 100% are possible by now – even at chokepoints such in the transition between backroom and sales floor, with large amounts of tags passing through the RF field simultaneously and differing angles between transponders and antennae.

Lesson 2: Price

The cost for near-field UHF transponder inlays, which can be integrated into existing barcode labels attached to the merchandise, reached a low of about 0.07€ in 2007. Despite this ongoing substantial drop in price level, transponder costs are still the dominant factor in return on investment (ROI) calculations regarding supply-chain-wide RFID implementation (i.e., in contrast to the use of closed-loop systems). However, according to Kaufhof's view, current transponder prices have decreased enough to make RFID-based EAS solutions economically feasible, particularly under the precondition that anti-theft functionality resides in the software and not on the tag. Although EAS systems alone do not constitute the one and only "killer application" that justifies the replacement of existing technology, they might, in the long run, become the hoped-for driver of widespread diffusion of item-level RFID in the industry that other applications can build upon.

Lesson 3: Processes

With the comparison between barcode-based and RFID-based inventory counting, Galeria Kaufhof received new insights into the factors that eventually determine the size of operational improvements (e.g., product type, position of the RFID label, and merchandise fixture, among others). The comparison has shown that RFID can accelerate the process significantly for the whole assortment, because employees do not have to take clothes from the shelves anymore in order to look for barcode labels. However, it became evident that additional activities would be necessary in order to ensure that all items would be equipped with transponders to prevent them from remaining undetected. Furthermore, the trial was the first time that Kaufhof implemented RFID-based mobile applications to support their sales staff. All search-related functionalities provided by these devices were highly appreciated and led to a reduction in non-value-adding activities. RFID was thus used not only as a cost-saving but also as a service-enhancing technology. Third, Kaufhof examined how to utilize RFID as a measurement instrument that generates more detailed information on in-store processes. The lesson that the company learned was that the reality of process execution in retail is often not entirely known unless an automatic data collection technology provides the necessary data. Hence, RFID may serve as a tool to detect existing inefficiencies and to continuously control the effectiveness of countermeasures.

Lesson 4: Performance

Besides the performance indicators "lead times", "labor cost reduction", and "data quality" (Loebbecke 2007), RFID was also used in this trial to generate several novel performance measures for operational processes, such as on-shelf service level and misplacement frequency, as well as detailed product histories, for single items and for whole categories. In addition, it was shown that the combination of RFID and POS data can be used not only to monitor but also to control activities on the sales floor – e.g., by triggering shelf replenishments.

Lesson 5: Privacy

In earlier trials, privacy risks and the actions of pressure groups had become a major issue in the U.S. and Europe. The fear among many observers was that retailers might link customer information to products for profiling or tracking even after purchase. A number of boycott campaigns on the Internet, protests against individual trials, and negative press reports affected various companies world-wide. In earlier trials, transponders were routinely cut off at the checkout counter in order to address the potential privacy concerns of Kaufhof customers. In this trial, Galeria Kaufhof decided to inform customers proactively of the presence of the RFID infrastructure with the help of highly visible adhesive labels, logos and instructions printed on transponder labels, as well as with freely available information brochures. Transponders were only removed from the products on request to allow for the identification of returned items. The company nevertheless experienced virtually no negative reactions during the whole trial period.

Lesson 6: Integration

Beyond the optimization of operational processes, the underlying case illustrates the benefit of RFID as an integrated data collection infrastructure whose whole exceeds the sum of its parts. The conclusions that might be drawn from such data have often been presumed – e.g., by Loebbecke (2007), who illustrates the idea using the following fictitious example of an intelligent clothes rack: "If a customer put a blouse back on the clothes rack after 5 s, she might not like the material. If 5 min went by, it is likely that the customer tried on the article." However, the Galeria Kaufhof trial was the first time that a retailer really explored the opportunities involved with the use of the technology. The case study has shown that the data quality and level of detail provided by RFID allow for the generation of a plethora of different analyses of product histories, customer behavior, in-store logistics, and merchandise performance, among others. Some of these finding support and refine what has already been determined from existing data, whereas others point to new and previously unobserved phenomena. Furthermore, the same data can be used to implement novel customer applications and to support sales processes, such as the search for products that meet specific criteria. In light of the manifold possibilities regarding how to make use of the gathered information, a future challenge for retailers will eventually be to develop capabilities that allow them to use these new options to increase sales and customer satisfaction.

Lesson 7: Interpretation

Another lesson from the Kaufhof project is that RFID data in themselves provide little value if they are not integrated with other operational systems to link tag IDs to product information and business events. The output generated by RFID reading devices should not be mistaken for meaningful business information, since it requires further filtering and enrichment before any interpretation makes sense.

Lesson 8: Interoperability

In earlier trials, no relevant RFID standards aside from the ISO 15693 air interface standard were available (Loebbecke and Huysken 2008). The lack of common numbering schemes for RFID-equipped items rendered the management of an entire store assortment on the item level nearly impossible. Hence, earlier trials were limited to products from only one supplier. Even today, RFID is still far from being a plug-and-play technology and requires substantial efforts with regard to configuration and adaptation to a company's requirements, as is necessary to achieve the desired level of performance and quality of data. However, it was not least of all the rise of the family of standards around the Electronic Product Code (EPC) that actually enabled the creation of Kaufhof's infrastructure. Standards, on the one hand increase the interoperability of single components. As a result, the need to possess extensive RFID knowledge to effectively cope with issues like RF interferences becomes less critical. On the other hand, standards lead to interoperable systems in and between organizations, which may lead to additional benefits in the future. Therefore, even companies that intend to start with isolated RFID artifacts should pursue an infrastructure-centered approach based on widely accepted industry standards.

Lesson 9: Involvement

The Kaufhof case also demonstrates the necessity of involving suppliers, as well as members of different organizational units in the project. In their survey of employees, for instance, Kaufhof gathered a number of valuable suggestions for improving the functionality of mobile devices. An interesting statement made by one employee, for example, was that mobile applications perhaps should not be used in front of the customer because the dependency on an electronic device could be regarded as a sign of incompetence. Applications should therefore be designed together with prospective users to achieve a good fit with their *individual* needs. Similarly, companies should be aware that the mere availability of data does not imply easy interpretation from a business perspective. It is therefore crucial to cultivate a project team that includes other individuals besides technology experts. Moreover, the Kaufhof project shows the need for tight collaboration between retailers and suppliers with regard to RFID event data, master data, and other product-related digital content. The desirable level of cooperation with suppliers, however, presumes that the retailer is willing to share data, too.

5 Summary

From a technical perspective, the project has been highly innovative regarding the use of the newest transponder technology (UHF) and interface standards for the realization of integrated RFID systems. Read rates of virtually 100% were possible even at chokepoints such as in the transition between backroom and sales floor. The Galeria Kaufhof RFID project in Essen offered Kaufhof a unique possibility to empirically investigate different relationships between RFID investments and the effects on operational and management processes, thus improving the understanding of the business value of RFID in retail.

Various possibilities for improvements of logistical in-store processes, as well as of customer processes, could be demonstrated in the course of the project. These include, for example, RFID-enabled inventory counting, accelerated goods receipt processes, or sales support through mobile devices. Besides automational effects on the operational level, however, the project surpassed earlier trials by illustrating the relevance of the technology for the department store management. RFID as a fully automated measurement and control instrument provides information about inventory level developments on the sales floor, the time cycle and the quality of replenishment activities, the number of try-ons (on the item level and on the level of the whole category), and the types of tried on articles, respectively. Additionally, the trial showed that by providing more transparency, RFID can help to close the "data void" which exists between the goods receipt and the POS of the department store. The technology can thus help to discover process inefficiencies, which were so far intransparent and to help to control effective countermeasures accordingly. Furthermore, RFID technology not only helps to cut costs, but also improves the service quality and the quality of service processes on the sales floor.

Despite the major price decrease of RFID transponders to about 0.07 Euros in the year 2007 (year when the trial started), tags are still representing the biggest cost driver. Nevertheless, the prices allow for an economic justification of the RFID-based electronic article surveillance, which could become in the future a decisive driver for the adoption of RFID in this industry.

While RFID transponder prices will further decrease, future challenges of RFID adoption in the apparel retail industry include technical, business case, and privacy-related aspects, once RFID transponders are permanently sewed-in clothing items. The technical challenges encompass the adjustment of the read range of the RFID readers and data filtering (cleansing) and aggregation with adequate algorithms for further data treatment. Regarding the business case challenges, retailers will have to find the right trade-off between data gathering and analysis and investments in RFID readers and their positioning. Business-case driven RFID installations will then replace innovation-driven infrastructures. Retailers will have to ask themselves what they want to find out or what to improve and install the RFID readers accordingly. Finally, privacy concerns might become a further challenge. In 2007/08, Kaufhof decided to inform customers proactively by indicating the presence of readers with the help of highly visible adhesive labels, logos and instructions printed on transponder labels, as well as with freely available information brochures. Transponders were only removed from the products on request to allow for the identification of returned items. The company nevertheless experienced virtually no negative reactions during the whole trial period. Whether this is a consequence of their open communications policy or the consumers' perception of RFID benefits cannot be ascertained for certain at this point in time. Another conceivable explanation could be that the privacy debate in recent years, despite its presentation in the media, never really resonated with the broader public. However, the implementation of integrated RFID systems might trigger a second wave of anti-RFID activities and re-intensify the discussion around privacy risks.

References

Al-Kassab J, Lehtonen M, Michahelles F (2008) D5.4. Anti-counterfeiting prototype report, Deliverable of the EU-BRIDGE project, Project no. 033546, pp 1–63

Amini M, Otondo RF, Janz BD, Pitts MG (2007) Simulation modeling and analysis: a collateral application and exposition of RFID technology. Prod Oper Manage 16(5):586–598

Angeles R (2005) RFID technologies: applications and implementation issues. Inf Syst Manage 22(1):51–65

Asif Z, Mandviwalla M (2005) Integrating the supply chain with RFID: a technical and business analysis. Commun Assoc Inf Syst (AIS) 15(24):1–57

Auto-ID Labs (2003) White paper: beyond the EPC^{TM} – making sense of the data proposals for engineering the next generation intelligent data network. http://www.autoidlabs.org/uploads/media/MIT-AUTOID-WH024.pdf. Accessed on 10 February 2010

Bamfield J (2004) Shrinkage, shoplifting and the cost of retail crimes in Europe: a cross-sectional analysis of major retailers in 16 European countries. Int J Retail Distrib Manage 32(5):235–241

Barney JB (1991) Firm resources and sustained competitive advantage. J Manage 17(1):99–120

Barratt M, Choi T (2007) Mandated RFID and institutional responses: cases of decentralized business units. Prod Oper Manage 16(5):569–585

Bharadwaj A (2000) A resource-based perspective on information capability and firm performance: an empirical investigation. MIS Q 24(1):169–196

Bhatt GD, Grover V (2005) Types of information technology capabilities and their role in competitive advantage: an empirical study. J Manage Inf Syst 22(2):253–277

BRIDGE (2007) D7.1 – supply chain management in the European textile industry, deliverable D7.1 of the EU-funded research project "Building Radio frequency Identification solutions for the Global Environment" (BRIDGE). Project no. 033546, August 2007

BRIDGE (2009) BRIDGE Overview brochure of the EU-funded research project "Building Radio frequency Identification solutions for the Global Environment" (BRIDGE). Project no. 033546, September 2009

Brown SA (1997) Revolution at the checkout counter. Harvard Business Review Press, Cambridge, MA

Cangialosi A, Monaly JE Jr, Yang SC (2007) Leveraging RFID in hospitals: patient life cycle and mobility perspectives. IEEE Appl Pract 45:18–23

Christopher M, Lowson R, Peck H (2004) Creating agile supply chains in the fashion industry. Int J Retail Distrib Manage 32(8):367–369

Cole PH, Ranasinghe DC (2008) Networked RFID systems and lightweight cryptography: raising barriers to counterfeiting. Springer, Heidelberg

Curran K, Porter M (2007) A primer on radio frequency identification for libraries. Libr Hi Tech 25(4):595–611

Curtin J, Kauffman RJ, Riggins FJ (2007) Making the "most" out of RFID technology: a research agenda for the study of the adoption, usage and impact of RFID. Inf Technol Manage 8(2):87–110

Delen D, Hardgrave BC, Sharda R (2007) RFID for better supply-chain management through enhanced information visibility. Prod Oper Manage 16(5):613–624

Deloitte (2008) 2008 Global powers of retailing, Deloitte Touche Tohmatsu LLC. http://www.deloitte.com/assets/Dcom-Global/LocalAssets/Documents/dtt_globalpowersretailing.pdf. Accessed on 9 November 2009

EPCglobal (2007) EPC information services (EPCIS) version 1.0.1 specification. http://www.epcglobalinc.org/standards/epcis/epcis_1_0_1-standard-20070921.pdf. Accessed on 9 November 2009

Ellram L, La Londe BJ, Weber MM (1999) Retail logistics. Int J Phys Distrib Logistics 29(7/8):477–494

Falk (2005) Technology corner, temple of the computer. Electron Libr 23(2):244–248

Gagnon JL, Chu JJ (2005) Retail in 2010: a world of extremes. Strategy Leadersh 33(5):13–23

Gaukler GM, Hausman WH (2008) RFID in mixed-model automotive assembly operations: process and quality cost savings. IIE Trans 40:1083–1096

Gaukler GM, Seifert RW (2007) Applications of RFID in supply chains. In: Jung H, Chen FF, Bongju J (eds) Trends in supply chain design and management: technologies and methodologies. Springer, Heidelberg, pp 29–48

Gigalis GM, Kourouthanassis P, Tsamakos A (2002a) Towards a classification framework for mobile location services. In: Mennecke BE, Strader TJ (eds) Mobile commerce: technology, theory, and applications. Idea Group, Hershey, PA, pp 64–81

Gigalis GM, Pateli A, Fouskas K, Kourouthanassis P, Tsamakos A (2002b) On the potential use of mobile positioning technologies in indoor environments. In: Proceedings of the 15th bled electronic commerce conference, Bled, Slovenia, pp 413–429

Haberman AL (2001) 17 billion reasons to say thanks. In: Habermann AL (ed) Twenty-five years behind bars. Harvard University Press, Cambridge, MA, pp 113–151

Heese HS (2007) Inventory record inaccuracy, double marginalization, and RFID adoption. Prod Oper Manage 16(5):542–553

Heim GR, William RW Jr, Peng DX (2009) The value to the customer of RFID in service applications. Decis Sci 40(3):477–512

Hicks LE (1975) The universal product code. AMACOM, New York, NY

Hou J-L, Huang C-H (2006) Quantitative performance evaluation of RFID applications in the supply chain of the printing industry. Ind Manage Data Syst 106(1):96–120

Hozak K, Collier DA (2008) RFID as an enabler of improved manufacturing performance. Decis Sci 39(4):859–881

IDTechEx (2005) RFID progress at Wal-Mart. http://www.idtechex.com/products/en/articles/00000161.asp. Accessed 5 February 2009

Karaer Ö, Lee HL (2007) Managing the reverse channel with RFID-enabled negative demand information. Prod Oper Manage 16(5):625–645

Kärkkäinen M (2003) Increasing efficiency in the supply chain fort short shelf life goods using RFID tagging. Int J Retail Distrib Manage 31(10):329–336

Kaufhof (2009) Kurzporträit. http://www.galeria-kaufhof.de/sales/unternehmen/unternehmen/daten-fakten.asp. Accessed on 9 November 2009

Kelly EP, Erickson GS (2005) RFID tags: commercial applications v. privacy rights. Ind Manage Data Syst 105(6):703–713

KPMG (2005) Vertikalisierung im Handel. Auswirkungen auf die zukünftige Absatzwegestruktur. http://www.ihkschleswigholstein.de/produktmarken/starthilfe/anhaengsel/handel/Vertikalisierung_im_Handel_de.pdf. Accessed 24 June 2009

Langer N, Forman C, Kekre S, Scheller-Wolf A (2007) Assessing the impact on RFID on return center logistics. Interfaces 37(6):501–514

Lee HL, Özer Ö (2007) Unlocking the value of RFID. Prod Oper Manage 16(1):40–64

Lehtonen M, Al-Kassab J, Michahelles F, Kasten O (2008) D5.3. Anti-counterfeiting business case report, Deliverable of the EU-BRIDGE project. Project no. 033546, pp 1–79

Lehtonen M, Michahelles F, Fleisch E (2007) Trust and security in RFID-based product authentication systems. IEEE Syst J, Special Issue RFID Technol Oppor Challenges 1(2):129–144

Li S, Visich JK (2006) Radio frequency identification: supply chain impact and implementation challenges. Int J Integr Supply Manage 2(4):407–424

Loebbecke C (2004) Modernizing retailing worldwide at the point of sale. MIS Q Exec 3(4):177–187

Loebbecke C (2007) Piloting RFID along the supply chain: a case analysis. Electron Mark 17(1):29–37

Loebbecke C, Huyskens C (2008) A competitive perspective on standard-making: Kaufhof's RFID project in fashion retailing. Electron Mark 18(1):30–38

Loebbecke C, Palmer JW (2006a) RFID in the fashion industry: Kaufhof department stores AG and Gerry weber international AG, Fashion Manufacturer. MIS Q Exec 5(2):15–25

Loebbecke C, Palmer JW (2006b) A real-world pilot of RFID in the fashion industry: Kaufhof and Gerry weber in Germany. Academy of Management Annual Meeting (OCIS Division), Atlanta, GA

Manthou V, Vlachopoulo M (2001) Bar-code technology for inventory and marketing management systems: a model for its development and implementation. Int J Prod Econ 71(1–3):157–164

Mata FJ, Fuerst WL, Barney JB (1995) Information technology and sustained competitive advantage: a resource-based analysis. MIS Q 19(4):487–505

Melville N, Kraemer K, Gurbaxani V (2004) Information technology and organizational performance: an integrative model of IT business value. MIS Q 28(2):283–322

Menor LJ, Tatikonda MV, Sampson SE (2002) New service development: areas for exploitation and exploration. J Oper Manage 20(2):135–157

Metro AG (2009) www.future-store.org. Accessed on 9 November 2009

Metro Group (2007) Annual report 2007. www.metrogroup.de/servlet/PB/menu/1152880_l2_ePRJ-METRODE-MICROSITE-MAINPAGE/index.html. Accessed on 9 February 2010

Metro AG (2005) RFID im Praxiseinsatz: Ein Pilotprojekt von Kaufhof und Gerry Weber. Metro Group Future Store Initiat 1–43

McAfee A, Sjoman A, Dessain V (2004) Zara: IT for fast fashion. Harvard Business, Harvard

Michael K, McCathie L (2005) The pros and cons of RFID in supply chain management. In: Proceedings of the international conference on mobile business (ICMB'05), Sydney, Australia, pp 623–629

Moon KL, Ngai EWT (2008) The adoption of RFID in fashion retailing: a business value-added framework. Ind Manage Data Syst 108(5):596–612

Nagy P, George I, Bernstein W, Caban J, Klein R, Mezrich R, Park A (2006) Radio frequency identification systems technology in the surgical setting. Surg Innov 13(1):61–67

Nelson JE (2001) Scanning's silver celebration. In: Haberman AL (ed) Twenty-five years behind bars. Harvard University Press, Cambridge, MA, pp 25–33

Ngai EWT, Suk FFC, Lo SYY (2008) Development of an RFID-based sushi management system: the case of a conveyor-belt sushi restaurant. Int J Prod Econ 112(2):630–645

Ngai EWT, Cheng TCE, Lai K-h, Chai PYF, Choi YS, Sin RKY (2007) Development of an RFID-based traceability system: experiences and lessons learned from an aircraft engineering company. Prod Oper Manage 16(5):554–568

Nikitin PV, Rao KVS, Lazar S (2007) An overview of near field UHF RFID. In: Proceedings of the IEEE RFID 2007 conference, Grapevine, TX, pp 1–8

Peppard J, Ward J (2005) Unlocking sustained business value from IT investments. Calif Manage Rev 48(1):52–69

Peppard J, Ward J, Daniel E (2007) Managing the realization of business benefits from IT investments. MIS Q Exec 6(1):1–11

Pommer MD, Berkowitz EN, Walton JR (1980) UPC scanning: an assessment of shopper response to technological change. J Retailing 56(2):25–44

Prater E, Frazier G, Reyes PM (2005) Future impacts of RFID on e-supply chains in grocery retailing. Supply Chain Manage Int J 10(2):134–142

Reyes PM, Frazier GV (2007) Radio frequency identification: past, present, and future business applications. Int J Integr Supply Chain Manage 3(2):125–134

Roth AV, Menor LJ (2003) Insights into service operations management: a research agenda. Prod Oper Manage 12(2):145–164

Samli C (2004) Up against the retail giants: targeting weakness, gaining an edge, 1st edn., Thomson, Mason, USA

Santhanam R, Hartono E (2003) Issues in linking information technology capability to form performance. MIS Q 27(1):125–153

Song J, Haas CT, Caldas C, Ergen E, Akinci B (2005) Automating the task of tracking the delivery and receipt of fabricated pipe spools in industrial projects. Automation Constr 15(2):166–177

Srivastava L (2007) Radio frequency identification: ubiquity for humanity. ITU 9(1):4–14

Tajima M (2007) Strategic value of RFID in supply chain management. J Purch Supply Manage 13(4):261–273

Thiesse F, Al-Kassab J, Fleisch E (2009) Understanding the value of integrated RFID systems: a case study from apparel retail. Eur J Inf Syst 18:592–614

Thiesse F, Fleisch E (2008) On the value of location information to lot scheduling in complex manufacturing processes. Int J Prod Econ 112(2):532–547

Tzeng SF, Chen WH, Pai FY (2008) Evaluating the business value of RFID: evidence from five case studies. Int J Prod Econ 112(2):601–613

Venables M (2005) Analysis: RFID – don't believe the hype. IEE Manuf Eng 84(5):8–9

Wang L-C (2007) Enhancing construction quality inspection and management using RFID technology. Automation Constr 17(4):467–479

Wang SW, Chen WH, Ong CS, Liu L, Chuang YW (2006) RFID applications in hospitals: a case study on a demonstration RFID project in a Taiwan hospital. In: Proceedings of the 39th Hawaii international conference on systems science, Los Alamitos, CA, pp 1–10

Wessel R (2007) Metro group's Galeria Kaufhof launches UHF item-level pilot, RFID Journal. www.rfidjournal.com/article/articleprint/3624/-1/1. Accessed on 10 February 2010

Wheatley M (2008) Learning to track ROI, Data collection technologies offer true business value – if you know where to look. Manuf Bus Technol 26(9):28–30

Wyld DC (2008) Death sticks and taxes: RFID tagging of cigarettes. Int J Retail Manage 35(7):571–582

Wyld DC (2006) RFID 101: the next big thing for management. Manage Res News 29(4):154–173

Yu S-C (2007) RFID implementation and benefits in libraries. Electron Libr 25(1):54–64

Zhu K (2004) The complementarity of information technology infrastructure and e-commerce capability: a resource-based assessment of their business value. J Manage Inf Syst 21(1): 167–202

The Potential of RFID Technology in the Textile and Clothing Industry: Opportunities, Requirements and Challenges

Elena Legnani, Sergio Cavalieri, Roberto Pinto, and Stefano Dotti

Abstract In the current competitive environment, companies need to extensively exploit the use of advanced technologies in order to develop a sustainable advantage, enhance their operational efficiency and better serve customers. In this context, RFID technology has emerged as a valid support for the company progress and its value is becoming more and more apparent. In particular, the textile and clothing industry, characterised by short life-cycles, quick response production, fast distribution, erratic customer preferences and impulsive purchasing, is one of the sectors which can extensively benefit from the RFID technology. However, actual applications are still very limited, especially in the upstream side of the supply network. This chapter provides an insight into the main benefits and potentials of this technology and highlights the main issues which are currently inhibiting its large scale development in the textile and clothing industry. The experience of two industry-academia projects and the relative fallouts are reported.

1 Introduction

The debate on RFID technology in the last few years has been characterised by some prevailing positions:

- the enthusiasm and expectations sparked by its potential, and somewhat boosted and drugged by mandatory RFID tagging projects undertaken for example by Wal-Mart, the US Defense Department, and European companies, such as Metro and Tesco (Ngai et al. 2008);
- the concern and implications in terms of possible consumer privacy violations (Ranasinghe and Cole 2008);

E. Legnani (✉)
Department of Industrial Engineering, CELS, Research CENTRE on Logistics and After-sales Service, University of Bergamo, Viale Marconi, 5, 24044 Dalmine, Italy
e-mail: elena.legnani@unibg.it

D.C. Ranasinghe et al. (eds.), *Unique Radio Innovation for the 21st Century*,
DOI 10.1007/978-3-642-03462-6_14, © Springer-Verlag Berlin Heidelberg 2010

- the prudent stance of many professionals on its widespread implementation timing which could be quite lengthy due to the numerous economic and organizational hurdles yet to be overcome;
- the current costs of the hardware, since it is still quite expensive to be compatible with large scale diffusion, especially if compared with the low costs of conventional bar codes.

Nevertheless, almost everyone agrees on the fact that the development process has been well and truly launched and this stands to generate profound changes above all in traditional sectors. As an example, with RFID collection of real-time data on individual items could become a reality, which was not possible with the use of bar codes (Tajima 2007).

A future vision of RFID even includes the "Internet of Things", a universal open network of computers and objects for identifying individual products and for tracking them as they flow through the global supply chain, anywhere in the world, instantly (Tajima 2007; Teresko 2003).

An example of a traditional sector that can benefit from an extensive and wise introduction of RFID technology is the textile and clothing industry. It is a diverse and heterogeneous industry which covers an important number of activities: from the transformation of fibres to yarns and fabrics to the production of a wide variety of products, such as hi-tech synthetic yarns, bed linen, industrial filters, geo-textiles for soil protection and reinforcement, apparel, etc.

Given the characteristics of the textile and clothing sector, RFID technology can surely be a valid and suitable support for sustaining the dynamics of the market competition and fostering the intra- and inter-organizational relationships between the supply tiers throughout the whole upstream and downstream network.

On these grounds, this chapter will provide an insight into the main benefits and potentials of the RFID technology and will highlight the main issues which are currently inhibiting its large scale development in this specific industry.

This book chapter is structured as follows: Sect. 2 provides an in-depth description of the textile and clothing supply network and its main requirements, in terms of information and knowledge exchange between the different tiers; Sect. 3 shows how RFID can be a valid support to the needs of the supply network and defines the opportunities and issues related to its adoption; Sect. 4 is about the industrial experience and fallouts coming from two industry-academia projects; Sect. 5 reports the main conclusions.

2 Structure, Processes and Main Requirements in the Textile and Clothing Supply Network

The textile and clothing sector represents an important part of the European manufacturing industry and plays a crucial role in the economy and social well-being in numerous European regions. According to the latest structural data available, collected by the European Commission, in 2006 there were 220,000 companies employing 2.5 million people and generated a turnover of 190 billion Euro, accounting for 3% of total manufacturing value added in Europe (European Commission 2009). It is mainly a SME (Small and Medium Enterprise)-based industry, as companies of less than 50 employees account for more than 90% of the workforce and produce almost 60% of the value added.

The largest producers are Italy, France, UK, Germany and Spain. They account for about three quarters of the European production of textiles and clothing. With regard to the external trade performance, the European textile and clothing sector is one of the two biggest players in the world market. It represents 29% of the world exports (not including trade between European member countries) after China which occupies the first place with 40% of world exports.

However, even though data are promising, the limited access to many third country markets, the low financial entry barriers and the easy access to technology, contribute to an increasing competition from other countries, with heavy ripples on the European employment rate. The great advances in the chemical-textile compound, moreover, favored chemical companies to the detriment of the traditional textile ones.

The way the European textile sector is facing this situation is by proposing a demand-driven production, providing high customization and responsiveness. This requires a thorough change of the managerial mindset in order to efficiently manage a huge number of references for different types of fabrics and complements. For instance, Zara, a Spanish-based chain owned by Inditex, has been able to achieve an excellent financial status thanks to its core competencies that provide the chain with a competitive advantage over traditional retailers in the industry. Zara is able to be so responsive through a competitor-crushing combination of technology-orchestrated coordination of suppliers, just-in-time manufacturing, and finely tuned logistics and of vertical integration (Gallaugher 2010).

By owning its in-house production, Zara is able to be flexible in the variety, amount, and frequency of the new styles it produces. Also, 85% of this production is performed through the season, which allows the clothing chain to constantly provide its customers with the latest designs in contrast to the traditional retailers who lack in this flexibility (Ferdows et al. 2004).

Regarding the Italian panorama, companies have been able to hold a leading position in some niche markets. In the luxury goods, for instance, Italian companies own 40% of the world market share. Their key success factors are related to: (i) the strong interconnections within the network, especially between the design, manufacturing and distribution processes; (ii) the introduction of very innovative and high quality products; and (iii) personalized answers to customers' needs.

In such a framework of global competition, research, innovation and new technologies are crucial elements to further develop the knowledge base of the textile and clothing industry.

In particular, information technologies are surely playing a prominent role in integrating people, information, and products across traditional supply chain boundaries, including management of engineering and manufacturing operations.

Moreover, the proliferation of smart materials represented by actuators and sensors, tagging represented by global positioning systems (GPS) and RFID tags, can further contribute to the advent of new manufacturing and logistics processes that provide a dynamically controlled supply chain network (Kumar 2007).

2.1 The Textile and Clothing Supply Network

The structure of the textile and clothing supply network is quite complex and composite. It encompasses several and very peculiar processes, especially during the manufacturing and processing cycles. As shown in Fig. 1, the network starts with the treatment of raw materials (chemical or natural fibres), it spans the production of semi-processed products (i.e. thread, yarn, fabric, knitted fabric) and ends with the tailoring of finished products and their distribution to the sales points where the final customers accomplish their purchases.

Complexity is also increased by the relation intricacies since the network is made up of several actors which interact with each other. In general, each process is managed by different companies which, in turn, outsource part of the work to other third parties during high peak seasons, or even on a regular basis. Furthermore, transportation and distribution are managed by external logistics operators.

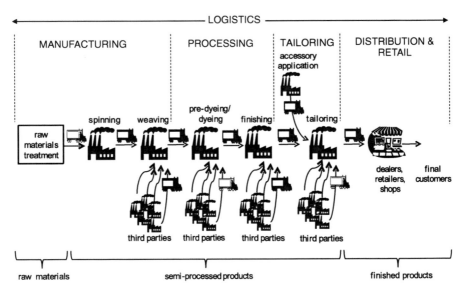

Fig. 1 The textile and clothing supply network

Within the textile and clothing supply network, four main macro-processes can be identified:

- *Manufacturing (M)* – it includes the production processes which range from raw material treatment to weaving (Table 1).

Table 1 Production sub-processes in the manufacturing phase

Raw material treatment	Extraction and treating of textile fibres, elements which require adequate sharpness, flexibility, high length/diameter ratio, to be workable by spinning and weaving machineries and appropriate to any textile use.
Spinning	Fibres working to obtain either threads or yarns.
Weaving	Interlacement of yarns or threads (called warp and filling) through a loom to form fabrics.

- *Processing (P)* – it encompasses some specific treatments on semi-processed products, such as pre- dyeing, dyeing and finishing (Table 2).

Table 2 Production sub-processes in the processing phase

Pre- dyeing	Cleaning of substances and surface fibres previously applied to the fabric during the weaving.
Dyeing	Imparting colours to a textile product by treatment with a dye. It draws on different technologies, determined by the characteristics of the fabric or by the effect that is required.
Finishing	Treatments aimed at giving fabrics the visual, physical and aesthetic properties which consumers ask for (e.g. foulard processing, coating, resin treatment, spraying etc.)

- *Tailoring (T)* – it refers to a set of specific hand and machine sewing and pressing techniques which are used to transform fabrics into products such as:

  - garments, knitted or woven (the so called "clothing" industry);
  - carpets and other textile floor covering;
  - home textiles (such as bed linen, table linen, toilet linen, kitchen linen, curtains, etc);
  - technical textiles (items with high technology content such as products with high tenacity yarns, special elastic or coated fabrics, etc.).

  In this phase, accessories (such as buttons, zips, buckles, etc.) are also applied.

- *Distribution and Retail (DR)* – it embraces both the distribution of finished products and the management of the sale points. Distribution can be either directed to the shops or mediated through the presence of dealers, retailers and stores. Sales points handling is about products assortment, products disposal and turnover on shelves, responsive products supply and management of on-site inventories.

Logistics also plays a fundamental role. It is a transversal process which encompasses the whole network and impacts on all the above mentioned macro-processes.

It involves activities related to transportation from/to third parties and companies, warehousing, material-handling and packaging.

2.2 Main Requirements

As aforementioned, the textile and clothing industry is characterised by a wide assortment of products, short life-cycles, high seasonality, high volatility, high-impulse purchasing and complicated distribution and logistics operations (Christopher et al. 2004). The supply network is extremely multifaceted with several intricate material and information flows, numerous supply chain tiers and a multitude of actors to be co-ordinated on a global level.

Moreover, for maintaining a sustainable level of competitiveness, companies need to address wider markets, to source supplies on a global scale, and to carry out a continuous and systematic development of new products and services. This relentless trend obviously contributes to raise the complexity of the textile and clothing organisations.

However, most of the companies, especially Small and Medium Enterprises (SMEs) are not aware of the utmost opportunities they could lever out from the adoption of appropriate logistics practices and technology systems. These tools could provide the real "holy grail" for managing their level of complexity and for surviving in the market.

Focusing on the impact that technology, such as RFID, can have on the development of the business, interviews conducted to sales and operations managers of numerous Italian textile and clothing firms contributed to draw up the following requirement list. This list addresses some of the most critical aspects that traditionally affect those companies operating in the textile area. Companies mainly call for unique product identification in order to provide:

1. tracking of the product along the network – ability to know exactly the position of an item within a network;
2. traceability of accurate and specific product information along the network – ability to build the "logistical history" of an item at any level of the network;
3. fast, automatic and efficient data management;
4. automatic and efficient management of manufacturing and operations processes to reduce human errors and inventory stock-outs;
5. Made-in-Italy brand safeguard to protect the creativity, research and innovation which enable a product to establish itself nationally and internationally;
6. theft and counterfeiting protection, especially for high-value goods; counterfeiting stands for an unauthorised use of a registered trade mark and/or the intentional copying of an external feature of a product;
7. containment and control of grey market – trade of products through distribution channels which, while legal, are unofficial and unauthorized. This implies: (i) the production and supply of textile goods that are not labelled with registered trade

marks which, however, may confuse final consumers owing to their appearance or the similarity of the trade marks they bear ("look-alike" goods); (ii) unlawful overproduction by textile products licensees;

8. involving and versatile shop experience for final customers.

Figure 2 classifies these requirements according to which macro-process of the network their resolution would have a major effect.

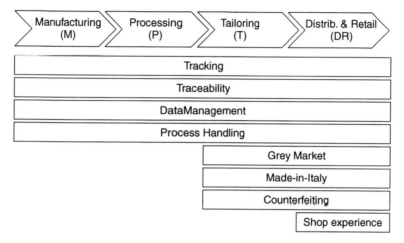

Fig. 2 Classification of the main requirements for the textile and clothing supply network

Data necessary to fulfil all these requirements are diverse. For example, the identification of a product at the manufacturing level might require technical data in order to assure quality compliance, whilst the identification at the retail level might require some commercial and advertising data in order to provide an exciting shop experience. Nevertheless, the exact identification of the necessary data depends upon the specific requirements of the company implementing such technology. The business specifications should drive the collection of RFID data.

An in-depth analysis of the specific data being collected through RFIDs in the textile and clothing network is out of the scope of this chapter. We suggest the reader to refer to the contribution by Niederman et al. (2007), who provide an interesting dissertation about the issues that practitioners are likely to face.

Nonetheless, it is important to mention the distinction between the data acquisition cycle and the data usage cycle (Redman 1996). While the former has mainly technological implications, the latter is more concerned with the meaningfulness and value of the information that can be extracted from raw data. The management of the data flows is a key issue for any organization deploying the RFID technology. The issues are not just limited to the implementation of a suitable IT infrastructure but also to the right interpretation of the captured data in order to elicit the proper information and knowledge. Concerning this, it is worth mentioning the recent development in the business intelligence field, which has led to the development of methods and technologies for ensuring an effective data exploitation.

3 Opportunities and Issues in the Adoption of RFID Technology in the Textile and Clothing Industry

Most of the requirements reported in the previous paragraph can be met through high coordination, information sharing, automation of some processes and transparency at each level of the network. In this context, RFID technology is an effective and valid tool for answering to these needs.

An analysis of the scientific literature about the applications of RFID systems in the textile and clothing industry revealed that, even though RFID is a relatively new topic and the main contributions have been lately published, an increasing number of companies are getting more and more aware of its potential. The evidence of this research is reported in Table 3.

Table 3 Literature review about the main RFID applications

Requirements	Literature contributions
Tracking	Metro (2003); Newstex (2007); Singh et al. (2008)
Traceability	Angeles (2005); Chappell et al. (2002); Choy et al. (2009); Loebbecke and Huyskens (2006); Sullivan (2005); Wilding and Delgado (2004)
Data management	Angeles (2005); Chappell et al. (2002)
Process handling	Angeles (2005); Collins (2006); Hogan (2003); Loebbecke and Palmer (2006); Pradhan et al. (2005); Sangani (2004); Shim (2003); Singh et al. (2008); Wilding and Delgado (2004)
Made-in-Italy	Guercini and Runfola (2009)
Counterfeiting	Jones (2005); Lehtonen et al. (2007); Rekik et al. (2009); Wong et al. (2005)
Grey market	Pradhan et al. (2005); Wilding and Delgado (2004)
Shop experience	Eckfeldt (2005); Want (2004)

In particular, through the automatic and unique item identification, relevant results have been achieved in tracking items, improving processes (in terms of higher efficiency, responsiveness and reductions in inventory) and in increasing the amount of information and reliable data along the network. Counterfeiting has also been significantly reduced.

Made-in-Italy brand protection has few contributions: this may depend on the fact that this issue is very distinctive of the Italian market. However, it is presumable that other countries are interested in protecting their own made-brand, and thus willing to apply RFID technology.

Experiences which significantly attest a control and a reduction of the grey/parallel market are still trivial, while there is a growing number of interesting applications which are addressed to boost the customer satisfaction and his/her shop experience.

Due to some issues still to overcome in the RFID technology (which will be more extensively reported in Sect. 3.2), the likely scenarios of RFID adoption regard the downstream part of the network, such as *tailoring*, *distribution* and *retail* processes. Minor relevance regards the processing and even less the *manufacturing* processes.

The literature analysis confirms that the main emphasis is given to applications in the downstream part of the network, while examples of applications more upstream in the network are almost absent. This is a clear sign that more experimentation is required in the upstream part and chances to make research are very broad. The Griva experience and the CLOTH project, reported in the next sections, are tangible examples of how also the upstream part of the network can benefit from the adoption of RFID applications.

3.1 Adopting RFID Technology – Opportunities

According to a research carried out by Politecnico di Milano (2007), benefits coming from the adoption of the RFID technology can be represented through "the tree of value", as reported in Fig. 3. The two main categories are:

- *tangible benefits*, which lead to an increase in efficiency and effectiveness;
- *intangible benefits*, which lead to an increase of "soft" elements (such as brand image, information, customer satisfaction and law compliance), harder to estimate quantitatively.

Regarding the first category, benefits related to efficiency come from improvements in the productivity of resources – mainly human – and/or in process quality. These enhancements allow either to reduce costs (maintaining the same production volume) or boost revenues (augmenting volumes due to the increased productivity of resources).

Benefits related to effectiveness, instead, depend on increases in external quality (higher value perceived by customers, higher availability of products or services, etc.) and in responsiveness. Customers price positively both results, thus the final

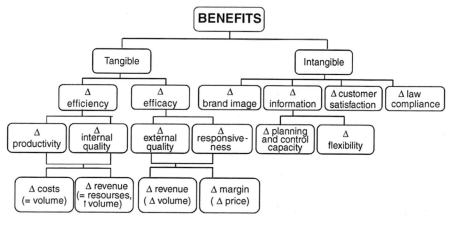

Fig. 3 "The tree of value" (Politecnico di Milano 2007)

effect is either an improvement in revenues (customers buy more) or a raise in margins (higher tariffs can be applied).

Concerning the intangible category, benefits may result in: (i) enhancements in brand image; (ii) higher number of available data and information which, in turn, can facilitate planning and controlling activities and favour flexibility to unplanned changes; (iii) higher customer satisfaction; and (iv) more compliance with law constraints.

However, it should be remarked that the application of RFID systems normally generates value and successful results when they are not applied as stand-alone technologies. In fact, as it happens for any relevant and innovative ICT-based project, the best outcomes generally do not merely stem just from the technology adoption but they depend on other activities like reengineering of processes, introduction of new functionalities or services, updates of information systems, and so forth.

To illustrate more in detail the results, which can derive from the implementation of the RFID technology, the main advantages have been grouped according to (Table 4):

- the macro-processes of the textile and clothing supply network which can get more benefits (Moon and Ngai 2008; Roberti 2006);
- the main categories identified in the "tree of value".

Table 4 Benefit of RFID technology in the textile processes

Effectiveness	M	P	T	DR
Sharply cut of database update times, almost close to real time	✓	✓	✓	✓
Management of simultaneous orders	✓	✓	✓	✓
Easier and faster picking process			✓	✓

Brand Image	M	P	T	DR
Guarantee of product authenticity – Made-in-Italy protection			✓	✓

Information	M	P	T	DR
Higher and more reliable information available for each item	✓	✓	✓	✓
Tracing of textiles movements from one department/third party/company to another	✓	✓	✓	✓
Automatic check of outbound loads and truck loading sequence	✓	✓	✓	✓
Collection of statistics about customers' habits and fashion tastes				✓

Efficiency	M	P	T	DR
Automatic identification of each item in any position along the network	✓	✓	✓	✓
Multiple and simultaneous identification of several items without visual contact needed	✓	✓	✓	✓
Association of the items to well-identified physical routes	✓	✓	✓	✓
Check of physical characteristics and conditions of use of the items	✓	✓	✓	✓
Associability of operator/process and material/process in case of controversy from customers	✓	✓	✓	✓
Robust cost-savings (working capital decrease and error margin reduction)	✓	✓	✓	✓
Higher precision in order fulfilment	✓	✓	✓	✓
Automatic control of inbound loads and receipt of goods without opening their packaging	✓	✓	✓	✓
Control of discrepancies in the length of fabric pieces	✓	✓		
Highly accurate inventory control through multiple treatment processes	✓	✓	✓	
Automatic control of item availability (size, colour) on the shelf and immediate re-assortment				✓
Shoplifting protection				✓

Customer Satisfaction	M	P	T	DR
Shop customer experience				✓
Enhancement of service levels				✓
Development of "smart" products like washing machines or wardrobes				✓

Law Compliance	M	P	T	DR
Control of grey/parallel market			✓	✓
Control of unauthorised uses of a registered trade mark			✓	✓
Preservation of reserved information and privacy maintaining	✓	✓	✓	✓

M manufacturing, *P* processing, *T* tailoring, *DR* distribution and retail

The example in the following section reports what Prada, a famous Italian fashion manufacturer, has carried out in its shop to increase its customer satisfaction and give them a unique shop experience.

3.1.1 The Smart Shop in New York City

RFID technology is being used at Prada's Epicenter store in New York City. The smart labels identify customers, merchandise, and link individual shoppers with information about their selections before and after they make a purchase.

RFID tags and readers are implemented at numerous touch points throughout the Prada Epicenter store to identify products, devices and staff. The technology creates a seamless shopping experience designed to enhance customer relationships. Prada sales personnel are equipped with a wireless RFID handheld reader that gives them up-to-date access to inventory and customer information stored in a centralized database. Sales personnel also use the device to read RFID-tagged products and identify staff wearing RFID "clips". The device also controls video screens throughout the store, which demonstrate products on the runway, show collection photographs and designer sketches, while providing more in-depth in-formation about the colour, cut, fabric and materials used to create Prada merchandise.

In the dressing rooms, RFID readers identify all merchandise a customer brings inside and displays information on the garment on the interactive video touch screen display. From the touch screen, customers can access product specifications as well as alternative and complementary items and accessories. Using RFID technology linked to customer information stored in a database, Prada ensures a high-quality customer experience across multiple sales associates and subsequent Epicenter locations (www.allbusiness.com).

3.2 Adopting RFID Technology – Issues

The main issues concern the application of RFID technology to the upstream part of the textile and clothing supply network. In fact, even though RFID systems would be of great support, their introduction is awkward for at least the following two reasons:

- cost of components and materials in the upper part of the network have a low unit value, often not so high to justify RFID implementation;
- aggressive processes (mechanical and chemical), which items are subjected to, can be devastating for RFID hardware (i.e. tags).

Regarding costs, the recent evolution of RFID technology and the continuous pressure exerted towards its improvement, point to tremendous advances in the future. It is presumable that production processes for making tags will be optimized and will allow strong cost reductions.

Regarding the resistance of tags, problems are related to strong mechanical stress and high temperatures, involved especially in the *processing* phase, that damage the RFID tags applied on the textile products. Normally, constraints are not due to RFID chips, but rather to antennas and the welding between the antenna and the chip. To overcome this issue, research laboratories have already come up with passive tags which can stand high temperatures (even up to 250°C) and low mechanical stress, for instance, able to bear painting and dyeing processes (Politecnico di Milano 2007).

However, what is still missing is a tag able to resist to strong mechanical stress and capable to get through the entire network, going across *manufacturing, processing* and *tailoring* processes. In this sense, researchers are considering the use of protective inlays even though, until now, there are no adequate ones. Most of the inlays are created to resist to very strong stress. Thus, they are very rigid and risk to damage the item or the processing machine and equipments; on the contrary, less rigid inlays do not guarantee the tag survival. Moreover, tag shape cannot be unique along all the processes, from upstream to downstream and, most of all, tag application must not cause defects to the items to which is applied.

However, these issues represent great challenges for researchers leading to new experimentations and investigations. If all the technological and physical constraints were overcome, RFID would add great and priceless value to the enhancement of all the textile and clothing supply network activities.

An interesting application comes from Griva, an Italian textile manufacturer. It represents a pioneer company which decided to: (i) adopt RFID for its upstream processes in order to efficiently track its fabric rolls and manage inventory through each phase of production and (ii) replace its inaccurate barcode system.

3.2.1 A Company that Looks to the Future

Griva S.p.A. is an Italian textile manufacturer, which produces more than 300,000 rolls of fabric each year. It distributes 500 pieces of fabric, equal to 20,000 m, each day. Its textiles are used for upholstery and drapery and sent to leading European retailers. With this diversity among its fabrics, Griva has to maintain highly accurate inventory control. It must trace the textiles starting with the initial raw thread material to woven fabric and finally to distribution and shipment of the fabric rolls. Adding challenge to the traceability is the fact that the fabric rolls undergo multiple treatment processes as they move through Griva's warehouse. These treatments include exposure of the fabric to harsh environmental factors, such as high temperatures, water and high humidity, and potent chemicals used in the dyeing process. In addition, the fabric – and subsequently the inventory tracking used on the rolls – has to withstand machine cutting and sizing, transporting, and shrink wrapping.

For years, Griva depended on traditional bar coding systems to track its fabric rolls through its manufacturing process. However, the bar code system was not accurate, mostly due to the mechanical and environmental stresses placed on the bar codes. So when Griva was ready to open its new automated warehouse in 2006, it decided to partner with an Italian RFID integrator. The technology provider developed a complete RFID textile tracking solution for Griva. The European standard ETSI 302-208 (866 MHz), using tags with data codified according to the EPC Global standard, was implemented. The software chosen was a middleware which interfaces the RFID reader firmware with ERP (Enterprise Resource Planning) software adopted by Griva to plan its resources.

This RFID solution was able to overcome all the problems Griva had previously faced with bar codes. With the RFID in place, Griva can guarantee traceability during all phases of fabric production and logistics. Moreover, tags passed the "plastic film" test that bar coding had failed. The plastic film used in shrink wrap-ping that protects the fabric during shipping frequently hid the old bar codes. With RFID, Griva can efficiently track rolls that are wrapped and ready for delivery, which saves time and also provides its customers with the most accurate information about the finished fabric. The RFID tags also allow Griva to sort its fabrics automatically in the warehouse. By associating ID information to each roll (such as product type, weight, and diameter), the roll can move efficiently through the warehouse to its next checkpoint. By having readers at various checkpoints, operators can decide instantly whether a roll should be moved into a storage facility or prepared for immediate shipping to a customer. In the ending stages, the automated system uses the RFID tag information to determine final packaging and the most cost-effective transportation options for each roll. Using the fabric's weight and destination information, warehouse operators can build customised shipping boxes and prepare each roll for its final distribution and transport by designated carriers (www.griva online.com).

4 Industrial Fall-Out: Experiences from Industrial Projects

This section reports the main evidence derived from two projects carried out by the authors: first on application of RFID to the downstream players of a textile network (named "TAGGIE" project), while the other is related to an experimental project for RFID application to the upstream part of a textile network (named "CLOTH" project).

4.1 The TAGGIE Project

"TAGGIE: identification and traceability using RFID to preserve Made-in-Italy products" is an industry-academia applied research project. Actors involved in its development were both from academic and industrial extraction: the working team was formed by two Universities (among which the University of Bergamo), an

RFID technology company and two companies in the fashion industry, one making accessories (zips) and the other clothing items. The project implied R&D and capital expenditure with the main purpose of embedding and integrating the RFID technology in the pilot companies' products.

The main reason why the TAGGIE project was carried out was to guarantee the identification and traceability of textiles in order to preserve their Made-in-Italy origin and to prevent them from counterfeiting threats.

The ever increasing problem of counterfeiting of garments and accessories, in fact, is very often related to little or lack of transparency throughout the manufacturing processes and it entails huge losses for companies. These faults are not just of economical nature, but more importantly, in terms of image and customers' trust in the brand. On the same grounds, the major associations in this sector, crucial to protect the true value of Made-in-Italy, are perfectly aware of the problem and claim that this situation urges for regulations that force manufacturers to clearly communicate the origin of their products.

Considering this background, the TAGGIE project team, after a thorough analysis of the manufacturing processes of the two textile pilot companies, decided to apply the RFID technology to the accessories, namely the zips. The main reason is that a zip can be exclusively inserted by the manufacturer during the production of an item, bags or garments, and it cannot be easily replaced without damaging the product. RFID on a zip can be a valid tool for identifying, tracing products (where, when and who made every production steps) and certifying their origin. The project has been framed around three different stages.

4.1.1 Process Mapping

This phase was a detailed analysis of the zip manufacturing processes and of the company layout. This was fundamental to identify when the tag had to be embedded in the zip through its manufacturing process and where the reader had to be placed at the company site to update tag information. Furthermore, a selection of the most crucial and essential information to be collected along the processes has also been carried out.

As reported in Fig. 4, the zip manufacturing process is made up of three different parts: the tape production, the slider production and the zip assembling. Tape and slider are two components that make the final zip and their production can be done at the same time, even though located in different areas. The final zip assembly starts when both the previous phases are completed.

The most suitable time to physically insert the tag is during the slider production, just after its colouring and before its possible exit to a third provider. Information is updated before the final zip enters the warehouse while waiting to be sold. Information is about technical parameters such as colours and length of the different components, production batch sizes, dates, used machines, etc.

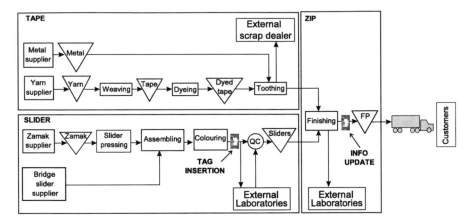

Fig. 4 An overview of the zip manufacturing processes

4.1.2 Technical Analysis

This phase was about research activities and experimental tests to find out the best tag and antennas able to meet the project requirements and make the "intelligent" zip. The team had to face some constraints, in particular the need of:

- very small tags (< 1 × 1 cm) that could fit the small zip size;
- a relatively small quantity of low cost tags;
- tags which do not interfere with the metal alloy (zamak in the analyzed case) that zip components are made up;
- customised antennas suitable for very small tags.

Different test have been performed using passive tags. Experiments have been made comparing:

- HF (High Frequency) vs. UHF (Ultra High Frequency) tags;
- copper vs. aluminium antennas;
- inlay vs. no inlay use.

According to the project requirements, the best solution was a HF tag (for its low interference with metal surfaces), with an aluminium antenna, inserted in a resin inlay. Some of the tag shapes utilized are depicted in Fig. 5.

4.1.3 Fashionable Design

The aim of the TAGGIE project was also to leverage the presence of the RFID on the product adding esthetical value to items, through a "fashionable" zip. For this reason, the team worked hardly in identifying the best material (resin) for the tag inlay and the best inlay shape. Moreover, to guarantee greater innovation and product authenticity, tags have been marked with the "brand names" of the two pilot companies. Different models have been proposed, as reported in Fig. 6.

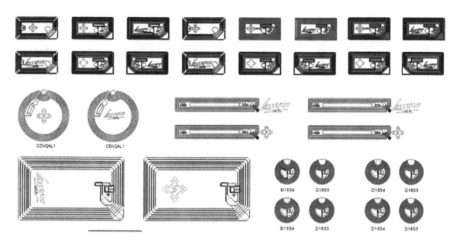

Fig. 5 HF tag varieties

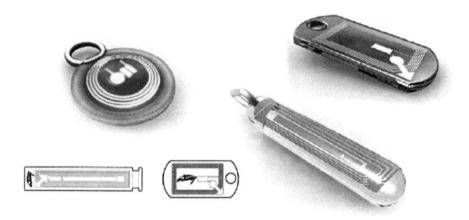

Fig. 6 Zip-wing models with resin inlays

As a result, the TAGGIE project team was able to equip a zip with an inlay which protects the RFID tag. Each item that wears this "intelligent and fashionable" zip is identified through a unique serial number impressed in the memory of the chip. To guarantee greater security of the information, the chip can be additionally protected by cryptographic keys and codes, much more difficult to copy.

Moreover, this new zip allows potentially numberless ways of personalisation, since the tag interacts through the web via dedicated software applications.

A core outcome of the project has been the registration of a patent concerning this innovative design of the zip slider which embeds the tag (Patent no. 08169048.9-1256).

Further developments of the TAGGIE project will be directed to a massive industrialization of these innovative zips and to the launch of a new line of "intelligent and fashionable" cloths, namely garments produced by the other pilot company with the patented zips. In this way, efforts will be addressed to create more coordination among the pilot companies, their suppliers and third party operators. This will hopefully allow them to share information and trace items along the entire network, being confident of the products authenticity and Made-in-Italy origin.

4.2 The CLOTH Project

CLOTH is an experimental project, still ongoing, which aims at introducing the RFID technology in a very upstream phase of a textile and clothing supply network. In particular, the challenge is to apply the RFID device at the fabric during the *processing* macro-process, which includes pre- dyeing, dyeing and finishing processes. Final goals of the CLOTH project are:

- to trace items along the network and retrace the sequence of the past processes they have undergone;
- to check and monitor process parameters and optimize data collection;
- to counter the threat of counterfeiting, thereby securing protection of brand names and of the Made-in-Italy label;
- to control discrepancies in the length of fabric pieces.

The project is being carried out by CELS – University of Bergamo, a company which makes special fabrics for clothing and footwear, and an RFID tag producer.

According to what reported in Sect. 3.2 about the main issues to face when adopting RFID, especially in the upstream part of the network, a decisive part of this project is to provide crucial evidence that the adopted tags can stand the *processing* macro-process (pre-dyeing, dyeing and finishing) and continue to be fully operational afterwards.

Thanks to the involvement in the project team of an RFID tag producer, several tests are ongoing to select the best tag, both in terms of performance and costs, which allows the use of the RFID technology. Attention is especially focused on the following activities:

- identification of a tag that is as small as possible and, in general, does not damage the fabric affecting its quality and limiting its use;
- identification of a tag that can be read/written by automatic systems and readable at a distance with wide ranges;
- assessment of the ability of the tag to withstand the mechanical and thermal stresses, as well as the action of chemical agents, that fabrics undergo during the pre- dyeing/ dyeing and finishing stages;
- identification of a low-cost, verifiable, robust, and simple process where the tag can be easily inserted/applied to the fabric piece and read/written during the different stages of the processing macro-process;

- drafting of the technical specifications that the hardware system for reading/writing the tag should have;
- assessment of virus problems and counterfeiting in relation to RFID and, therefore, the identification of safety arrangements, such as tag blockers and cryptography.

The results of the first tests are reported below. Regarding pre-dyeing, 12 UHF tags have been treated at 94°C for 60 min. More than 50% of treated tags (8 over 12) successfully passed the experiments. Those tags that failed the tests revealed problems at the antennas, which melted in most of the cases.

Regarding dyeing, 8 UHF tags have been enveloped in a textile inlay and sewed onto the fabric itself. They have been treated at 105°C for 90 min. Just 2 tags withstood the experiments. A microscope analysis revealed that the chip either melted or moved from its original position and the antenna regularly broke in some parts.

Once these experimentations will be ended, the team project aims at proposing and industrializing tags able to trace and identify each fabric during the main stages of its production and while it is stocked pending to be acquired by the customers.

5 Conclusions

The value of RFID technology is becoming more and more apparent in several businesses: in such a competing society, in fact, firms need to extensively exploit the use of advanced technology in order to develop competitive advantage and succeed in today's market. The adoption of such technology can be of priceless worth to enhance company operational efficiency, differentiate from competitors and better serve customers.

The textile and clothing industry, characterised by short life-cycles, quick response production, fast distribution, erratic customer preferences and impulsive purchasing, could extensively benefit from the RFID introduction. However, actual utilizations are still quite limited and they are mainly research and pilot projects. The main emphasis is given to applications in the downstream part of the network, such as tailoring, distribution and retail, while examples of applications more upstream in the network are almost absent.

The goal of this chapter has been to point out both the potential benefits that the textile and clothing industry can get through the application of RFID and the main issues which still obstacle its wide spread. Having an in-depth understanding of the impact of the RFID on a business, in fact, is fundamental to the prediction of a firm's adoption decisions and, in turn, to provide a better understanding of technology transfer and enhancement in a less technologically advanced industry, like the textile and clothing one (Moon and Ngai 2008).

From the book chapter, it emerges that the main obstacles to the RFID application, and which call for major research and experimentations, regard the upstream part of the textile and clothing supply network. In fact, even though RFID systems would be of great support, their introduction results difficult both for economical reasons (cost of components is very low to justify RFID introduction) and the chemical/mechanical aggression of the involved production processes.

The industry-academia projects reported in this chapter confirm how applications of RFID are more accessible in the downstream part of the network (TAGGIE project) while are still under research and of experimental nature in the upstream side (CLOTH project).

Other issues that obstacle the RFID expansion and that are current challenges to overcome are related to standards, infrastructure, cost and privacy (Davison and Smith 2005). Even though great progress has been made in last years, especially in the establishment of European standards, much work has still to be carried out as RFID is not a fully mature technology yet. Furthermore, a seamless end-to-end RFID-based supply network vision is still lacking and requires more efforts and attention for the next future.

Acknowledgments The authors wish to thank Regione Lombardia for having funded the TAGGIE and the CLOTH projects. These projects have been backed to sustain the development of SMEs and local industrial zones.

References

Angeles R (2005) RFID technologies: supply-chain applications and implementation issues. Inf Syst Manage 22(1):51–65

Chappell G, Durdan D, Gilbert G, Ginsburg L, Smith J, Tobolski J (2002) Auto-ID on delivery: the value of Auto-ID technology in the retail supply chain. Auto-ID Center. http://www.autoidlabs.org/single-view/dir/article/6/104/page.html. Accessed 18 November 2009

Choy KL, Chow KH, Moon KL, Zeng X, Lau HCW, Chan FTS, Ho GTS (2009) A RFID-case-based sample management system for fashion product development. Eng Appl Artif Intell 22:882–896

Christopher M, Lowson R, Peck H (2004) Creating agile supply chains in the fashion industry. Int J Retail Distrib Manage 32(8):367–376

Collins J (2006) Marks and spencer to extend trial to 53 stores. RFID Journal. http://www.rfidjournal.com/article/articleview/1412/1/1/. Accessed 18 November 2009

Davison J, Smith SE (2005) Retail RFID stirs process change and improves product availability. Gartner, Stamford, CT

Eckfeldt B (2005) What does RFID do for the consumer? Commun ACM 48(9):77–79

European Commission (2009) Enterprise and industry – textiles and clothing http://ec.europa.eu/enterprise/sectors/textiles/index_en.htm. Accessed 18 November 2009

Ferdows K, Lewis MA, Machuca JAD (2004) Rapid-fire fulfilment. Harv Bus Rev 82(11):104–110

Gallaugher J (2010) Zara, in Information systems: a manager's guide to harnessing technology. FlatWorld Knowledge online book, http://www.flatworldknowledge.com/pub/gallaugher?ion=true#book-undefined. Accessed 10 February 2010.

Guercini S, Runfola A (2009) The integration between marketing and purchasing in the traceability process. Ind Mark Manage 38(8):883–891

Hogan J (2003) Fashion firm denies plan to track customers. New Sci 178(2391):11

Jones MA, Wyld DC, Totten JW (2005) The adoption of RFID technology in the retail supply chain. Coastal Bus J 4(1):29–42

Kumar S (2007) Connective technologies in the supply chain. Auerbach Publications, Boca Raton, FL

Lehtonen M, Staake T, Michahelles F, Fleisch E (2007) The potential of RFID and NFC in anti-counterfeiting. In: Cole PH, Ranasinge DC (ed.) Networked RFID Systems and

Lightweight Cryptography: Raising Barriers to Product Counterfeiting, Springer, Germany, DOI: 10.1007/978-3-540-71641-9_11

Loebbecke C, Huyskens C (2006) Weaving the RFID yarn in the fashion industry: the Kaufhof case. Manage Inf Syst Q Exec (MISQE) 5(4):169–179

Loebbecke C, Palmer JW (2006) RFID in the fashion industry: Kaufhof department stores AG and Gerry weber international AG fashion manufacturer. Manage Inf Syst Q Exec (MISQE) 5(2):15–25

Metro AG (2003) Metro group future store initiative website. www.future-store.org. Accessed 18 November 2009

Moon KL, Ngai EWT (2008) The adoption of RFID in fashion retailing: a business value-added framework. Ind Manage Data Syst 108(5):596–612

Newstex (2007) Now your clothes will be washed under watchful eyes of RFID. RFID Weblog, 17 October 2007

Ngai EWT, Moonb KKL, Rigginsc FJ, Yib CY (2008) RFID research: an academic literature review (1995–2005) and future research directions. Int J Prod Econ 112:510–520

Niederman F, Mathieu RG, Morley R, Kwon I-W (2007) Examining RFID applications in supply chain management. Commun ACM 50(7):92–101

Politecnico di Milano (2007) RFID: alla ricerca del valore – Rapporto 2007 Osservatorio RFID. www.osservatori.net (in Italian) (2007). Accessed 18 November 2009

Pradhan S et al (2005) RFID and sensing in the supply chain: challenges and opportunities. http://www.ship2save.com/page_images/wp_hp_rfid_supplychain.pdf. Accessed 18 November 2009

Ranasinghe DC, Cole PH (2008) Addressing insecurities and violations of privacy, Networked RFID system and lightweight cryptography: raising barriers to product counterfeiting. Springer, Heidelberg

Redman TC (1996) Data quality for the information age. Artech House, Norwood, MA

Rekik Y, Sahin E, Dallery Y (2009) Inventory inaccuracy in retail stores due to theft: an analysis of the benefits of RFID. Int J Prod Econ 118:189–198

Roberti M (2006) RFID is fit to track clothes. RFID Journal. http://www.rfidjournal.com/article/articleview/2195/1/1/. Accessed 18 November 2009.

Sangani K (2004) RFID sees all. IEE Rev 50(4):22–24

Shim R (2003) Benetton to track clothing with ID chips. CNET, 11 March 2003. http://news.com.com/2100-1019-992131.html. Accessed 18 November 2009

Singh SP, McCartney M, Singh J, Clarke R (2008) RFID research and testing for packages of apparel, Consumer goods and fresh produce in the retail distribution environment. Packaging Technol Sci 21:91–102

Sullivan L (2005) Apparel maker tags RFID for kids' sleepwear. Information Week. http://www.informationweek.com/news/mobility/RFID/showArticle.jhtml?articleID=165701942. Accessed 18 November 2009

Tajima M (2007) Strategic value of RFID in supply chain management. J Purch Supply Manage 13:261–273

Teresko J (2003) Winning with wireless. Ind Week. http://www.industryweek.com/articles/winning_with_wireless_1262.aspx. Accessed 5 August 2009

Want R (2004) The magic of RFID, Intel research, www.acmqueue.com, October (2004). Accessed 18 November 2009

Wilding R, Delgado T (2004) RFID – applications within the supply chain. Supply Chain Pract 6(2):30–43

Wong KHM, Hui PCL, Chan ACK (2005) Cryptography and authentication on RFID passive tags for apparel products. Comput Ind 57:342–349

www.allbusiness.com. Accessed 18 November 2009

www.grivaonline.com. Accessed 18 November 2009

A New Security Paradigm for Anti-Counterfeiting: Guidelines and an Implementation Roadmap

Mikko Lehtonen

Abstract Product counterfeiting and piracy continue to plague brand and trademark owners across industry sectors. This chapter analyses the reasons for ineffectiveness of past technical anti-counterfeiting strategies and formulates managerial guidelines for effective use of RFID in anti-counterfeiting. An implementation roadmap toward secure authentication of products tagged with EPC Gen-2 tags is proposed and possible supply chain locations for product checks are discussed.

1 Introduction

Since the beginning of the history of manufacturing, man has always copied the creations of other men. For example, amphorae used to transport wine from Italy to Gaul around 200BC used to bear stoppers with the wine merchant's mark, but these stoppers got faked by Roman counterfeiters.[1] Accordingly, the need to protect trademarks and inventions has been long acknowledged. In 1544, Charles V of France passed a verdict saying that anyone who falsified marks of authenticity of the famous Flemish tapestry would have his right hand chopped off (Hopkins et al. 2003).

To fight counterfeiting today, corporate security officers apply what is essentially the very same mitigation strategy as their ancestors used to protect amphora stoppers and Flemish tapestry: mark genuine articles with high-tech labels (e.g. holograms, security inks, color-shifting films, special printings, or microscopic taggants) and let corporate lawyers prosecute the infringers. However, despite the abundance of high-tech security features since 20 years, customs statistics (Taxation and Customs

---

M. Lehtonen (✉)

ETH Zürich, Information Management, Scheuchzerstrasse 7, 8092 Zürich, Switzerland

e-mail: mlehtonen@ethz.ch

[1] http://www.wipo.int/wipo_magazine/en/2009/01/article_0009.html

D.C. Ranasinghe et al. (eds.), *Unique Radio Innovation for the 21st Century*,
DOI 10.1007/978-3-642-03462-6_15, © Springer-Verlag Berlin Heidelberg 2010

Union 2008) show that the extent of product counterfeiting has been nothing but skyrocketing.

Brand owners might receive relief to counterfeiting from RFID which is currently being adopted as a new Auto-ID technology across industry sectors. It has especially strong advocates in the consumer goods and retail industry where many potential benefits are asserted (Leimeister et al. 2007). Since many products will be tagged with low-cost RFID[2] anyway, it makes practical sense to use the technology also to fight product counterfeiting. However, it is not yet known if and how low-cost RFID can achieve a major impact in anti-counterfeiting, instead of becoming "just another security feature" among hundreds of others. The central thesis of this chapter is that RFID can make a major impact in anti-counterfeiting, but this requires a paradigm shift in the way managers think about security.

This chapter is organized as follows. The following section gives a brief view to the status quo of RFID security. Section 3 presents general guidelines about security in anti-counterfeiting and Sect. 4 argues why past high-tech labeling strategies have failed to be effective and how a new security paradigm could increase the effectiveness. Last, Sects. 5 and 6 provide guidelines for effective use of RFID in anti-counterfeiting by presenting an implementation roadmap toward secure product authentication and by analyzing different supply chain locations for authenticity checks.

2 RFID Security – Where we are Today

RFID is being used to identify pallets, cases, single articles, venue tickets, transport cards, animals, skis, fixed assets etc. In many RFID applications cloning and impersonation of tags could be attractive to hackers and criminals while being harmful for licit businesses' reputation and revenue. This gives rise to requirements to mitigate the threat of tag cloning and impersonation.

Security engineers do not consider today's low-cost RFID tags (e.g. EPC Gen-2) secure against tag cloning attacks, though also simple tags can provide practical hurdles that make cloning harder. One option is to use the chip's static ACCESS and KILL passwords, but these schemes are vulnerable to eavesdropping and pose some delicate implementation challenges (Juels 2006; Koscher et al. 2008). Another option is to use the chip's unique hardware number, such as the transponder ID (TID) number, to verify the chip's identity. This method has been used by Pfizer to fight fake Viagra,[3] but it is not bullet proof since it can be fooled by an adversary who has access to chip manufacturing technology or a $10 tag impersonation device (Lehtonen et al. 2009d).

RFID chips can be secured against cloning attacks also with chipmakers' proprietary cryptographic protocols. However, a series of severe attacks against

---

[2]This article focuses on EPC Class-1 Gen-2 tags
[3]http://www.rfidjournal.com/article/articleview/2075/1/9/

proprietary protocols between 2005 and 2008 (namely: SpeedPass™ (Bono et al. 2005), MiFare Classic (Courtois et al. 2008), and Keeloq (Bogdanov 2007)) demonstrated the risks of proprietary protocols and evoked the need for strong standard cryptography.

Strong standard cryptography is available in RFID tags in the HF frequency band (13.56 MHz), but not yet in tags in the UHF frequency band (868–956 MHz) that are often used in logistic applications. However, advanced encryption standard (AES) (Feldhofer et al. 2005) and elliptic curve cryptography (ECC) (Hein et al. 2008) have been demonstrated on silicon in a way that complies with the rigid chip size and energy consumption requirements of passive UHF tags, though these have not yet been elaborated into market-ready products that also address side-channel attacks.

In addition to cryptography, RFID tags can be authenticated with physical unclonable functions (PUFs) that provide the functionality of a crypto unit on the tag with a small hardware overhead. This approach has already been commercialized on HF tags (Devadas et al. 2008; Tuyls and Batina 2006) and it will compete against cryptographic approaches in the RFID security market – with an apparent cost advantage.

In addition to the above described approaches that aim at making cloning of tags hard, tag cloning attack can also be mitigated by a system that reliably detects cloned tags. These techniques correspond to intrusion detection systems. The starting point is checking the validity of serialized ID numbers (e.g. SGTIN) (Koh et al. 2006) and the second step is reasoning whether any of the read events are generated by cloned tags (Lehtonen et al. 2009b; Mirowski et al. 2008; Staake et al. 2005). In addition, the RFID tags' user memory can be used to store renewable random numbers, so called *synchronized secrets*, that a back-end server can verify to detect tag cloning attacks (Lehtonen et al. 2009c).

3 What Managers Need to Know About Security

We now put RFID security aside for a while and approach security engineering from a holistic point of view. The presented guidelines provide a background for understanding technical anti-counterfeiting measures as security applications.

Though security is easy to understand as a synonym for protection in everyday language, security cannot be analysed without formal definitions that make underlying assumptions explicit. Security always involves an *asset* ("anything that has value to the organization[4]") that is protected against *threats* ("potential cause of an unwanted incident, which may result in harm to a system or organization") caused by adversaries. In addition to the direct protection, security *deters* the adversary by decreasing his or her expected payoff. This is demonstrated in Fig. 1.

---

[4]This can be information, software, physical, service, people, or intangible

Fig. 1 General security concepts

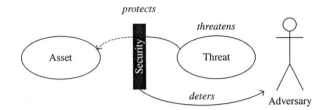

3.1 Security is About Trade-Offs and Practice

An optimal product authentication system would have only one security feature that is highly secure, not expensive, and that can be easily verified. However, security never comes for free and a security application always needs to balance between the cost, level of security, and performance and usability of the solution (Eisenbarth et al. 2007; Schneier 2003). These fundamental trade-offs are also strongly present in the design decisions of anti-counterfeiting applications (cf. Fig. 2). For instance, machine-readable security features (high usability) often fail in practice because of the high cost of the feature and the reader, high complexity of reader control and distribution, and because the system can be compromised if readers get fall into the wrong hands (Hopkins et al. 2003).

These trade-offs explain why the challenges behind the research on cryptographic RFID tags revolve around balancing between tag cost, level of security, and performance in terms of reading distance and speed (electricity consumption and clock cycles). Though the research community always provides incremental improvements to the aforementioned trade-offs, advances in technology also make cracking of devices easier for potential adversaries, for instance by decreasing the cost of computing power.

Managers also need to understand that the level of protection that a countermeasure provides is determined by the way the countermeasure is used in practice. The importance of correct usage has been recognized in several scholarly papers (Bishop 2003; Schneier 2000, 2003) and it is the Achilles' heel of many real-life security systems. In general, security mechanisms (e.g., passwords, encryption,

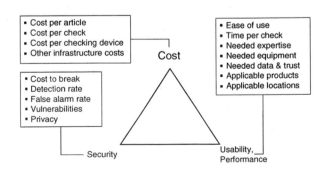

Fig. 2 Major trade-offs in technical anti-counterfeiting measures

firewalls, antivirus software, intrusion detection systems, etc.) have an "intrinsic level of security" that is characterized by its cost-to-break or detection rate, but they need to be used in a correct way to be effective in practice (e.g. passwords need to be kept safe, doors must be kept locked, etc.). In technical anti-counterfeiting systems this means that the check not only needs to be reliable (high intrinsic security), but it also needs to be applied to potential counterfeit products (high check rate).

3.2 Product Authentication is a Process

Product authentication is the process of verifying that a product under study has the claimed identity. This process inherently deals with uncertainty and different product authentication techniques lead to different levels of confidence to the result. This principle is employed in practice, for example, by first verifying an overt security feature (e.g. a hologram), then a semi-covert or a covert feature (e.g. color-shifting ink), and last a forensic feature (e.g. molecular markers) (CACP 2009), illustrated in Fig. 3.

Product authentication has four possible outcomes. Favorable outcomes for the brand owner are that a counterfeit product raises an alarm (true positive) and that a genuine product passes the check without alarm (true negative). On the flip side, a counterfeit product can pass the check without raising an alarm (false negative) and a genuine product can raise an alarm (false alarm). The probability that a counterfeit product is detected characterizes the level of security of a check and the false alarm rate represents a cost factor invoking unnecessary inspections.

Fig. 3 Level of confidence and verification steps in a product authentication process

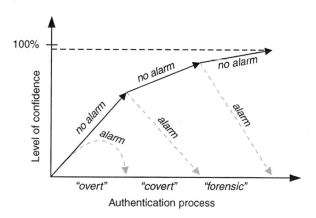

3.3 Securing a Supply Chain is a Process

Securing a supply chain from counterfeits is not only about product authentication, but a process that combines technical, organizational, and legal countermeasures.

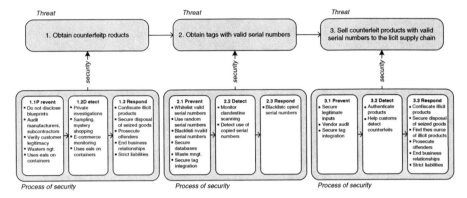

Fig. 4 The integrated process of securing a supply chain against counterfeit products

This process can be constructed by identifying the counterfeiter's course of action and then identifying which preventive, detective, and responsive countermeasures illicit actors can apply to mitigate these actions.

The counterfeiter's course of action includes (i) obtaining counterfeit products, (ii) obtaining tags with valid serial numbers, and (iii) selling the counterfeit product to the licit supply chain. Obtaining valid serial numbers is in fact not a mandatory step for the counterfeiter, but not doing it enables the counterfeit product to be easily detected based on invalid serial numbers. Current best practices regarding organizational and legal countermeasures are published by Staake (2007) and the Coalition Against Counterfeiting and Piracy (CACP 2009).

Figure 4 illustrates the resulting integrated process of securing a supply chain that combines mass serialization-based technical countermeasures, as well as organizational and legal measures. This model shows how a supply chain is secured through three multiple prevent-detect-respond processes that each make counterfeiting hard. By revealing a broad set of possible points of intervention for licit actors, the model gives a comprehensive view of the available measures that licit actors can apply to secure a supply chain from counterfeits.

4 Paradigm Shift for Security in Anti-Counterfeiting

An investment in a technical anti-counterfeiting system is an investment in security. The traditional way to think of security in anti-counterfeiting – so called *traditional security paradigm* – is to equate security with the effort to clone or forge the security feature. Thus, affected companies that invest in technical countermeasures pay for cost-to-break and, indeed, modern anti-counterfeiting technologies (e.g. laser surface analysis, color-shifting inks, taggants, microscopic printings, serialized 3D holograms, micro wires, etc.) provide a high cost-to-break.

A market pull for a high cost-to-break is also echoed by affected companies' reasons for investing in technical countermeasures. These include removal of the uncertainty about goods' origins, reliable evidence for legal cases, and decrease liability by demonstrating actions. Though these reasons are valuable for affected companies, they only deal with the symptoms of the problem instead of the actual problem itself. This reveals a shortcoming in the traditional security paradigm for anti-counterfeiting: it focus on making products "copy-proof" but it does not sufficiently address usability and performance aspects which contribute to high check rate and make a technical solution effective in practice.

Anecdotal evidence of low check rates support the argument that many existing technical countermeasures fail regarding effectiveness. According to statements of affected brand owners, in today's supply chains only a very small ratio of products are checked for authenticity, the checks are not systematic but sporadic, and counterfeit products are rather detected accidentally than as a result of explicit authenticity checks. Even customs inspect only less than one per cent of consignments.

Counterfeiters' reactions to security features represent further evidence of low check rates. The argument goes that *if the risk that a counterfeit product is inspected is low, counterfeiters do not need to clone or imitate the security features of genuine products*. Indeed, empirical evidence suggests that many counterfeit products can be distinguished even without security features. Staake (2007) analysed the characteristics of 128 counterfeit articles by interviewing experts in over ten product categories. The results suggest that up to 87.4% of fake products could be recognized by trained inspectors based on visual quality (cf. Fig. 5).

Shortcomings in the traditional way to think of security in anti-counterfeiting can be overcome with a paradigm shift. Since the functional goal of a product authentication system is not to make products "copy-proof" but to detect counterfeit products, security of a technical anti-counterfeiting solution should rather be equated with the system's capability to detect counterfeit products. This represents a profound paradigm shift for what security means in anti-counterfeiting.

The new paradigm focuses on securing a supply chain instead of a product, and it can be applied both to licit and illicit supply chains that have different possible points of intervention (cf. Sect. 6). The qualitative goal is a high counterfeit product detection rate (P_{det}) which is equal to the probability that a counterfeit product is inspected (P_{check}) times the probability that a counterfeit product is detected in an inspection ($P_{reliability}$). This metric represents both the direct effect of security and the impact on a counterfeiter's business model. Moreover, the new paradigm concentrates the practical level of protection provided by the security measure since

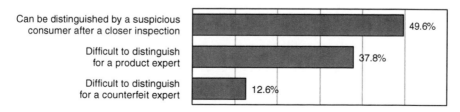

Fig. 5 Visual quality of counterfeit products, $N = 128$ (Staake 2007)

Table 1 Traditional and new paradigm for security in anti-counterfeiting

	Traditional security paradigm	New security paradigm
Secured assed	• Product	• Supply chain
Threat	• Cloning of the genuine product	• Injection of a counterfeit product to the secured supply chain
Qualitative goals	• Make genuine products "copy-proof" • Remove uncertainty of goods' origins • Reliable evidence for legal cases • Decrease liability • Visual value-added feature	• Effective detection of counterfeits • Make counterfeiting financially unattractive
Quantitative goals	• $P_{reliability} = 100\%$ • High cost-to-break	• $P_{reliability} =$ "good enough" • $P_{check} = 100\%$

it also takes into account the usability and performance aspects. The differences between these two paradigms are summarized in Table 1.

4.1 Low-Cost RFID and the New Security Paradigm

Operationalizing the new security paradigm shift requires decreasing cost and effort to conduct the checks, which is why RFID appears a particularly effective tool for anti-counterfeiting. Moreover, a solution based on standard RFID tags and readers could overcome the problems of dedicated checking equipment and hardware vendors. Combined with the ability to check multiple products at once without line of sight, this can lead to a dramatic increase in check rate.

Integration of authentication to identification is a potentially powerful tool to achieve the augmented check. It means authenticating every product that is identified. Figure 6 illustrates the required protocol-level changes that include an initialization phase, where a secure connection is established to a back-end server,

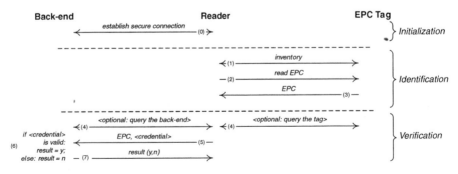

Fig. 6 Illustration of protocol-level changes when integrating authentication and identification

and a verification phase, where optional credential is gathered from the tag and sent to the back-end. Measurements on a prototype implementation suggest that a verification phase that involves both back-end-to-reader and reader-to-tag communication (step 4 in Fig. 6) can increase a tag's processing time by 180% (Lehtonen et al. 2009c). This time overhead can decrease the applicability of certain approaches in processes where bulk reading is needed.

By moving from static hard-to-copy features to detection-based authentication (e.g. track and trace checks), brand owners can steer away from the cat and mouse game of updating security features as they get forged by counterfeiters. The downside of detection-based measures is the uncertainty caused by incomplete visibility in cases where the trace data is not complete. Our simulator studies demonstrate that this uncertainty causes false negatives ($P_{reliability} < 100\%$) and false alarms when detecting cloned tags from track and trace data, though most of the false alarms can be eliminated with a filtering algorithm that detects missing reads (Lehtonen et al. 2009b).

The practical implication is that an alarm needs to be preceded by another (e.g. manual) inspection to ascertain the origin of the product. Though this represents additional effort, the number of needed inspections can be minimized using the synchronized secrets method that triggers only as many alarms as there are cloned tags that have entered the RFID system (Lehtonen et al. 2009c).

Regarding the cost of the solution, RFID is a relatively expensive anti-counterfeiting technology. Even low-cost RFID tags still cost ten times more than typical security features (e.g. about 10 cents instead of 1–3 cents), without considering other infrastructure cost. However, since RFID is a platform technology that enables multiple Auto-ID applications, only a part of the tag cost needs to be accounted to anti-counterfeiting applications. Brand protection can thus be perceived as an additional way to depreciate the overall RFID investment.

5 Roadmap Toward Secure Product Authentication with RFID

Researchers have come up with several approaches how RFID-tagged products can be authenticated (Juels 2006; Lehtonen 2008), providing different levels of protection at different costs. The needed level of protection partially depends on how much resources adversaries invest in breaking or bypassing the security measure for that particular product; a too strong security measure (e.g. a cryptographic RFID tag on a pack of chewing gum) is a waste of money, while too weak protection represents a security risk. However, it is hard to know which measures are needed to enable secure product authentication (i.e. a high-enough $P_{reliability}$) for different products.

To support brand owners in choosing suitable authentication approaches for different kinds of products, we have constructed a roadmap toward secure product authentication with EPC Gen-2 technology. The starting point of all approaches is identification of the tagged object and verification that the object's serialized identifier (e.g. SGTIN) is *valid*. This approach only requires a *white list* (Koh et al. 2006) where the valid identifiers are stored.

The basic measure is not secure as such, but it can be turned into secure product authentication by addressing tag cloning and tag removal and reapplying attacks. The former is addressed by security goals of increasing tag cloning resistance and detecting cloned tags and the latter by increasing tag-product integrity. For products where the risk due to counterfeiting is low, such as some non-branded fast moving consumer goods, the basic measure provides a good starting point. For products where the risk due to counterfeiting is augmented, such as patented life-saving drugs or airplane spare parts, the need for security is higher and the first technical countermeasure should already include some more advanced security measures, such as track and trace checks or cryptographic tags.

The aforementioned three security goals correspond to three groups of security measures which are presented as different axes in Fig. 7. The graph on left presents the available security measures for UHF RFID tags (e.g. EPC Gen-2). Since some of the security measures can be implemented also using serialized barcodes (e.g. data matrix), the corresponding graph for serialized barcodes is included to the right to provide a comparison.

Overall, UHF RFID provides a broader set of available approaches to enable secure product authentication. In particular, cloning resistance of barcodes is restricted to copy detection patterns that are computer generated images whose copies can be detected based on changes in noise patterns, but they can hardly be used reliably without a deep understanding of printing processes so they are not plug-and-play solutions (Picard 2001). This means that UHF RFID has a greater potential to enable secure product authentication than serialized barcodes. The following sections detail the security measures available for UHF RFID.

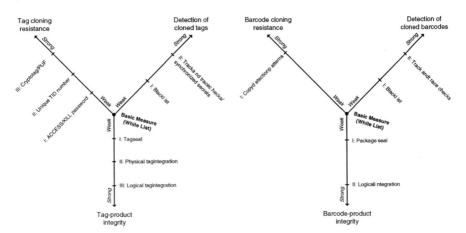

Fig. 7 Implementation roadmap toward secure product authentication with UHF RFID tags (*left*) and serialized barcodes (*right*)

5.1 Tag Cloning Resistance

Traditional security measures mitigate tag cloning attacks by increasing the tag cloning resistance. Two password-protected commands of Gen-2 tags, KILL and ACCESS, can be used as basic tag authentication techniques, though they are vulnerable to eavesdropping (Juels 2006). Implementation of KILL-password based authentication is feasible in deployed tags, but presents some delicate technical challenges (Koscher et al. 2008). Overall, these approaches are vulnerable to brute force attacks of the 32-bit passwords and eavesdropping.

Also unique factory programmed read-only Transponder ID (TID) numbers can increase the cloning resistance of Gen-2 tags, but the TID only represents a practical hurdle against tag cloning and does not provide secure authentication in the long-term (Lehtonen et al. 2009d). The highest level of tag cloning resistance is achieved with cryptographic tags (Feldhofer et al. 2005; Hein et al. 2008) or physical unclonable functions (PUF) (Devadas et al. 2008; Tuyls and Batina 2006) that have been demonstrated on silicon and HF tags, and that can become available on UHF tags (e.g. EPC Gen-2) in the future.

5.2 Detection of Cloned Tags

The first detection-based measure is about *black listing* identifiers of those products who have been sold, consumed, or otherwise become invalid and thus cannot pass the check. This measure restricts the time window when an identifier is valid and thus can be exploited by a counterfeiter to fool the basic measure. More advanced detection-based measures include track and trace checks (Lehtonen et al. 2009b) and so called synchronized secrets approach that can detect cloned tags based on visibility (Lehtonen 2008).

Compared to simple black listing, these techniques can also detect cloned tags before the corresponding identifier is black-listed, e.g. when the genuine product is still in the supply chain. While track and trace checks do not increase a product's processing time, the synchronized secrets protocol introduces approximately a 180% time overhead compared to simple identification.

5.3 Tag-Product Integrity

Tag-product integrity guarantees that a tag is attached to the right product. Sealing of a tag to the product or its packaging is a straightforward way to improve tag-product integrity because it makes removing and reapplying genuine tags harder.

In addition, state-of-the-art techniques allow secure integration of RFID tags to various physical products depending on the characteristics of the product, for example inside watches that are fully made of metal (though not with standard Gen-2 tags) (Cook et al. 2008). Physical tag integration can make the tag hard to find, hard to remove without breaking the tag and/or the product, and hard to reapply to a counterfeit product.

The strongest tag-product integrity is achieved through logical tag integration where the product's unique features are stored in the tag and verified to establish that the tag is attached to the right product (Nochta et al. 2006).

5.4 Risk Management

This section summarizes the risks relating the aforementioned RFID-based security measures and how they can be mitigated with preventive and responsive actions. The results are summarized in Table 2.

6 Supply Chain Locations for Product Authentication

An RFID-based product authentication system can be used in various supply chain locations covering check points both in licit and illicit supply chains. The choice these locations has a major influence on which part of the problem the technical solution can address.

We have identified and analysed possible supply chain locations for product authentication by gathering and clustering possible usage scenarios from SToP and BRIDGE projects and real-life examples (cf. Fig. 8). The identified locations are mapped to a generic model of licit and illicit supply chains spanning from manufacturer to consumer/end-user, excluding parts' and components' suppliers. The analysis covers the following dimensions: (i) product handling in current business processes, (ii) integration of authenticity checks to existing Auto-ID processes, and (iii) how checks in this location affect counterfeiting.

6.1 Inside Distribution

Counterfeit products can enter the licit supply chain in the distribution level between manufacturing and retail (Pfizer 2007). Counterfeits can appear either as complete consignments of faked goods or co-mingled with genuine goods (CACP 2009). Authenticity checks in the distribution level, e.g. in distribution centers, help detecting these counterfeits. Since logistic units are identified using Auto-ID inside the distribution level today, the existing Auto-ID processes provide an opportunity for integration of automatic and systematic authenticity checks.

When products are handled in known lot sizes or one by one (e.g. luxury goods), untagged counterfeit articles can be detected without additional effort to count the verified products. Another important efficiency factor is the relatively small number of distributors (e.g. compared to the number of retailers); when all genuine products flow through a relatively small amount of supply chain locations, the whole

Table 2 Risk matrix of RFID-based security measures

Security measure	Risks	Actions
Basic measure (white list)	∎ Product ID number copied	∎ Detect cloned tags and/or increase tag cloning resistance
Black list	∎ List not updated, counterfeit product not detected	∎ Systematic updates e.g. always when product is sold/consumed
Track and trace checks	∎ False alarms ∎ False negatives	∎ Maintain an accurate model of how genuine products generate events ∎ Maintain a high level of visibility, eliminate location uncertainty
Synchronized secrets	∎ Cloned tag passes the check without raising an alarm, alarm is raised by the genuine tag ∎ Manipulation of a genuine tag's synchronized secret (denial of service)	∎ Have a high tag read rate compared to the time an adversary needs to clone and inject a tag ∎ Secure the synchronized secret from reading and writing using an ACCESS password
ACCESS/KILL password	∎ Eavesdropping ∎ Brute force attack	∎ Secure the reader environment ∎ Use tags secured against brute force attacks with an internal timer and counter
Unique TID number	∎ Tag impersonation device ∎ Manufacturing of programmable tags	∎ Inspect tags physically ensure no impersonation device is used ∎ Survey the market if programmable tags become available
Crypto tag	∎ Cryptanalysis ∎ Side-channel attacks ∎ Physical attacks	∎ Use standard cryptography and long-enough keys ∎ Use tags secured against side-channel- and physical attacks ∎ Use tag seals and/or physical tag integration
PUF	∎ Removal and reapplying	∎ Use tag seals and/or physical tag integration
Tag seal	∎ Removal and reapplying ∎ Seal is broken ∎ Seal is not verified	∎ Use strong-enough seals ∎ Systematic verifications of seals
Physical tag integration	∎ Tag is detected and removed and reapplied	∎ Hide tag inside product ∎ Secure tag integration (tag is destroyed upon attempted removal) ∎ Customize tags and products to make reapplication hard
Logical tag integration	∎ Object-specific feature can be imitated	∎ Use truly unique and random object-specific features with high-granularity measurements

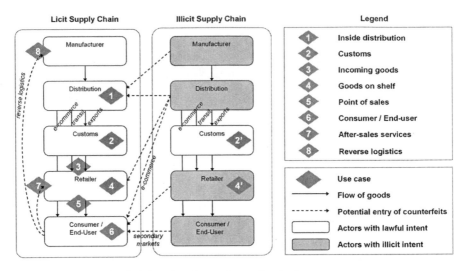

Fig. 8 Possible supply chain locations for product authentication

product population can be authenticated with a smaller number of check locations. Furthermore, authenticity checks inside distribution can detect the counterfeit products as soon as they enter the licit supply chain, increasing the chances of detecting infringers. However, counterfeit products can enter the supply chain also further downstream.

When the brand owner or manufacturer does not have its own distribution network but relies on external parties (i.e. external licit supply chain), active collaboration of the external distributors is required. Obtaining the needed commitment from external distributors can be challenging since distributors might not get direct business benefits from conducting authenticity checks.

6.2 Customs

Customs conduct most counterfeit seizures in the world and thus it is a key stakeholder in any anti-counterfeiting strategy. Though anti-counterfeiting is not the number one priority for customs, customs is often the best locations to interfere the illicit supply chain. Therefore supporting customs in anti-counterfeiting not only protects the licit supply chain from counterfeit products but also affects the illicit supply chain, inflicting a broad effect on counterfeiters' business.

Brand owners can collaborate with customs by offering training and tools to detect counterfeit products. However, customs are reluctant to adopt multiple devices to authenticate different kinds of products. Though many brand owners provide customs with different devices to authenticate their devices, customs officers rarely know or remember how to use them. Rather, a simple standard solution that can handle different kinds of products is strongly preferred. Such a standard solution does not exist today and currently hundreds of different product authentication solutions are being used by affected companies, but integration of authentication to Auto-ID technologies such as EPC/RFID has potential to change it.

Today customs verify less than one per cent of imported consignments. These checks are sporadic and not coupled with regular goods handling. As a result, a system that is able to authenticate one good at a time is sufficient. In addition, customs need mobile or hand-held verification devices since inspections are conducted not only in customs warehouses, but also on highways, in company warehouses, and other remote locations. Sporadic checks of single samples helps customs identify counterfeit consignments faster and easier, but they are not effective in detecting small quantities of counterfeit articles that are co-mingled among genuine products. Last, a solution adopted by customs does not need to provide a 100% level of confidence since the affected brand owner is in the end responsible of proving the clandestine origins of seized products.

6.3 Incoming Goods

Retailers are in a critical position to engage in countermeasures against product counterfeiting. The retail level comprises typical consumer good retailers and other end-points such as pharmacies, hospitals, as well as small boutiques and garages. Authenticity checks in the retail level can be integrated to the process where incoming goods are scanned in to the inventory before placing them to the back room storage or shop-floor. Authentication of incoming goods in the retail level is potentially a very effective way to secure the licit supply chain.

When incoming goods are subject to verifications in the existing processes, such as expiry data verification and order completeness verification, the overhead of integrating an authenticity check to the existing process can be done with a small overhead. This has been demonstrated in the pharmacy trial of the SToP project (Lehtonen et al. 2009a). A small overhead is also a requirement when the process of scanning in incoming goods is time-critical. Furthermore, when the lot sizes of incoming goods are fixed or otherwise known, detection of untagged counterfeit products can be automated.

A general downside with authenticity checks in the retail level is that the counterfeit products are detected in a late point in the supply chain, which makes tracing the source of counterfeit goods harder. Also, more check points are needed to cover the same product flow than in upstream locations.

6.4 Goods Available for Sale

Authenticity checks can secure the retail level from counterfeits also through verifications of goods available for sale on shelves and on shop-floor. This can be done either with the consent of the retailer, for instance as an audit by the brand owner, or without the consent of the retailer, for instance as mystery shopping by a private investigator to gather evidence of the counterfeit penetration within a market. A prerequisite for these checks is that the verified products are openly displayed, which restricts application of this scenario mostly to consumer goods. This restriction, however, can be overcome by conducting test purchases and authenticating the bought samples afterward. Therefore this usage scenario also covers test purchases through e-commerce. In principle, also consumers can authenticate goods on retail shelves if they are empowered with the needed technology and have an incentive to do so.

Checks of goods on shelves are sporadic and can be targeted to suspicious or high-risk targets for an increased effectiveness. It is not likely that these checks can be conducted as a part of other processes where goods are identified or otherwise verified, and therefore they represent additional effort. But this effort might be worth while since, together with checks in customs, authentication of goods on shelves is the only way to interfere with the illicit supply chain (excluding infiltrating private investigators among the illicit supply chain).

An RFID-based solution with a large read range and a bulk reading mode suits this usage scenario especially well since it enables quick and imperceptible verifications of goods on retail shelves. In order to detect untagged counterfeit items, however, the number of verified items must be manually counted. In addition, since this check is conducted at a late stage of the supply chain, tracking down the sources of detected counterfeit goods can be hard.

6.5 Point of Sales

Authenticating products at the point of sales or at the point of consumption (e.g. a drug that is consumed in a hospital) secures the last link of the licit supply chain. At this final step of the distribution channel products are handled one by one so no additional effort is needed to count the verified articles so as to detect untagged counterfeit articles. Furthermore, point of sales systems already identify products with Auto-ID (e.g. to scan the price, to verify the expiration date of drugs). These two conditions can minimize the additional effort needed to integrate authenticity checks to the process how products are handled.

On the other hand, introducing systematic authenticity checks in the point of sales level is challenging. Foremost, authenticating products in front of the consumer, patient, or end-user interferes with the customer relationship. For example, authentication of pharmaceutical products in front of the patient can decrease the trust toward the doctor or pharmacist, and authentication of luxury goods can "*break the romance*" of the carefully designed buying experience. In general, retailers do not want to deal with product counterfeiting issues in front of their customers since it can generate negative associations among customers. In particular, this usage scenario can be the first time when customers learn that counterfeit products could appear in the retail shop. The retailer's dilemma is that the associations are perceived as negative, though the authenticity checks arc conducted for the customers' own good.

Also other factors make these checks challenging. The checks take place in a time-critical process where additional delays are not welcome and they take place far from the sources of counterfeits. Last, the vast number of possible point of sales locations makes diffusion of the technology and process changes burdensome and probably possible only with standards, mandates, and regulations.[5]

6.6 Consumer/End-User

Enforcing consumers and end-users with the capability to authenticate products has a big potential in enabling secure supply of genuine goods and countering deceptive counterfeiting in regions where counterfeit products might appear in the retail level. For instance, drugs sold in Ghana have a unique numeric code that can be scratched and send with an SMS to verify their authenticity.[6] Nokia uses similar numbers in batteries,[7] and a similar system is in use in the Hong Kong Airport.[8]

Involving consumers through mobile phones could potentially empower masses of people with the ability to authenticate products in locations where brand owner cannot access otherwise, including secondary markets (c.g. flea markets, C2C sales) and new geographic areas. In addition to relying on consumers' own devices, this usage scenario can also be enabled by installing reader kiosks where consumers can authenticate products themselves. However, in this approach the consumer has no real means of assuring whether the reader kiosk is trustworthy and yields right results or not.

Despite the vast potential of involving consumers, currently this usage scenario is only very rarely utilized due to an atmosphere of denial and secrecy. In general, many brand owners believe that *"you should not involve your customers in your dirty laundry"*. Indeed, there are arguments for not involving consumers in anti-counterfeiting efforts. First, there might be a sales drop due to bad publicity and admittance of the problem. Second, the required effort and cost of empowering consumers might be too high to justify the benefits. Third, by giving consumers the capability to recognize counterfeits more consumers might turn to the secondary markets to buy second-hand products instead of new ones. And fourth, possible false alarms of the authentication technology could lead to liability claims. In addition, consumers also buy some counterfeit products intentionally, which limits this usage

---

[5]For instance, the European Federation of Pharmaceutical Industries and Associations (EFPIA) is trialing methods to meet the European Commission's new traceability requirements. http://www.efpia.org/Content/Default.asp?PageID=566

[6]http://www.mpedigree.org/home

[7]http://www.nokia.co.id/nokia/0,,82227,00.html

[8]http://www.rfidjournal.com/article/view/5022/1

scenario to those product categories where consumers have good incentives to buy the genuine product.

6.7 After-Sales Services

In some cases counterfeit goods can enter the licit supply chain in after sales services when customers return goods that are already bought. This can be a relevant scenario for example in the luxury goods industry where products are used during long periods of times and sometimes they need to be returned for repair, polishing or refurbishment. Though authentication of products in after-sales services does not prevent a consumer from getting a counterfeit product, it enables easy detection of counterfeits in an early phase of the service.

From the process point of view, authentication of these products is relatively easy since these products are handled one by one or in small quantities, within the premises of a retailer or brand owner (e.g. a luxury goods boutique). Due to the interference with the customer relationship (cf. point of sales scenario above) it might be preferable not to authenticate these products in front of the customer but rather in the back room or service level. This is also a preferable practice in those cases where the customers knowingly bring counterfeit goods to after-sales services with the hope of getting them replaced by genuine goods since a face-to-face conflict with these fraudulent customers is avoided.

From the technical point of view, lack of complete trace data limits the use of location-based authentication approaches in this usage scenario. Regarding the migration of serialization labels, this usage scenario also needs to handle non-tagged genuine products, including those product categories that are not tagged as well as older articles that were not yet tagged. Last, tracing the source of the counterfeit products detected in this usage scenario can be very hard.

6.8 Reverse Logistics

Similar to the after-sales services scenario, counterfeit products can enter the licit supply chain also through reverse logistics of products that are returned to the manufacturer under warranty. This can be an issue in particular with electronics, batteries, computer chips and mechanical components or accessories, where manufacturers are seeing an increase in counterfeit parts being returned to manufacturers under warranty and claiming replacement. Manufacturers of these products are therefore having a problem authenticating returned articles and, without appropriate technology and processes, have found themselves forced to replace a counterfeit chapter with a genuine chapter. In this case an authenticity check can be integrated in the service process on the manufacturer's side.

Compared to checks in the lowest levels of the supply chains, only a very small number checking locations is needed. The downside of this usage scenario is that it is very far from the source of counterfeits and its benefits are limited to eliminating the losses due to replaced or fixed counterfeit products.

6.9 Implications to a Technical Solution

The choice of supply chain locations where the authenticity checks are conducted restricts the applicability of certain product authentication approaches. In principle, track and trace data can be gathered only until the point of sales – though in most cases the manufacturer loses the trace of the product much earlier. This limits the applicability of location-based authentication approaches. Table 3 presents the conceptual limitations of the product authentication approaches considered in different supply chain locations.

Table 3 Feasibility of different RFID-based product authentication approaches

Location	Basic measure	Black list	Track & trace, sync. secrets	Password/TID/ Crypto tag/PUF
1. Inside distribution	Ok	Ok	Ok	Ok
2. Customs	Ok	Ok	Ok	Ok
3. Incoming goods	Ok	Ok	Ok	Ok
4. Goods available for sale	Ok	Ok	Ok	Ok
5. Point of sales/consumption	Ok	Ok	Ok	Ok
6. Consumer/end-user	Ok	Limited[9]	Limited[10]	Ok[11]
7. After-sales services	Ok	Limited[9]	Limited[10]	Ok
8. Reverse logistics	Ok	Limited[9]	Limited[10]	Ok

7 Conclusions

Though technology is ready to turn the tide in the fight against counterfeiting, its adoption is still subject to a profound paradigm shift in the way executives think about security. The current security paradigm in anti-counterfeiting concentrates on a high cost-to-break but as a result it somewhat neglects the veritable goal of a technical anti-counterfeiting system: detection of counterfeit products. By giving up some of the check reliability, brand owners could gain orders of magnitude increases in the check rate with RFID, to enable effective detection of counterfeit products in protected, licit or illicit, supply chains. Since the check reliability depends on

---

[9] In addition to copied tags, also the genuine tag will raise an alarm after blacklisting the ID

[10] Single cloned tags cannot be reliably detected once the genuine product is no longer traced, but the existence of multiple cloned tags can still be detected, especially if the number of copied tags with same ID number is high

[11] Can be made available only to trustworthy parties if the verifier learns the password/secret

the used security measures, a roadmap is presented suggesting how secure product authentication can be achieved in a war of escalation by employing more and more reliable security measures as they become necessary. Last, eight supply chain locations are identified and analysed for conducting these checks.

Acknowledgments This work has been partly funded by the Auto-ID Lab of University of St. Gallen/ETH Zürich.

References

Bishop M (2003) What is computer security? IEEE Secur Privacy Mag 1(1):67–69

Bogdanov A (2007) Attacks on the KeeLoq block cipher and authentication systems. In: 3rd conference on RFID security, volume 2007, Budaperst, Hongria

Bono S, Green M, Stubblefield A, Juels A, Rubin A, Szydlo M (2005) Security analysis of a cryptographically-enabled RFID device. In: USENIX security symposium, Baltimore, MD, USA, pp 1–16, USENIX

CACP (2009) Intellectual property protection and enforcement manual: a practical and legal guide for protecting your intellectual property rights. The Coalition Against Counterfeiting and Piracy (CACP). http://www.ipr-policy.eu/media/pts/1/Brand_Enforcement_Manual_FINAL.pdf. Accessed 24 June 2009

Cook C, Vogt H, Muller J, Dada A, Pfletschinger M, Ortel N, Molan M, Naraks A, Gourmanel F (2008) Report on integration of smart/intelligent tags in products. Deliverable D4.3 of the SToP project

Courtois NT, Nohl K, O'Neil S (2008) Algebraic attacks on the crypto-1 stream cipher in MiFare classic and oyster cards. Cryptology ePrint Archive, Report 2008/166

Devadas S, Suh E, Paral S, Sowell R, Ziola T, Khandelwal V (2008) Design and implementation of PUF-based "Unclonable" RFID ICs for anti-counterfeiting and security applications. In: IEEE International Conference on RFID, Las Vegas, Nevada, USA, 16–17 April pp 58–64

Eisenbarth T, Kumar S, Paar C, Poschmann A, Uhsadel L (2007) A survey of lightweight-cryptography implementations. IEEE Design Test Comput 24(6):522–533

Feldhofer M, Wolkerstorfer J, Rijmen V (2005) AES implementation on a grain of sand. IEE Proc Inf Secur 152(1):13–20

Hein D, Wolkerstorfer J, Felber N (2008) ECC is ready for RFID – a proof in silicon. In: Conference on RFID security, Budaperst, Hongria

Hopkins D, Kontnik L, Turnage M (2003) Counterfeiting exposed: protecting your brand and customers, 1 edn. Wiley, Hoboken, NJ

Juels A (2006) RFID security and privacy: a research survey. IEEE J Selected Areas Commun 24(2):381–394

Koh R, Schuster E, Chackrabarti I, Bellman A (2006) Securing the pharmaceutical supply chain. Auto-ID Labs White Paper, Massachusetts Institute of Technology. http://www.autoidlabs.org/single-view/dir/article/6/160/page.html. Accessed 24 June 2009.

Koscher K, Juels A, Kohno T, Brajkovic V (2008) EPC RFID tags in security applications: passport cards, enhanced drivers licenses, and beyond. Manuscript. ftp://ftp.cs.washington.edu/tr/2008/10/UW-CSE-08-10-02.PDF. Accessed 24 June 2009.

Lehtonen M (2008) From Identification to Authentication – A Review of RFID Product Authentication Techniques. In: Cole PH, Ranasinghe DC (eds) Networked RFID Systems and Lightweight Cryptography:Raising Barriers to Product Counterfeiting, Springer, Germany

Lehtonen M, Boos D, von Reischach F, Magerkurth C, Müller J, Bogataj K, Gout E, Gourmanel F, Ippisch T, Oertel N, Dada A (2009a) Final evaluation of project results accordingly to the identified requirements. Deliverable D5.4 of the EU-SToP Project, project number IST-034144

Lehtonen M, Michahelles F, Fleisch E (2009b) How to detect cloned tags in a reliable way from incomplete RFID traces. In: 3rd IEEE international conference on RFID – IEEE RFID 09, Orlando, FL, pp 257–264

Lehtonen M, Ostojic D, Ilic A, Michahelles F (2009c) Securing RFID systems by detecting tag cloning. In: Tokuda H, Beigl M, Friday A, Brush A, Tobe Y (eds) 7th international conference on pervasive computing – pervasive09. Lecture notes in computer science, vol 5538. Springer, Heidelberg, pp 291–308

Lehtonen M, Ruhanen A, Michahelles F, Fleisch E (2009d) Serialized TID numbers – a headache or a blessing for RFID crackers? In: 3rd IEEE international conference on RFID – IEEE RFID 09, Orlando, FL, pp 233–240

Leimeister J, Knebel U, Krcmar H (2007) RFID as enabler for the boundless real-time organisation: empirical insights from Germany. Int J Netw Virtual Organ 3(1):45–64

Mirowski L, Hartnett J, Williams R, Gray T (2008) A RFID proximity card data set. Technical Report of University of Tasmania. http://eprints.utas.edu.au/6903/1/a_rfid_proximity_card_data_set.pdf. Accessed on 3 March 2009

Nochta Z, Staake T, Fleisch E (2006) Product specific security features based on RFID technology. In: Saint-workshop, international symposium on applications and the internet workshops (SAINTW'06), Phoenix, AZ, USA, 23–27 January, pp 72–75

Pfizer (2007) Case study: lipitor US recall. http://media.pfizer.com/files/products/Lipitor USRecall.pdf

Picard J (2001) Copy detectable images: from theory to practice. In: NIP24: international conference on digital printing technologies and digital fabrication 2008, Pittsburgh, PA, pp 796–798

Schneier B (2000) Computer security: will we ever learn? Crypto Gram Newsletter, May 15, 2000

Schneier B (2003) Beyond fear. Thinking sensibly about security in an uncertain world. Copernicus Books, Springer, New York, NY

Staake T (2007) Counterfeit trade – economics and countermeasures. PhD thesis, University of St. Gallen. Dissertation no. 3362

Staake T, Thiesse F, Fleisch E (2005) Extending the EPC network – the potential of RFID in anti-counterfeiting. In: ACM symposium on Applied computing, Santa Fe, New Mexico, USA, 13–17 March, pp 1607–1612

Taxation and Customs Union (2008) Results at the European border – 2007. Report on community activities on counterfeiting and piracy

Tuyls P, Batina L (2006) RFID-tags for anti-counterfeiting. In: Pointcheval D (ed) Topics in cryptology – CT-RSA 2006, The cryptographers' track at the RSA conference 2006. Lecture notes in computer science. Springer, San Jose, CA, USA, pp 115–131

Green Logistics Management

Yoon S. Chang and Chang H. Oh

Abstract Nowadays, environmental management becomes a critical business consideration for companies to survive from many regulations and tough business requirements. Most of world-leading companies are now aware that environment friendly technology and management are critical to the sustainable growth of the company. The environment market has seen continuous growth marking $532B in 2000, and $590B in 2004. This growth rate is expected to grow to $700B in 2010. It is not hard to see the environment-friendly efforts in almost all aspects of business operations. Such trends can be easily found in logistics area. Green logistics aims to make environmental friendly decisions throughout a product lifecycle. Therefore for the success of green logistics, it is critical to have real time tracking capability on the product throughout the product lifecycle and smart solution service architecture. In this chapter, we introduce an RFID based green logistics solution and service.

1 Green Logistics and RFID

Green logistics applies green principles to all the stages of traditional forward and reverse logistics, i.e. throughout product lifecycle, starting from product design, material sourcing, manufacturing processes, delivery of the final product to the consumers, after sales, product return, remanufacturing/reuse and recycling.

In spite of the fact that there are few researches, many companies come to have interests in reverse logistics and green logistics (Starik and Marcus 2000). Current research suggests that only considering reverse logistical processes may not be enough to reduce the impact of business practices on the environment resulting from operations in the entire supply chain (Van Hoek 1999). Instead it is suggested that

Y.S. Chang (✉)
School of Air Transport, Transportation and Logistics, Ubiquitous Technology Application Research Center, Korea Aerospace University, Goyang, Republic of Korea
e-mail: yoonchang@kau.ac.kr

D.C. Ranasinghe et al. (eds.), *Unique Radio Innovation for the 21st Century*,
DOI 10.1007/978-3-642-03462-6_16, © Springer-Verlag Berlin Heidelberg 2010

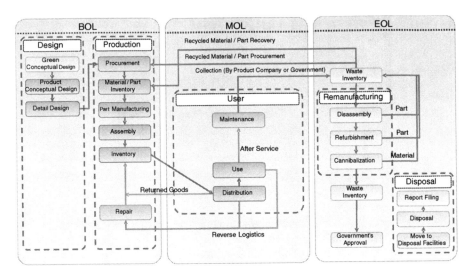

Fig. 1 Green logistics processes considering the lifecycle of a product

research emphasize the need to extend the concept of reverse logistics to green supply chains to reduce the impact of business practices on the environment. There is also research which expands the concept of the traditional one-way supply chain (e.g. forward supply chain) to a closed loop supply chain by adding end-of-life product and packaging recovery, collection, re-use, recycling and remanufacturing process (Beamon 1999; Bintrup et al. 2008; Ouertani et al. 2009).

Product Lifecycle Management (PLM) is the process of managing the entire lifecycle of a product from its conception, through design and manufacture, to service and disposal (Bintrup et al. 2008; Parlikad and McFarlane 2007). Since green logistics deals with all stages of a product, it is better to apply a complete lifecycle approach for managing and understanding green logistics.

Thus we believe that green logistics should address all the stages of a product's lifecycle which includes Beginning of Life-Cycle (BOL), Middle of Life-Cycle (MOL), End of Life-Cycle (EOL) as shown in Fig. 1.

Since the scope of green logistics includes the whole life of a product, accurate and real time provision of information along the product life span will critically impact the performance of green logistics. In such respects, Radio Frequency Identification (RFID) devices are a key aid for this new logistics paradigm.

2 Green Logistics Oriented Business Environment (GLOBE)

2.1 Overview

Green Logistics cover not only reverse logistics but also forward logistics with an environmental perspective. There has been a growing consensus on the need for intelligent green logistics in South Korea. A number of significant motivations for moving towards green logistics in Korea are as follows:

- *Importance of green initiative*: EU's Directive for Waste Electrical and Electronic Equipment (WEEE) requires that every electrical and electronic equipment must have a minimum revival rate of 80% and a minimum recycle rate of 75% based on their total weight (WEEE).
- *Track and trace capability of RFID for green logistics*: Most previous RFID applications in Korea have focused on forward logistics. Keeping product pedigree in real time is critical for end-of-life applications and RFID can play a very important role in such applications.
- *Opportunities for new solutions, services, and market*: According to recent research by Aberdeen Group (2006), many companies reported utilizing spreadsheet programs for at least some aspect of reverse logistics.

There has been a government funded RFID based green logistics research consortium called "Green Logistics Oriented Business Environment (GLOBE)" in South Korea (GLOBE Research Group). Currently there are four universities and nine companies involved. The aim of this project is to construct an RFID based intelligent green logistics solution and service which will create a new market sector and bring new processes and value-added services. Figure 2 shows an overview of research areas and projects to be developed in the consortium.

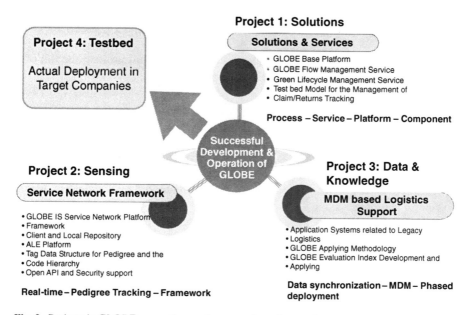

Fig. 2 Projects in GLOBE consortium and an overview of research areas

2.2 Research

In many green logistics related research efforts, the focus has been given to the optimisation of transportation, modelling, modal shift, CO_2 reduction etc. (Green Logistics 2009). However, in GLOBE, we are focusing more on the improvement of traceability and the creation of RFID supported business models in the forward and reverse supply chains.

In the planning stage of GLOBE, we have collected existing issues and requirements from industry partners for designing RFID based green logistics solutions and services.

The following is a summary of their primary issues and concerns.

- Issues in traceability both in forward logistics and reverse logistics.
- Need for solution platforms which address both forward logistics and reverse logistics: existing solution platform generally covers either forward logistics or reverse logistics. There are almost no solution which covers both forward and reverse.
- Issues in return management: insufficient visibility, long processing times for return management and the need for moving away from paper based return processes.
- Issues in after-sales service management: There are cases for insufficient visibility in service history and service part inventory.

  - Currently, product warranty is given at the entire product level even though each component within the product might have different usage and lifecycle.
  - There are illegal/counterfeit service parts for some valued replacement components.
  - Successful and efficient after-sales service process will prolong the life of a product and can improve the value of a product.

- Issues in traceability after collection process: According to the association of electronics environment of Korea, there are losses of end of lifecycle (EOL) consumer electronics products because of insufficient visibility from collection process to recycling process. Typical issues are: differences in quantities reported between collection points and recycling centres, loss of valued recycled parts (motor, compressor, circuit board etc.), illegal sale to 2nd market dealers, illegal landfill and incineration.
- Issues in reuse: there are many products disposed even though they can be reused if they can be donated or sold.
- Issues in warning: no alarming messaging service in case of urgent service situation or emergencies.
- Issues for unsold product management: unsold products can be sold by dynamic pricing or dynamic promotions suggested by decision support S/W.

2.3 Solutions and Services

Considering issues including example issues in Sect. 2.2, the GLOBE solution aim to address both forward logistics and reverse logistics by providing components and services in Fig. 3. Figure 3 shows brief technical architecture of GLOBE.

As illustrated in Fig. 3, RFID based real time events are collected from the service network platform layer. In designing the service network platform, we used the EPC Network (EPC Global Inc 2007). In Fig. 3, U-GLIS stands for Ubiquitous Green Logistics Information Service and it is identical to EPC Information Service (EPCIS) of EPC Network. A web 2.0-based platform has been developed to run

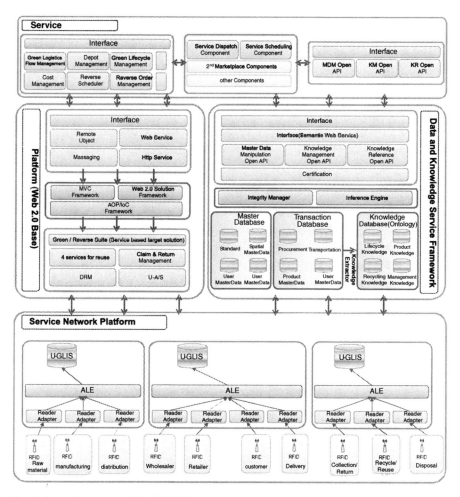

Fig. 3 Overall structure of the GLOBE solution

Fig. 4 Architecture of the
green logistics platform

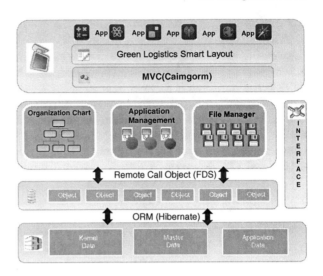

various applications. The platform applied the MVC (model-view-controller) framework and provides a suitable layout for a logistics environment (see Fig. 4). Flex
Data Service (FDS) or Life Cycle Data Service provides a programming model for
automatically managing data sets that have been downloaded to the Flex client. Once
data is loaded from the server, changes are automatically tracked and can be syn-
chronised with the server at the request of the application. We also applied Hibernate
to addresses object-relational impedance mismatch problems.

After reviewing industrial issues for several months, we designed 8 service
components for green logistics. Their brief descriptions are as follows:

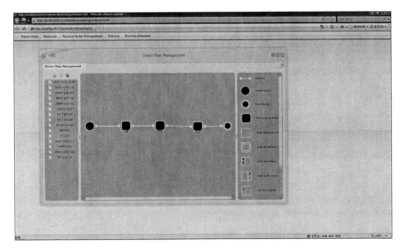

Fig. 5 Snapshot of green workflow management

- *Reverse order management service*: As in most of other business solutions, green logistics solutions start from order management. It manages different reverse logistics orders such as claim orders, after-sales service orders, recycling orders etc. (for forward logistics orders, we applied existing concepts along with most of the existing business applications).
- *Green logistics flow management service*: It defines activities and attributes of workflow in green logistics chains (Fig. 5). The service provides traceability in both forward and reverse logistics chains.
- *Green lifecycle management service*: It manages the life-cycle of a product after its manufacture. It considers activity based lifecycle management and is linked with cost management service. As in the Fig. 6, lifecycle management itself is closely related to the real time activities of a product. By this service, one can extend the life of a product with healthy condition (e.g. smart maintenance decisions such as suggestions made considering life-gauge value, reduction of disposal considering life-gauge, etc.)
- *Depot management service*: It covers functionality of general warehouse management but also considers the management of products during the after-sales service (A/S) process (e.g. temporary management of a product during its repair). It also considers management of service parts. Better management of depot will help to extend the life of a product.
- *Reverse scheduler service*: It generates work schedules based on the information from reverse order management and green logistics flow management.
- *Cost management service*: It calculates cost of a product considering lifecycle information and activity (i.e. lifecycle value).

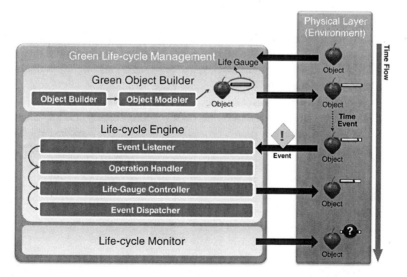

Fig. 6 Concept of green lifecycle management

- *Warranty management service*: It tracks warranty information and provide a review of warranty considering information on the cost and lifecycle information of a product such as after-sales service.
- *Notification service*: E-Mail notification service, SMS notification service and web site notification service. Using this service, an organization can respond to the event by various modes (e.g. real time service, pseudo real time service, etc).

3 Pilot Case Study: RFID Based After Service Process

We have done a pilot test on the RFID based after-sales service system for a water purifier company. The followings are summary of key user requirements:

1. Currently the test bed company has limited visibility in forward and reverse logistics process. It is hard to have overall visibility throughout the lifecycle of a product: for example, no information link between forward logistics activities (e.g. sales, delivery, etc.) and reverse logistics activities (e.g. service history, disposal information).
2. The company wants to find a greener way of service dispatch. Currently they try to satisfy customer request (e.g. Just in time service).
3. The company wants to prevent illegal distribution/use of service parts such as filters. Generally a filter costs around US$200 and they have identified the use of illegal parts from customers as a real issue.

The company uses a barcode based system to identify product types but do not use the system for product tracking (product tracking is done manually). Figure 7 shows the RFID enabled logistics flow of the test bed company. In order to achieve overall visibility, we have identified data collection points and data required. "J A/S center" stands for its after-sales service (A/S) center located at "J" city and "H logistics center" stands for its logistics center which is close to the capital city, Seoul. "Planner" is the field service force who visits customers on site and performs simple scheduled maintenance tasks such as cleaning or changing filters, etc. Service team represents a team of technicians employed by the company.

The goal of after-sales service (A/S) in the test bed company is responsiveness to customer requests (e.g. just in time service). From the user requirements, we have studied A/S types from different perspectives (i.e. not from the viewpoint of just in time) then classified A/S types as the following three:

1. remote service;
2. customer oriented service; and
3. customer support team oriented service.

As in Fig. 8, the most cost effective service is remote support. In the case of customer oriented service (see Fig. 8), the test bed company is able to achieve higher levels of customer satisfaction but it can often result in greater aggregate travelling

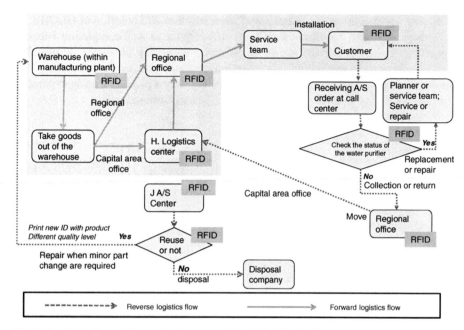

Fig. 7 Logistics flow of the test bed company (the *dashed line* indicates the reverse logistics flow and the other represents the forward logistics flow)

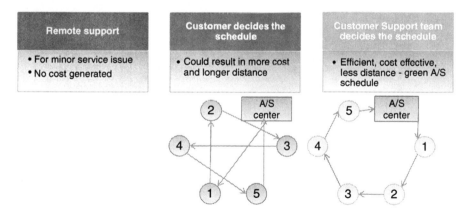

Fig. 8 Types of customer support

distances by the service team, and consequently higher costs. In the case of customer oriented support (see Fig. 8), it is easy to support just in time philosophy but it may not be easy to make efficient use of the available resources (such as the number of spares being carried in trucks, truck space not utilized completely, etc.). The last case, which we named as green A/S, results in less travel distances and cost, and more importantly allows the service company to schedule their A/S force more effectively.

Figure 9 shows the A/S processes and the relation of A/S solution of GLOBE. In A/S, there is a need for traceability from A/S request to A/S completion because sometimes customer service teams should trace necessary historical information back to purchasing (including customer information) and to manufacturing. By implementing an RFID system, they now have not only a product tracing capability but also the ability to track products individually and automatically.

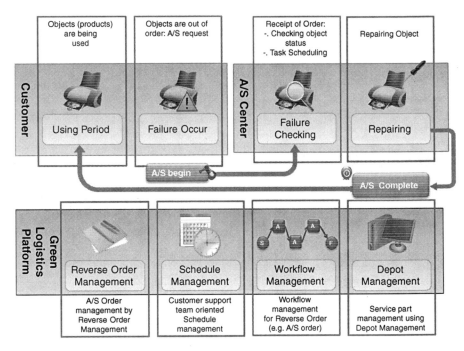

Fig. 9 GLOBE components in A/S case

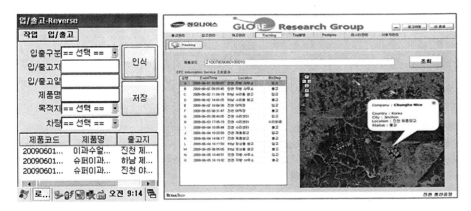

Fig. 10 Tracking of logistics (forward and reverse and both directions)

Figure 10 shows examples of the user interfaces. On the left, in Figure 10, is a PDA screen which shows reverse logistics information and on the right is GUI (graphical user interface) for providing traceability information on both the forward and reverse logistics.

Another critical issue in A/S is service part management and the distribution of counterfeit/fake service parts. In water purifier business, the cost of the filter is over US$200 and provides an adequate financial incentive for counterfeiters. Depot management solution is used for service part management and for the management of a returned product for A/S. Figure 11 shows the concept of RFID based illegal part management solution. Currently there are two ways to prevent illegal part use.

In the first method, an RFID reader is installed into the water purifier and reads the part number of filter when a planner installs a filter. The information is sent to the server at headquarter, then it gives confirmation on the part number from database. In the second method, a planner reads the part number of the filter using an RFID enabled PDA and then sends the part number to headquarters or transfers the information to a USB device for offline verification.

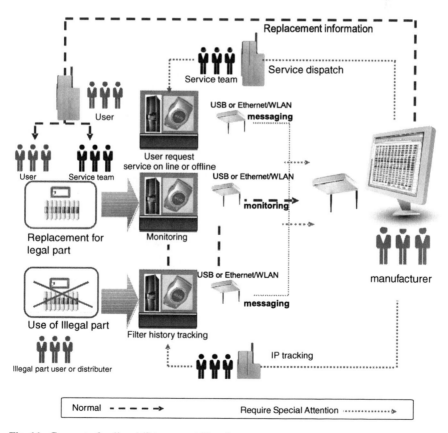

Fig. 11 Concept of online A/S to prevent illegal part use

During the case study, the test bed company found the following potential benefits from the RFID based A/S process.

• Elimination of manual process errors (for example, data entry errors).
• Possibility to provide greener A/S dispatch.
• Overall visibility of a product (i.e. in both forward and reverse logistic processes).
• Prevent distribution and use of counterfeit spare parts.

Among the methods we introduced to prevent counterfeit parts, the company prefers the RFID enabled water purifier because they can check the health of water purifier remotely as well as identify illegal parts. However due to the current cost of RFID readers, it will take more time to be in the market.

4 Conclusion

In this chapter, we have presented an RFID based green logistics solution along with number of service provisions. In designing the solution and the services, we embraced industry standards such as the EPC network architecture and Web 2.0. We have presented a brief pilot case study at a water purifier manufacturing company in Korea. From the pilot study, we confirmed that RFID could address existing issues of the company by improving traceability.

Acknowledgments This work was supported in part by Ministry of Knowledge Economy of Korea under Grant Number 10029935. The Authors would like to acknowledge this support.

References

Aberdeen Group (2006) Revisiting Reverse Logistics in the Customer-Centric Service Chain, September 2006. http://www.aberdeen.com/Aberdeen-Library/3475/RA_RevLogReport_RG_3475.aspx. Accessed on 24 June 2009
Beamon BM (1999) Designing the green supply chain. Logistics Inf Manage 12:332–342
Bintrup A, Ranasinghe DC, McFarlane D, Parlikad AK (2008) A review of the intelligent product across the product lifecycle. In: International product lifecycle management conference (PLM08), Seoul, Korea
EPC Global Inc (2007) The EPCglobal architecture framework, Final version 1.2. http://www.epcglobalinc.org/standards/. Accessed 20 October 2010
GLOBE Research Group (2009) http://www.globe.re.kr. Accessed 20 October 2009
Green Logistics (2009) http http://www.greenlogistics.org/. Accessed 20 Oct 2009
Ouertani MZ, Srinivasan R, Parlikad AK, McFarlane D, Ranasinghe DC, Kelepouris T, Lopez TS, Thorne A, Harrison M, Brintrup A, Cuthbert R (2009) Integrated asset maintenance: a lab-based demonstrator. In: 16th CIRP international conference on life cycle engineering (LCE09), Cairo, Egypt
Parlikad AK, McFarlane D (2007) RFID-based product information in end-of-life decision making. Contr Eng Pract 15(11):1348–1363
Starik M, Marcus A (2000) Special research forum on the management of organization in the natural environment: a field emerging from multiple paths, with many challenges ahead. Acad Manage J 43(4):539–546

Van Hoek RI (1999) From reversed logistics to green supply chains. Supply Chain Manage 4(3):129–134

WEEE (2009) European Commission's WEEE Directive. http://ec.europa.eu/environment/waste/weee/index_en.htm. Last Accessed October 20, 2009

Object Oriented Business Process Modelling in RFID Applied Computing Environments

Xiaohui Zhao, Chengfei Liu, and Tao Lin

Abstract As a tracking technology, Radio Frequency Identification (RFID) is now widely applied to enhance the context awareness of enterprise information systems. Such awareness provides great opportunities to facilitate business process automation and thereby improve operation efficiency and accuracy. With the aim to incorporate business logics into RFID-enabled applications, this book chapter addresses how RFID technologies impact current business process management and the characteristics of object-oriented business process modelling. This chapter first discusses the rationality and advantages of applying object-oriented process modelling in RFID applications, then addresses the requirements and guidelines for RFID data management and process modelling. Two typical solutions are introduced to further illustrate the modelling and incorporation of business logics/business processes into RFID edge systems. To demonstrate the applicability of these two approaches, a detailed case study is conducted within a distribution centre scenario.

1 Introduction

Radio Frequency Identification (RFID) is a re-emerging technology that is intended to identify, track, and trace items automatically. Nowadays, business globalisation and commoditisation advocates wide adoption of RFID technology in retailing, manufacturing, supply chain, military use, health care, etc. Through readers to RFID middleware systems, the information and movement of tagged objects can be used to trigger business transactions, and thereby enable business automation to deal with object-level information without human involvements. These features change the way of dealing with the physical world from a quantity-based mode to an object-based mode. Under this background, both research communities and

X. Zhao (✉)
Information Systems Group, School of Industrial Engineering and Innovation Sciences, Eindhoven University of Technology, Eindhoven, The Netherlands
e-mail: x.zhao@tue.nl

D.C. Ranasinghe et al. (eds.), *Unique Radio Innovation for the 21st Century*,
DOI 10.1007/978-3-642-03462-6_17, © Springer-Verlag Berlin Heidelberg 2010

industry companies are putting great efforts to integrate business logics and real-time item-level awareness together, by fusing RFID edge systems and application systems.

As a powerful data sensing/collecting technology, RFID brings a lot of operational benefits to the business applications that require accurate data or more collection points, as indicated by Overby (2008) of Forrester Research. With regard to general business process management, RFID provides the ability of tracking goods moving through a supply chain. This enables organisations to shorten business cycles, detect and resolve delivery exceptions, prevent out-of-stock situations, and pinpoint affected products in a product recall, while minimising inventory and safety stock levels.

This real-time visibility is expected to benefit business process management in following aspects, where new business values are highly sought after:

- Possibility to handle business on site. This can significantly enhance the operational efficiency by reducing the response time.
- Improved productivity of business processes. For example, RFID might be used to simultaneously read all of the cartons on a pallet as the pallet passes through a portal, or read all of the serial numbers virtually at once as a pallet of goods leaves a production cell.
- Improved sensitivity of business intelligence. Real-time visibility supports vendor-managed inventory programs, helps prevent repository shrinkage and diversion (untraceable loss and change of stock), and discourages counterfeiting by making it easier to identify fake products.

By integrating business processes into RFID data management, we can effectively facilitate business process automation, and thereby improve the agility and efficiency of current business operations in the end. Towards this ultimate goal, this book chapter introduces the business process modelling for RFID-enabled applications. The content of this chapter is based on our previous work on RFID event/data handling (Zhao et al. 2009) with particular emphasises and discussions on object-oriented business process modelling.

2 The State of the Art

RFID community has put a lot of efforts on RFID event processing, particularly in aspects of data cleansing and filtering. Work "Stream-based And Shared Event" (SASE) processing has defined an SQL like complex event language to aggregate RFID events (Gyllstrom et al. 2007; Wu et al. 2006). The implemented SASE system uses a persistent storage component to support queries over historical data and allows the query results from the stream processor to be joined with stored data. In addition, extended sliding window control and indexing techniques have been adopted by Bai et al. (2007) and Park et al. (2007) to improve the performance of continuous query processing over RFID event flows. Hu et al. (2006) have addressed the query issue from the perspective of energy efficiency. Wang and Liu (2005) have investigated the temporal management of RFID data. They have adapted traditional database query techniques to the temporal relationships of RFID data, and defined a set of temporal complex event constructors in their follow-up work (Wang et al. 2006). Two partitioning mechanisms have also been proposed in their work to support efficient queries. However, none of the mentioned works have provided an explicit solution on how to handle the delayed effects in event management, or how to integrate business process automation into RFID event management.

After data cleansing and filtering, RFID data/events will be sent to middleware systems for semantic elicitation. Ranganathan and Campbell (2003) have proposed a middleware to allow heterogeneous agents to acquire contextual information and uniform such knowledge using ontologies. Yau et al. (2002) have built a reconfigurable context-sensitive middleware for developing context sensitive pervasive computing softwares and their runtime operations. In service computing area, Gu et al. (2005) have developed a service-oriented middleware to support the acquisition, discovery, interpretation and access of various contexts to build context-aware services. To sum up, these middleware systems are mainly designed for general sensors, and therefore they fail to fully address the various characteristics of RFID technology. For RFID, Kim et al. (2007) have reported a business aware framework for business processes in EPC network. This framework allocates its business aware layer between traditional RFID data middleware and upper applications, and the framework mainly focuses on the conversion of RFID raw events to business events, and the invocations to upper level business services.

For business process management, to facilitate business process modelling in data-intensive scenarios, the object-oriented (or artifact-oriented, artifacts delegate the main business objects or documents) perspective has been proposed recently as a new business modelling method (Bhattacharya et al. 2007; Hull 2008; Küster et al. 2007; Liu et al. 2007; Nigam and Caswell 2003; Wahler and Küster 2008). This perspective uses objects to denote the information entities that capture process goals. Business rules are used to assemble related services together, and define how the entities respond to real-time dynamics. Compared with traditional business process modelling approaches, object-oriented modelling approaches focus on business contexture and behaviours, rather than activity sequences. Therefore, the object-oriented modelling advocates a complete data-driven execution mechanism, and thereby enables business actors to be aware of what can be done instead of what should be done.

3 Incorporating Business Process Modelling into RFID-enabled Applications

Many researchers have investigated RFID application in typical supply chain scenarios, like distribution centres, etc., (Bottani 2008; Zhao et al. 2009). These works have identified the following distinct characteristics of RFID-enabled applications from traditional applications:

1. Activities are triggered by RFID data rather than humans.
2. RFID systems tend to generate a huge amount of event data, as the readers continuously report all pass-by objects.
3. Movements of some RFID tagged objects reflect swarming phenomena, as many RFID tagged objects act with similar behaviours, particularly in packaging and transportation stages.
4. Products of the same type and the same batch may participate in different business processes, yet it is hard to pre-define the correlation between products and their involved business processes.

These characteristics pose challenges to the deployment of business process modelling. In regard to facilitating business process automation with RFID technologies, Mark Palmer has identified "Digest the Data Close to the Source", "Turn Simple Events into Meaningful Events", and "Cache Context" as three principles for RFID data management in work (Palmer 2004). Basically, these principles emphasize the following aspects of data management:

- The raw RFID data should pass the cleansing, consolidation and summarisation processes at edge systems to ensure better reliability and protect central IT systems from data flood.
- Turn simple RFID read events into meaningful events to derive actionable knowledge from discrete events.
- Further understand and process RFID event data in a specific business context with cached reference data and related scenario context.

These principles emphasise the edge-side event processing according to contextual business process logics. Further, the concurrence of multiple different business processes within the same edge infrastructure system highlights system reusability and independence. Traditional activity-based workflow models architect business processes with a main focus on control flow dependencies, where activities execute in accordance to pre-defined activity sequences rather than contextual dynamics. To adapt to event-based communications and effectively utilise real-time object information, this chapter intends to model business processes from an object-oriented perspective, and drive business process according to contextual dynamics.

Object-oriented methodology has been widely applied to model real world applications. Features of polymorphism, inheritance and encapsulation are its three most typical distinctions. Table 1 lists the comparison between these features and their reflections in RFID-applied environment.

From this table, we can see the strong similarity and high potential benefits of applying the object-oriented perspective in business process modelling in the RFID-applied environments. To further justify the rationality of object-oriented modelling perspective in the RFID-applied environments, we compare the object-oriented business process modelling with the traditional process-centric modelling.

Table 1 Comparison between OO features and their reflections in RFID-applied environments

	Polymorphism	Inheritance	Encapsulation
Meaning in OO paradigm	The ability of one type to appear as and be used like another type.	Form new classes using classes that have already been defined. The new classes, known as derived classes, inherit attributes and behaviour of the pre-existing classes.	A language construct that facilitates the bundling of data with the methods operating on that data.
Reflection in RFID-applied environment	An entity type may own multiple definitions and its method may own multiple implementations. Once deployed in a practical scenario, it will choose proper definition or implementation according to the actual situation.	An involved entity type can be easily extended or specialised by adding new methods or attributes. This feature is particularly useful when deploying the defined classes in a practical scenario.	Each class can encapsulate the implementation of its operations/methods. In addition, a composite class can encapsulate the related classes, involved event patterns and business rules into a new class.
Examples	An assembly line in a packing station may pack goods from the centre of the pallet to balance the weight distribution if the goods are of one type; while it may pack goods in proportional spacing to match different product sizes with goods of different types. This reflects the function level polymorphism.	Class of forklift car can be specialised with methods of loading and unloading pallets in the distribution centre case; while it may be specialised with attributes of latest position, latest speed, etc., in the case of monitoring the locations of cars.	Related class, rules, event patterns can be encapsulated into deployable RFID application contexts. Once such RFID application contexts are deployed to the data-driven middleware systems, the business logics can be pushed down to RFID edge systems.

Table 2 Comparison between process-centric approaches and object-oriented approaches

	Process-centric approaches	Object-oriented approaches
Modelling focus	Function oriented	Goal oriented
Business logics are embedded in ...	Pre-defined process models	Business contexture and behaviours
Main content	Activity sequences	Objects and declarative rules
Execution mechanism	Relatively complex	Lightweight
Reusability	No explicit process inheritance support	High reusability via class inheritance
Human readability	Easy to read	Hard to read

In process-centric approaches, each process is designed to describe the procedure of fulfilling a business function. According to embedded business logic, such a process may involve many related organisational units, staff, services as well as items, and thereby create a complex model. Normally, such models can be represented as easily readable diagrams. In comparison, object-oriented approaches do not follow the activity control flow. Instead, they model business contexture and behaviours with objects and related rules. A group of objects collaborate together to achieve a given goal, and such collaboration creates a business contexture. Table 2 lists the comparison between these two kinds of approaches.

This comparison illustrates that object-oriented approaches possess powerful capability for expressing item-level behaviours. Unlike process models, rule-based approaches treat items individually, and therefore provide a finer control over business transactions. Compared to process-centric approaches, rules own higher automation and efficiency for processing large volumes of items, as rules explicitly define how the system responses to certain events and conditions. All these features prove that object-oriented approaches better fit into the business process modelling in RFID-applied environment.

In the following section, we will further illustrate how object-oriented business process modelling incorporates into the RFID environment by introducing two models.

4 Introduction to Two Models for Business Process Modelling in the RFID-Applied Environments

4.1 Object-Oriented Business Scenario Model for RFID-enabled Applications

Following the object-oriented paradigm, Zhao et al. (2009) have proposed a model to characterise business processes in RFID-applied environments.

This model abstracts the entities involved in a business scenario, such as products, equipments, staff, tools, related documents, etc., into *classes*. Continuous

RFID tag reading *event series* reflect the movements of objects, while these movements can be interpreted into business meaningful events using *event patterns*. According to *business rules*, *objects* will respond to these elicited events by updating their internal status or invoking external operations. Thereby, a business process is fulfilled through the interactions between these objects. By setting up multiple rule sets, it can support objects to serve multiple business processes at the same time. By redefining business rules, an *RFID application context* can be customised, and thereby RFID edge systems can be reconfigured to adapt to new requirements.

Figure 1 illustrates the relationships between key notions of this model. The shadowed area represents the static part of the RFID application context, which consists of classes, rules and event patterns. A composite class may contain both base class(es) and composite class(es), and a class can inherit the characteristics of another class by extending the latter. Event patterns can extract business meanings from tag reading events, and with event patterns the rules can define the conditions of state transitions or operation invocations. An RFID application context defines a self-contained, self-acting and encapsulated entity which can invoke the operations of edge systems or external systems in response to real-time events. The unshadowed area represents the run time part, which describes the interactions between objects. The status and attribute values of an object indicate the progress stage of its lifecycle. An RFID reader sends reading events when it observes a pass-by RFID tagged object.

Particularly, this modelling approach emphasises the encapsulation and inheritance features of the object-oriented perspective. Each class encapsulates related attributes, operations and the rules for state transitions without regard to external events. These features benefit the reusability of the modelled business scenario, and can assist the system configurability by reusing the pre-defined classes. More details of this model can be found in work (Zhao et al. 2009).

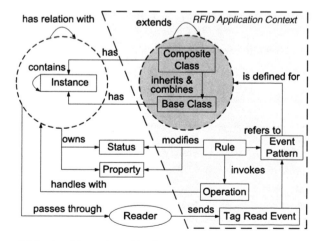

Fig. 1 Meta-model for the object-oriented business model

4.2 Event-Calculus Based Modelling Approach

As discussed before, event handling plays a very important role in RFID data management, and thereby the efficiency of event representation and handling is highly sought after by many business process modelling approaches.

The model proposed by Zhao et al. (2009) inherited the object-oriented modelling perspective and deployed event calculus as the tool for event and rule modelling. Event calculus is a logic-based formalism that infers what is true when given what happens, when and what actions do, and it is based on the supposition that "all change must be due to a cause, while spontaneous changes do not occur" (Shanahan 1999). Event calculus particularly fits into event-based rule design and analysis in the event-rich environment of RFID-enabled applications. Compared with other event/state modelling approaches, such as UML state diagram and Event Process Chain (EPC) (Scheer 1994), event calculus has advantages in modelling actions with indirect or non-deterministic effects, concurrent actions, and continuous changes, in a much concise representation.

The main components of event calculus include events (or action), *fluents* and *time points*. A fluent is anything whose numerical or boolean value is subject to change over time (Shanahan 1999). In this approach, fluents are confined to be propositional fluents, i.e., Boolean fluents, for simplification (note, this can be easily extended to situational fluents, i.e., the range of fluents can be extended to be situations). A scenario modelled by event calculus constitutes predicates and axioms, which may refer to fluents, events, and time points as parameters.

Table 3 lists the primary event calculus predicates.

Rules and queries construct the main skeleton of the model. The rules regulate the dynamic behaviours of the system in terms of fluents' value changes. Syntactically, a rule r is defined as $P \leftarrow \overset{n}{\underset{i=0}{\vee}} [(\overset{m}{\underset{j=0}{\wedge}} \exp_{ij}) \vee \exp_i]$, where

- $P \in \{Initiates(e, f, t), Terminates(e, f, t), HoldsAt(f, t)\} \cup \{$domain-specific predicates$\}$;
- $\exp_{ij}$ and $\exp_i \in \{Happens(e, t), HoldsAt(f, t)\} \cup \{$domain-specific predicates$\}$.

Table 3 Event calculus predicates

Predicates	Explanation
$Initiates(e, f, t)$	Fluent f starts to hold after event e at time t
$Terminates(e, f, t)$	Fluent f ceases to hold after event e at time t
$Initially_P(f)$	Fluent f holds from time 0, i.e., the initial time point
$Initially_N(f)$	Fluent f does not hold from time 0
$t_1 < t_2$	Time point t_1 is before time point t_2
$Happens(e, t)$	Event e occurs at time t
$HoldsAt(f, t)$	Fluent f holds at time t
$Clipped(t_1, f, t_2)$	Fluent f is terminated between time points t_1 and t_2
$Declipped(t_1, f, t_2)$	Fluent f is initiated between time points t_1 and t_2

Queries are used to retrieve the fluent values at a given time point, and the query results can trigger proper operations as the system's responses change. Syntactically, a query q_t at a given time point t is defined as $\leftarrow_{\Delta(t_0, t) \wedge I_{t0}} \rho_t$, where

- ρ_t denotes the target statement for the query in form of a conjunction of several $HoldsAt(f, t)$ predicates, i.e., $\rho_t = \bigwedge_i HoldsAt(f_i, t)$;
- $\Delta(t_0, t)$ denotes the set of events that are occurred from the beginning time point t_0 to t;
- I_{t0} denotes the initial settings at time t_0, i.e., the values of fluents at time t_0;

Figure 2 illustrates the relationship between aforementioned notions. An RFID schema consists of a set of RFID classes, which specify the fluents and the attributes for describing domain-specific rules in an RFID scenario. Such an RFID scenario represents a self-contained system at build time, while an RFID environment represents the run time dynamics including fluent values at the beginning time and the continuous event flow. Queries are used to retrieve real-time fluent values, and the edge system can invoke proper operations via operation triggers in response to the query results.

Based on event calculus, this approach is specialised in event aggregating and elicitation. Due to the dependency between fluent values and events at different time points, the fluent values at a time point are subject to the fluent values or events at previous time points. This phenomenon results in an awkward trouble that each query execution may need to calculate the events occurred from t_0 up to current time point. As time goes, the number of events increases towards infinity, and in turn this will reluctantly increase the query execution time towards infinite. Since

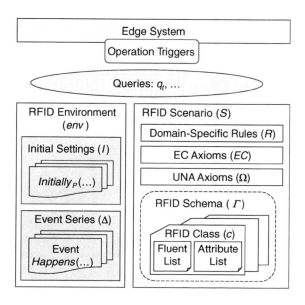

Fig. 2 RFID data management model architecture

RFID queries are always running continuously/periodically, it is possible to optimise RFID query performance by cutting off event series to a limited scope and reusing the fluent values at previous time points from previous query results. Zhao et al. (2009) have investigated this issue and proposed a 2- block buffering mechanism to shorten necessary event series for query execution.

This 2-block buffering mechanism runs two buffers to record the data with delayed effects to later queries. The one with an earlier time point is named *back buffer* (*bb*), while the other is *fore buffer* (*fb*). Each buffer stores the fluent snapshot at a certain time and the events that have occurred during a certain period of time.

Figure 3 illustrates how the buffering mechanism serves query execution. Suppose query q_t runs at time t ($t>fb.t$) and all the required historical data occurred after $fb.t$, these historical data can be obtained by calculating from time $fb.t$ with the events buffered in fb and the events from time $fb.t$ to t. Suppose another query q_t' runs at time t' ($t'>fb.t$) and q_t' requires some historical data of the time earlier than $fb.t$, it indicates that some past events and fluent values of time period x, as shown in Fig. 3, have delayed effects to q_t'. For the fluent values at time points other than $fb.t$ in period x, they cannot be contained by calculating the events buffered in fb, as the calculation may result in queries over earlier historical data. In this case, we calculate these fluent values using bb's fluent snapshot and its buffered events.

A set of experiments has been conducted to test the performance enhancement. Compared to the naive mechanism, the 2-block buffering mechanism reduces the query execution time from exponential time to an approximate flat time. Further, the 2-block mechanism outperforms the periodical mechanism by nearly 50%. More details about the buffer setting and query execution can be found in work (Zhao et al. 2009).

Fig. 3 2-Block buffering mechanism

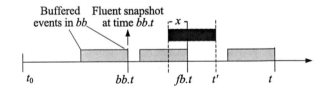

5 Case Study

To better illustrate the feasibility and advantage of the introduced models, we take a distribution centre for assembling shipments as an example in this section.

The business process shown in Fig. 4 depicts the procedure of pallet packing at the distribution centre. The process is drawn in the Business Process Modelling Notation (BPMN) diagram format, where the legend of symbols is given below the diagram.

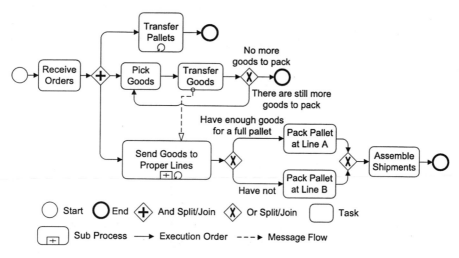

Fig. 4 Simplified business process diagram for the goods packing process

When receiving an order from customers, the process concurrently handles three tasks, i.e., transferring pallets to the distribution centre, picking the ordered goods from inventory, and transferring the picked goods to the distribution centre. When these goods arrive, the packing station periodically packs them at two packing lines. Line A can only do full pallet packing with the goods of the same type, while line B can do partial packing and mixed packing. When all the ordered goods are packed, they will be sent for shipment.

5.1 Modelling with the Object-Oriented Business scenario model

In this scenario, several kinds of entities are involved, such as assembly lines, forklift cars, packing stations, etc. To sort out the interaction behaviours between these entities using the object-oriented business scenario model, we first need to abstract these entities into proper classes.

Figure 5 shows an example of an "Assembly Line" class. This class abstracts the attributes and operations of assembly lines at a distribution centre. An object of the class denotes a concrete assembly line, and technically the instance is materialised with actual business data at run time. In the object-oriented business scenario model, the status of a class object indicates the internal stage of the delegated entity, and will help navigate the behaviours of the object. The state transition diagram in Fig. 5 illustrates the states of the class and the transitions between these states under proper conditions.

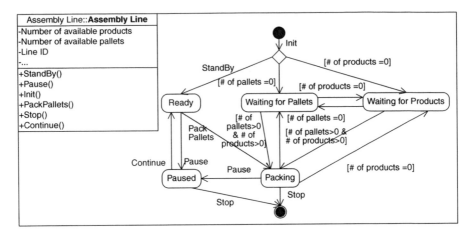

Fig. 5 Class "Assembly Line" and its state transition diagram

With this class, the pallet packing scenario can be defined as shown in Example 1. In this scenario, two events are imported, i.e., *Arrives* and *sentOff*, which can be elicited from RFID raw events, and indicate product arrival and sending off a pallet, respectively.

Example 1. Partial content of the pallet packing scenario.

Class
   AL – Assembly line.

Events
   Arrives – a product arrives to the assembly line;
   sentOff – a pallet of packed products are sent off.

Counter
   emptyPallets – this counter records the number of empty pallets at the assembly line.

Rules
   (1.1) *changesTo(AL, sentOff, "wait for pallets", t)* ← *occurs(sentOff, AL, t)∧emptyPallets=0*;
   (1.2) *invokes("call for pallets")* ← *holdsState(AL, "wait for pallets", t)∧occurs(Arrives, AL, t)*;
   (1.3) *changesTo(AL, Arrives , "ready", t)* ← *holdsState(AL, "wait for pallets", t) ∧occurs(Arrives, AL, t)∧¬emptyPallets=0*;
   . . .

Rule (1.1) specifies that the assembly line will change to state "wait for pallets", if it has no empty pallets after sending off the packed pallets. Rule (1.2) specifies that the assembly line will request for new pallets, if it is in state "wait for pallets", when new products arrive. Similarly, rule (1.3) specifies the condition for the assembly line to change from state "wait for pallets" to state "ready". These rules specify how external events influence the state transitions of class "Assembly Line". Due to the space limit, we do not list the full set of rules.

So far, the given example shows partial content of the pallet packing scenario, while it certainly can be enriched with more classes, events and rules. For example, assembly lines and related pickers, forklift cars, drivers and operators can join into a packing station. Such a packing station can handle the packing process from picking products, transferring pallets, and packing pallets. At conceptual level, we define a composite class to represent such combined entities. Figure 6 shows the content of composite class "Packing Station" and its state transition diagram, which composes the state transition diagrams of its component classes.

Such a composite class builds up a self-contained unit, which owns better reusability by hiding the internal complexity, and thereby enhances the scalability of RFID system integration.

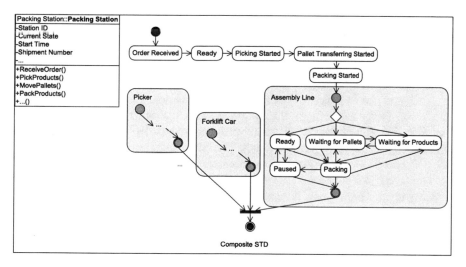

Fig. 6 Composite class "packing station" and its state transition diagram

5.2 Modelling with the Event-Calculus based approach

Based on this example, we further investigate sub process "Send Goods to Proper Lines". Figure 7 shows the details of this sub process in a BPMN diagram. In this scenario, the packing station has a temporary repository to store the received products. Once a product of type 1 comes, RFID readers will send out a "g1Arr" event. This event will be captured by task "Add the Number of Product Type 1", which is

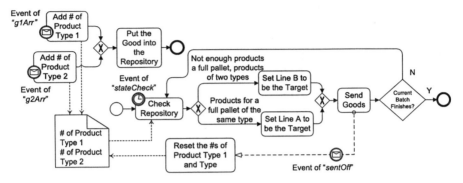

Fig. 7 Sub process "Send Goods to Proper Lines"

responsible for counting the products of type 1 in the repository, and recording the number in a document, denoted by "# of Product Type 1…". Similarly, task "Add the Number of Product Type 2" is responsible for counting the products of type 2 in the repository. The dashed lines with tilted arrows represent the data reading/writing operations.

Event "stateCheck" is a periodical event, which triggers the execution of task "Check Repository". In this task, a dispatcher machine will check the number and types of the received products in the repository, and determine the assemble line for packing these products. Here, line A can only pack full pallets of products of the same type, and line B can do partial or mixed pallet packing.

If no more than a pallet of goods arrive in a 4-s period, the current batch of products is considered to be completed, and the packing station will empty the repository and wait for the next batch. The main process covered by RFID technology was the tracking of the movement of goods from the band conveyer to the dispatcher. Tagging was done at individual product level.

This scenario represents a specific case, where the system should react in accordance with the events and product status in the repository. To well specify the dependency and interactions between events and product status, here we model this scenario with the event-calculus based approach.

In this approach, the corresponding RFID scenario $S=(\Gamma, R, EC, \Omega)$ constitutes the RFID class schema Γ including the fluents listed below, axioms EC and Ω, and the domain-dependent rule set R, where R comprises the following events, fluents and rules:

Example 2. Content of an RFID scenario sample and related queries and operation triggers.

Events
    $g1Arr$ – a product of type 1 arrives to the dispatcher;

g2Arr – a product of type 2 arrives to the dispatcher;
sentOff – the deposited products are sent to a packing line;
stateCheck – a periodical event to initiate the state checking of the dispatcher.

Fluents

Mixed – the deposited products are of two types;
Full – the repository has enough products for a full pallet;
Finish – the current batch has been handled;
NotEmpty – the repository is occupied;
Idle – the dispatcher is standing by, rather than working.

Rules

(R1) $[num_1++, num_2++] \leftarrow Happens([g1Arr, g2Arr], t)$;

(R2) $Initiates([g1Arr, g2Arr], NotEmpty, t) \leftarrow Happens([g1Arr, g2Arr], t) \wedge \neg HoldsAt(NotEmpty, t)$;

(R3) $Initiates(Mixed, t) \leftarrow Happens([g1Arr, g2Arr], t) \wedge (num_1 \geq 0) \wedge (num_2 \geq 0) \wedge \neg HoldsAt(Mixed, t)$;

(R4) $Initiates(Full, t) \leftarrow Happens([g1Arr, g2Arr], t) \wedge (num_1 + num_2 = MAX) \wedge \neg HoldsAt(Full, t)$;

(R5) $Terminates(Idle, t) \leftarrow Happens([g1Arr, g2Arr], t) \wedge HoldsAt(Idle, t)$;

(R6) $Terminates([Mixed, Full, NotEmpty], t) \wedge num_1 = 0 \wedge num_2 = 0 \leftarrow Happens(sentOff, t) \wedge HoldsAt([Mixed, Full, NotEmpty], t)$;

(R7) $Initiates(Idle, stateCheck, t) \leftarrow Happens(stateCheck, t) \wedge \neg HoldsAt(Idle, t) \wedge NoSentOff(t-4, t)$;

(R8) $Initiates(Finish, stateCheck, t) \wedge Terminates(Idle, stateCheck, t) \leftarrow Happens(stateCheck, t) \wedge \neg HoldsAt(Finish, t)) \wedge HoldsAt(Idle, t) \wedge KeepsIdle(t-4, t)$;

Queries

q_{1t}: $\leftarrow_{\Delta(t_0, t) \wedge I_{t0}} HoldsAt(Full, t) \wedge \neg HoldsAt(Mixed, t)$;

q_{2t}: $\leftarrow_{\Delta(t_0, t) \wedge I_{t0}} HoldsAt(Full, t) \wedge HoldsAt(Mixed, t)$;

q_{3t}: $\leftarrow_{\Delta(t_0, t) \wedge I_{t0}} HoldsAt(Idle, t) \wedge \neg HoldsAt(Full, t) \wedge HoldsAt(NotEmpty, t)$.

Triggers

Trigger 1: $| q_{1t} | \Rightarrow$ invoke operation "send to Line A";
Trigger 2: $| q_{2t} | \vee | q_{3t} | \Rightarrow$ invoke operation "send to Line B".

In the rule set, (R1) uses num_1 and num_2 to record the numbers of arrived products of *Product Type* 1 and *Product Type* 2, respectively. Square brackets represent a selective relation between the contained elements. (R2-5) adjust the values of fluents *NotEmpty*, *Mixed*, *Full* and *Idle* when a product arrives. (R6) resets the values of fluents *Mixed*, *Full* and *NotEmpty* to be false once a "sentOff" event occurs. (R7-8) handle with "*stateCheck*" events, where (R7) turns Dispatcher into "Idle" mode if no "sentOff" events have occurred in last 4 s before the latest "state-Check" event, and (R8) turns Dispatcher into "Finish" mode if it has been in "Idle" mode for the last 4 s before the latest "stateCheck" event. The referenced predicates *NoSentOff* and *KeepsIdle* are defined as follows,

$$NoSentOff(t_1,\ t_2) = \bigwedge_{t_i \in [t_1,t_2]} \neg Happens(sentOff,\ t_i);$$

$$KeepsIdle(t_1,\ t_2) = \bigwedge_{t_i \in [t_1,t_2]} HoldsAt(Idle,\ t_i).$$

Once scenario S is defined, it can be input into RFID edge system, i.e., the dispatcher machine. Thus, the dispatcher machine is empowered with the awareness of products' arrivals and the business logics on where to send products for packing.

Further, to guide the edge system's responses to these events, we deploy the listed queries and operation triggers. Here, query q_{1t} checks whether there are enough products of the same type for a full pallet at time point t; q_{2t} checks whether there are enough products of the different types for a full pallet of at time point t; q_{1t} checks whether there are not enough products for a full pallet at time point t. The two operation triggers will invoke the operation of sending the products in the temporary repository to proper assembly lines. These queries and operation triggers enable the dispatcher machine to intelligently react to real-time dynamics.

6 Conclusions

This book chapter advocated the object-oriented business process modelling in RFID-applied environments. The rationality of applying such object-oriented modelling perspective was analysed in terms of the suitability between data-intensive RFID event handling and object-oriented modelling features. The advantages of object-oriented modelling perspective were discussed in comparison with the traditional process-centric modelling approaches. Two modelling approaches were introduced to illustrate the application of object-oriented perspective in business process modelling in the RFID-applied environments. The first approach focused on the migration from the traditional control flow oriented modelling perspective towards the object-oriented modelling perspective, with the following features:

- Business abstraction: A class encapsulates the internal details of a type of entities, while a class instance delegates the runtime status of the corresponding entity.
- Behaviour specification: State transition diagrams specify the object behaviours and their interactions.

- Class composition: Classes can be combined together to delegate a comprehensive assembly of related entities. This improves system reusability and re-configurability.

The second approach emphasised the event handling and event-based business logic modelling on the basis of event calculus. The logics are represented as propositional logic clauses with the following features:

- Propositional fluents are used to represent the status of objects and interaction status.
- The business logics are represented in a concise event calculus format, which well expresses the dependency between fluents and their value changes according to raised events.
- Query optimisation mechanism to shorten the querying time, and enhance query efficiency by buffering historical evens and fluent snapshots.

The case study demonstrated how the approaches can be applied to model practical scenarios. The object-oriented approach modelled the scenario by abstracting involved entities into classes with proper state transition diagrams. By means of composing related classes together, we can create a pre-configured and self-contained component, which can be reused to other business scenarios. The event-calculus based approach modelled the scenario by setting the proper fluents. Instead of state transition diagrams, rules are used to specify how these fluents change their values according to the dependency relations and contextual dynamics.

References

Bai Y, Wang F, Liu P, Zaniolo C, Liu S (2007) RFID data processing with a data stream query language. In: Proceedings of the 23rd international conference on data engineering, Istanbul, Turkey, pp 1184–1193

Bhattacharya K, Gerede CE, Hull R, Liu R, Su J (2007) Towards formal analysis of artifact-centric business process models. In: Proceedings of the 5th international conference on business process management, Brisbane, QLD, Australia,pp 288–304

Bottani E (2008) Reengineering, simulation and data analysis of an RFID system. J Theor Appl Electron Com Res 3(1):12–29

Gu T, Pung HK, Zhang D (2005) A service-oriented middleware for bbuilding context-aware services. J Netw Comput Appl 28(1):1–18

Gyllstrom D, Wu E, Chae H-J, Diao Y, Stahlberg P, Anderson G (2007) SASE: complex event processing over streams. In: Proceedings of the 3rd biennial conference on innovative data systems research, Asilomar, CA, USA, pp 407–411

Hu W, Misra A, Shorey R (2006) CAPS: energy-efficient processing of continuous aggregate queries in sensor networks. In: Proceedings of the 4th annual IEEE international conference on pervasive computing and communications, Pisa, Italy, pp 190–199

Hull R (2008) Artifact-centric business process models: brief survey of research results and challenges. In: Proceedings of the OTM 2008 confederated international conferences, Monterrey, Mexico, pp 1152–1163

Kim S, Moon M, Kim S, Yu S, Yeom K (2007) RFID business aware framework for business process in the EPC network. In: Proceedings of the 5th ACIS international conference on software engineering research, Management and Applications, Busan, Korea, pp 468–475

Küster JM, Ryndina K, Gall H (2007) Generation of business process models for object life cycle compliance. In: Proceedings of the 5th international conference business process management, Brisbane, QLD, Australia, pp 165–181

Liu R, Bhattacharya K, Wu FY (2007) Modeling business contexture and behavior using business artifacts. In: Proceedings of the 19th international conference on advanced information systems engineering, Trondheim, Norway, pp 324–339

Nigam A, Caswell NS (2003) Business artifacts: an approach to operational specification. IBM Syst J 42(3):428–445

Overby C (2008) Forrester research http://www.forrester.com/rb/analyst/christine_overby. Accessed 24 June 2009

Palmer M (2004) Seven principles of effective RFID data management. Technical primer. http://citeseerx.ist.psu.edu/viewdoc/download?doi=10.1.1.119.5952&rep=rep1&type=pdf. Accessed 24 June 2009

Park J, Hong B, Ban C (2007) A continuous query index for processing queries on RFID data stream. In: Proceedings of the 13th IEEE international conference on embedded and real-time computing systems and applications, Daegu, Korea, pp 138–145

Ranganathan A, Campbell RH (2003) A middleware for context-aware agents in ubiquitous computing environments. In: Proceedings of ACM/IFIP/USENIX international middleware conference, Rio de Janeiro, Brazil, pp 143–161

Scheer WA (1994) Business process engineering. ARIS-navigator for reference models for industrial enterprises, Springer, Berlin

Shanahan M (1999) Book section: the event calculus explained. In artificial intelligence today. Springer, Berlin, pp 409–430

Wahler K, Küster JM (2008) Predicting coupling of object-centric business process implementations. In: Proceedings of the 6th international conference business process management, Milano, Italy, pp 148–163

Wang F, Liu P (2005) Temporal management of RFID data. In: Proceedings of the 31st international conference on very large data bases, Trondheim, Norway, pp 1128–1139

Wang F, Liu S, Liu P, Bai Y (2006) Bridging physical and virtual worlds: complex event processing for RFID data streams. In: Proceedings of the 10th international conference on extending database technology, Munich, Germany, pp 588–607

Wu E, Diao Y, Rizvi S (2006) High-performance complex event processing over streams. In: Proceedings of the ACM SIGMOD international conference on management of data, Chicago, IL, USA, pp 407–418

Yau SS, Karim F, Wang Y, Wang B, Gupta SKS (2002) Reconfigurable context-sensitive middleware for pervasive computing. IEEE Pervasive Comput 1(3):33–42

Zhao, X., Liu, C., Lin, T. (2009) Incorporating Business Process Management into RFID-enabled Application Systems (accepted on Oct. 18, 2009). Business Process Management Journal.

Zhao, X., Liu, C., Lin, T. (2009) Enhancing Business Process Automation by Integrating RFID Data and Events. In Proceedings of the 17th International Conference on Cooperative Information Systems, Algarve, Portugal. 255-272.

Part V
Business and Investment

Legal Regulation and Consumers: The RFID Industry's Perspective

Daniel Ronzani

Abstract Many journal articles have presented research on the adoption and diffusion of Radio Frequency Identification (RFID) from a regulatory or consumer perspective. This research takes a reverse viewpoint. It researches the industry's experience with regulation by law and its experience with consumers. First, semi-structured interviews with RFID industry stakeholders are conducted on the topics of (UHF) frequency law, database law, and privacy law. Second, the industry's experience with (i) regulation by law and (ii) the consumers is collected in a worldwide online survey with companies and organisations that research, produce, sell, and consult on RFID technology. Third, empirical data is evaluated by different territories and industries to discuss four observations about legal regulation and consumers made by the authors with four feedback observations from the online survey. Given the evaluation of the empirical data, this article recommends that the RFID industry engage in better constructive dialogue with the legal regulator, strengthen its knowledge on applicable legislation, and re-evaluate its information policy to the consumer.

1 Introduction

In the past few years there have been various scholarly articles presenting empirical data on Radio Frequency Identification (RFID), for instance: a survey of librarians in the USA to research the public perception of RFID technology (Strickland and Hunt 2005). That survey shows, inter alia, that even among highly educated members of the public there is more fear than knowledge of information collecting technologies; or the European Commission's survey on RFID titled "The RFID Revolution: Your voice on the Challenges, Opportunities and Threats" (Commission of the European Communities 2006). That survey is a public consultation with the key finding that

D. Ronzani (✉)
Copenhagen Business School, Centre for Applies ICT, 2000 Frederiksberg, Denmark
e-mail: dan@zurich.ibm.com

D.C. Ranasinghe et al. (eds.), *Unique Radio Innovation for the 21st Century*,
DOI 10.1007/978-3-642-03462-6_18, © Springer-Verlag Berlin Heidelberg 2010

there is insufficient information available to make an informed analysis of RFID technologies; or an empirical study of anticipated consumer response to RFID presented in Canada (Angeles 2007). That study evaluates the consumers' willingness to purchase RFID-tagged products; or consumer reactions to RFID-based information systems in retail in Germany (Rothensee and Spiekerman 2008). That research shows that people are moderately privacy aware and that their privacy awareness is negatively related to their acceptance of the service.

It is claimed here that researching primarily the view by legal experts or consumers makes the debate encompassing RFID technology lop-sided. The point is that if there should be a balance between regulation by law, consumers, and industry in the adoption and diffusion of RFID technology (Kelly and Erickson 2005), then the industry's opinion must be considered as well. The RFID industry's view on regulation and its perception of consumer awareness is necessary for an economically, legally, and ethically acceptable account of RFID. This article will focus on evaluating the RFID industry's perspective in individual interviews and in a worldwide online survey. The purpose of this research is to show the RFID industry's perspective of selected regulatory- and consumer-oriented issues with RFID, and to reveal the shortcomings with the current perspective.

This research is executed in three phases. In the first phase individual in-person telephone interviews with RFID experts from internationally operating companies and organisations are conducted. These interviews serve as an exploratory tool and survey question testing. In a second phase a survey is submitted to a list-based population of international companies and organisations engaged in consulting, system integration, retail, and provision of software and hardware of RFID technology. The survey invitations rolled out in two batches of 3680 e-mail invitations and 1283 manually completed Internet web forms. In the third phase the empirical data of 111 survey respondents is presented and discussed.

The structure of this article is as follows: The following Sect. 2 provides insight to the interviews that were conducted with the RFID experts. Section 3 introduces the survey research method. Section 4 presents the survey data of selected questions and discusses the findings of eight observations (O1–O8) (Shank and Cunningham 1996). Section 5 concludes with three recommendations for the RFID industry to support the adoption and diffusion of RFID.

2 Industry Interviews

This section outlines the first phase of the research. Three 1-h interviews with RFID experts at international companies and organisations engaged in consulting, system integration, retail, software and hardware are conducted: The first interviewee is the global RFID services leader at Company A, which is an international RFID integrator; the second interviewee is the communications & public affairs RFID expert at Company B, which is an international retailer; the third interviewee is senior researcher and architect at Company C, which provides supply chain management solutions for RFID at international level. The methodology used for the in-depth interviews is similar to that used by Johnson (2001). They are semi-structured and conducted partly for exploration purposes, partly for testing of the survey questions. The authors committed to keeping the names of the entities and interviewees confidential, which is why the names have been anonymised to A, B, and C.

Two main subjects are addressed in the interviews: The industry's views and opinions on regulation by law (radio spectrum, database, and privacy) and the RFID industry's relationship to consumers. The topics addressed herein relate mainly to passive ultra high frequency band (UHF) RFID technology.

2.1 Industry's Experience with Regulation by Law

Three fields of regulation by law are addressed in the interviews: radio spectrum, database regulation, and privacy protection. First, all interviewed RFID experts at the three companies are aware of the current radio spectrum regulations. They welcome the harmonisation of radio spectrum for RFID but note that further engagement in this respect is necessary. Radio spectrum incompatibility is not only an issue between the different spectrum regions 1, 2 and 3 (ITU 2005). There are, for instance, also radio spectrum issues within the European member states (region 1) despite regulatory provisions by the European Commission to adopt the harmonisation of radio spectrum for RFID devices operating in the UHF band by the end of May 2007 (European Commission 2006). In many EU member states the conditions for the availability and efficient use of radio spectrum for RFID devices in the spectrum bandwidth between 865 and 868 MHz have been adopted. But this still does not rule out compatibility problems between different countries. The use of a certain bandwidth in one country is not also necessarily possible in the next country. Regulation can become an inhibitor.

> So [...] even though the EU regulation or proposal was there, you still had countries like Italy and Spain that had not freed up that spectrum because it was used by the military. So discussions that we had were an issue: that we had to get a specific site license to do a pilot. So if we wanted to run a pilot with an EPC-oriented tag or UHF tag specifically we had to get a license. So that actually had helped *break the market, to slow down the market* in places like Italy. Two things happened there: a) it did not accelerate as quickly as it should have, there; and b) they started looking at other technologies, high spectrum instead. (Interview with Company A; authors' emphasis).

Indeed, the Commission of the European Communities (2006) revealed that slightly more than a third of the survey respondents do not believe that harmonisation of the radio spectrum between 865 and 868 MHz is sufficient to accelerate the establishment of a fully functioning internal market.

Second, the interview partners also all confirm that for several years the debate on RFID regulation has moved away from radio spectrum towards privacy issues. Two interview partners are more critical of the way the issues around privacy have evolved. On the one hand, one industry partner acknowledges that the way the industry appeased initial consumer concerns a few years ago has left a wrong impression of the RFID industry. On the other hand, in the industry's perspective, consumer organisations were just hyping scenarios that were technically impossible:

> [...] a lot of the things we have today are not really relevant to privacy. And there really has been a hype by some organizations that pushed this privacy problem. I do not say there is no problem at all. But it was pushed in a way which was sometimes *not really realistic*. I mean if you look at [...] the problem in retail [it is] often not that you have RFID; but that you have a [loyalty card]. And that gives you all information about the consumer already. Nobody cares about that. If there is an RFID on that is, in my opinion, not that important. (Interview with Company A; authors' emphasis).

This view of technically non-feasible claims by RFID opponents is replicated by the RFID expert at Company B:

> So I pick up those things [by the consumer groups and RFID opponents] and go back to my engineering colleagues and say 'well, this is what they claim, this is what they throw against us'. And they say: 'It's rubbish. This can never happen because it cannot work. You cannot do this. It is technically impossible, physics is not right. You cannot do this.'

Another partner argues similarly but more distinctly insofar as the privacy debate evolved because the regulator focussed more on consumer organisations. It is understood here that the miscommunication a few years ago between the legal regulator, consumer groups, and the RFID industry fostered the distrust. At that early experimental RFID phase a few years ago the industry could not yet participate in the debate because

> [...] the privacy debate specifically focused on the end result of an RFID enabled world, basically. And tries to emphasize *what could be done as if this could be done already today*, leading to a situation where you have on the one hand pressure groups [...] who singled out RFID as very heavily threatening technology for privacy and data protection. And regulators who came in to look at this issue [...] and had practically two choices to inform themselves: one were those pressure groups and the others were businesses who were still in very early experimental phases of RFID and could not say much or could only talk about the end vision. [...] And so it somehow flawed impression of what is possible if RFID developed, among non-technical people, lawyers and regulators. [This leads] to a situation where now [...] perceived threats are actually what we are talking about. And perceived threats are what regulators actually try to address in their regulation. Not real threats. (Interview with Company B; authors' emphasis).

Such perceived threats seem to form today's political agenda. It is argued that privacy is a thankful political agenda. It involves all people and every interested citizen can participate in the debate. Privacy groups have been more successful in convincing the public of lasting negative effects on privacy rights (Xiao et al. 2007). This has catered distrust in the adoption and diffusion of RFID technology. Whether the perception meets the technical possibilities and legal provisions is doubted.

> Privacy is a very interesting part because many, many people have very diffused concerns about privacy. Many of the privacy arguments are intuitively understandable to a whole lot of people. I mean I would not rule myself out there. And I think nobody really can. And so you can really play ball with those fears in the way that you say: 'Well, I am the guardian, I protect you. And there is a new technology, some people say it is scary, I do not necessarily

believe that, but I will take care that it does not become scary, and I will make sure that privacy is protected with that new technology'. (Interview with Company B).

Finally, aside from these radio spectrum and privacy regulations the third regulation-related topic addressed to the interview partners – database regulation – does not seem to be an issue with which the industry is yet much concerned. The next section on consumers details on this perception regarding database regulation.

2.2 Industry's Experience with Consumers (General Public)

Either there is not much awareness on database regulation, or there is awareness but the applicability is seen somewhere in the future because current deployment of passive UHF RFID is mainly (still) case and pallet level oriented. This stands in contrast to, for instance, passive HF RFID, which is currently deployed on a broader basis in items of daily use and therefore could be viewed as more human-centric from a database perspective.

[T]he data that is in the database is not consumer oriented. It's at case and pallet level and it is supply chain oriented. However, when RFID starts to adopt across the supply chain from the manufacturer all the way down to the retailer and on to the shop floor, you are going to have a lot of data sharing. And you are going to have multiple databases [but] we are not there yet. You know, we are far from item level [tagging]. But at least we can work at a regulation now to put it in place rather than inhibiting it. (Interview with Company A).

A full scale item level situation in retail, for instance, is likely to be realised only in the next ten to fifteen years. The read reliability needs to increase and the tag cost needs to decrease.

[T]echnically speaking we can do item level already now. [...] It is more that customers currently don't accept some high tech issue. I mean in the consumer product goods area there is just no market need yet for item level tagging, mainly because it is still too expensive. [...] And secondly there are still severe technical issues to realize in order for the technology to be actually reliable enough. So that is why (i) reliability of technology at that level, plus (ii) cost is still a hindrance at item level tagging. (Interview with Company C).

This means that

[t]oday we are not tagging at item level yet. So it is really not touching the consumer in any way today, at all. So it is good we start to raise this discussion and we need to start to set a framework. But in actual fact there are only very few [...] pilots at item level but it is not really touching the consumer today. So I think that is a key point. (Interview with Company A).

But in any case, despite the consumer not being in the RFID loop yet (except for pilot deployments), the industry, especially the retail industry, is dependent on the consumers. Privacy problems influence people's very decision whether or not to use a service. Emotional reactions to, and distrust of, shopping are negatively related to how much people value their privacy and that privacy must be an essential element of RFID roll-out (Rothensee and Spiekerman 2008). The RFID industry is aware of this and takes the debate seriously:

> If you are exposed to consumer decision on a level like retailers are, you are very sensitive. It is almost foolish to think that we would on a technology issue say 'well, we do not care about the people, we just do it'. It cannot go. We certainly fully agree with [taking consumer concerns seriously]. We are retailers, right? People need to trust us to enter our stores. We depend on the trust of our customers. An RFID business case is a two-part thing: (i) 50% does the technology work and does it bring benefits to us, and (ii) 50% customers. Is there any customer that would not shop [in a] store anymore because you use RFID? Then that business case would certainly deteriorate quite a lot. (Interview with Company B).

This view of the interviewee of Company B and Rothensee and Spiekerman's (2008) research seem to counter-evidence the considerable scepticism of Kelly and Erickson (2005) whether retailers' economic self-interest will in fact really take a back seat to ethical concerns over customer privacy. Competitiveness plays a fair role for the RFID business case. Hence, it is reasonable to assume that the industry is not deploying RFID for purely altruistic reasons. But since trust is difficult to generate, easily shaken, and once shaken, extremely difficult to rebuild (Ben Shneiderman in: Rothensee and Spiekerman 2008), the industry is not likely to ignore the opinion of its valued customers.

The problem seems to be, however, that there is a discrepancy between the industry's and the consumers' awareness of RFID. Whereas the RFID industry – as RFID driver – is obviously knowledgeable about the topic, people in general are not. There is not a substantive understanding of the technology by the general public (Strickland and Hunt 2005): the results of respondents are not better than random guessing. The education level of the sample in the survey of librarians in the USA was high. Thus, it is anticipated that the overall understanding of the technology by general public is actually even worse. This correlates with the experience of one interview partner:

> [...] well, the people are not necessarily not sensible, they are just not deep enough into the topic knowing the technology good enough to understand which is just a nice, although threatening, idea of what the potential future is, and what the reality of the technology is today and for the foreseeable future. (Interview with Company B).

3 Survey Research Methodology

This section outlines the second phase of the research. A survey was prepared based on the findings of the three interviews. The survey sample, the survey design, and the precautions for the interpretation of the survey data are discussed in this section.

3.1 Survey Design

Certain RFID opponents claim that the use of RFID is unethical (Albrecht and McIntyre 2005). Notwithstanding such claims, RFID as topic is neither ethically nor politically inappropriate for a survey research (Sapsford 2007). Therefore it is justified to collect the missing data by engaging RFID stakeholders in a survey.

To the best of the authors' knowledge, an accurate worldwide and publicly available list of companies and organizations interested and involved in RFID technology, be it as suppliers, implementers, operators, or users does not exist (but see for instance IDTechEx[1] or Gartner[2] where lists can be purchased). The population of companies or organisations engaged in RFID technology is unidentified. Because the count of certain populations will always remain unknown (Couper 2000), it is suggested that with certain limitations, a subset of a population of companies and organizations engaging in RFID technology can be compiled using a web-based survey. This population will obviously not be a complete worldwide list of companies and organizations interested and involved in RFID technology. But it provides usable input nonetheless. Whereas it is acknowledged that many different online databases with companies and organisations engaged in RFID technology exist, it is maintained here that companies and organisations engaging in the emerging field of RFID technology will likely participate in either of the two online platforms of EPCglobal (EPCglobal 2009) for standards, and RFID Journal (see http://www.rfidjournal.com) for news.

The survey comprises 6 parts with sixteen questions, of which ten are presented herein. All material questions of this survey on RFID are close-ended questions (Nardi, 2003; Rea and Parker 2005), with the possibility of adding comments. Most rating questions use a 5 point Likert rating scale (Rea and Parker 2005)

3.2 Survey Sample

To compile a subset of a population it is argued here that the databases of EPCglobal and RFID Journal are a good starting point. EPCglobal is the leading subscriber-driven organisation comprised of industry leaders and organisations for industry-driven standards for the Electronic Product Code (EPC) to support the use of RFID. RFID Journal claims to be the world's first independent and world's leading media source of RFID news and insights. Both of these sources provide lists of companies and organisations engaging or interested in RFID.

For this particular survey, the unit of analysis is defined as a company or organisation, for which an employed individual completes the questionnaire acting as proxy (Rea and Parker 2005; Yin 2003). As this survey is conducted on a worldwide basis, two questions regarding the unit of analysis need to be answered: First, what happens with national companies or organisations that have several entities within a country? For national companies or organisations with several entities or representations in a country only one entity or representation (preferably the headquarter) is included in the population. Second, what happens with international companies or organisations? For international companies and organisations every country, in which such company or organisation is located, is included in the population. Within

[1] http://www.idtechex.com/
[2] http://www.gartner.com/technology/home.jsp

each of these countries the same rule applies as for national companies or organisations: an international company or organisation is surveyed only once per country, even if such company or organisation has several entities or representations in a country.

A list of 1321 companies and organisations engaged or interested in RFID is compiled from the EPCglobal and RFID Journal vendor databases. Based on 1144 of these companies and organisations an extensive manual Internet search generates 4963 contact addresses worldwide in 224 countries and geographical regions (depending on how the companies and organisations are structured globally). These 4963 contacts receive an online invitation to participate in the authors' online survey. In total 724 survey invitations are undeliverable (e.g. post mail error). 4239 survey invitations are delivered. A total of 185 invitees respond to the survey. Of these 185 survey respondents 74 are unusable (e.g. no answers), 11 survey respondents answer the questionnaire partially, and 100 survey respondents complete the questionnaire. For the remainder of this article only the 111 full and partial survey replies are relevant.

3.3 Precautions for Interpretation of the Survey Data

Among the many different types of survey (Rea and Parker 2005), it is noted that nowadays most quantitative analyses are done using computers (Vogt 2007). Hence, the survey conducted for this RFID research is in the so called technological mode (Schaeffer and Presser 2003). Web-based surveys have many advantages compared to mail out or telephone surveys, such as, cost-efficiency, speed, flexible response schedule, negligible marginal costs for additional survey submissions. The disadvantages of a web-based survey are that it is limited to populations that have access to e-mail and computers and thus also computer literacy. Furthermore the advantages enumerated here can work against web-based surveys: since logistics of e-mail and web-based surveys are cheap(er) there is a flood of surveys submitted by researchers (not ruling out the authors of this survey).

3.3.1 Threats to Validity

First, there are certain threats to validity in research that must be taken into consideration (Schaeffer and Presser 2003; Vogt 2007), such as:

1. Self-selection effects (Heckman 1979): Subjects are not randomly assigned to the interest groups of the researcher. They assign themselves and do not do so randomly. For instance, some members can choose not to respond.
2. Attrition effects: Also often called mortality, attrition can occur when subjects drop out of a study. It is another form of self-selection effect, but involves self-selecting out, not in.
3. Volunteer effects: People who consent to being studied can often differ in important ways from those who do not consent to being studied.

4. Interpretation effects: Respondents "construct a 'pragmatic meaning'" that can include their interpretation, the reason it is being asked, and what an acceptable answer would be when they hear or read a survey question.

Hence, in a survey a proportion of the sample will not be traceable, will be refused by the respondent, or will yield incomplete and unusable questionnaires. This would be irrelevant if the lost respondents were themselves a random sample. But this is very unlikely (Sapsford 2007).

Second, the survey is sent to the entire population, leaving the respondents a certain self-bias to reply. There are numerous out of the office replies as well as direct replies by the respondents stating (mostly) that participation in any kind of survey violated their corporate communication policies. No assessment is done as to why many survey invitees did not answer the survey (or why the invitees that did answer, answered).

Third, within the selected company or organisation it is open who answers the survey. Is it a subject matter expert or is it the apprentice?

Fourth, a few web pages in Eastern Europe and Asia Pacific do not provide any English webpage translation and the appropriate e-mail contact cannot not be found. These companies and institutions are out of scope (population) and are not contacted.

Fifth, a few web pages do not provide any online contact, neither e-mail nor web forms. These companies and institutions are out of scope (population) and are not contacted.

Finally, the statistical analysis of the data in this survey is conducted under the two assumptions that (i) the sample is fully unbiased; and (ii) the distribution is normal according to the commonly understood bell-shaped curve (Rea and Parker 2005).

Despite these possible sources of error it is suggested that the analysis of the survey data by the respondents is a reasonable method for conducting the research for this article. Schaeffer and Presser (2003) note that the researcher must determine the level of accuracy he or she will try to achieve with the analytical goals and resources at hand.

3.3.2 Non-probability and Non-response

A key question of survey sampling is how large the research sample should be (Vogt 2007). On the one hand, there is support for large sample sizes as being more accurate (Nardi 2003); on the other hand, it is argued that that in Internet research other rules might apply (Couper 2000) because the Internet population is different than the general population in several respects (accessibility, speed, cost, etc.). Couper (2000) seems to contradict Vogt (2007) in that he states that there is a misguided assumption that in web-based surveys large samples necessarily mean more valid responses. However, taking a self-critical stance, one must regard the relative response rate of 4.36% (185 replies from 4329 submitted and received survey invitations) as insufficient; even by Couper's (2000) standard. There is clearly a large non-response bias (Vogt 2007). This makes a generalization about the target populations quite suspect. Clearly, a larger response rate would have been desirable and more favourable to the result.

3.3.3 Level of Confidence and Error Margin

At any rate, in case the 111 responses were a random sample, here is what the level of statistical significance would be:

The correct sample size depends mostly on the tolerance for uncertainty and how much risk one is willing to take of drawing a false conclusion. Ultimately one needs to decide whether the risk or error is worth accepting (Vogt 2007). Two interrelated factors must be addressed before proceeding with the selection of the sample size: level of confidence and confidence interval, to which the finding must conform (Rea and Parker 2005). A typical level of confidence is normally either 95% (i.e. 5% risk of error) or 99% (1% risk of error). The sampling accuracy indicated as error margin is typically set at 10, 5 or 3%.

In total, 185 participants of the 4239 delivered survey invitations responded. Of these 185 respondents 58 were empty; 11 respondents answered the questionnaire partially; 100 respondents completed the questionnaire. For the remainder of this paper only the 111 full and partial survey replies are relevant.

The conservative calculation approach includes only the 100 full replies in the computation. It does not include the partial replies. This will ensure that, if and when the partial replies are included, the error margin will become smaller, not larger. This approach is favourable for purposes of accuracy. The quantitative data can therefore be analysed with a 95% level of confidence at an error margin of 9.68%, i.e. rounded 10%. The margin of error of 9.68% is an upper bond computed. It does not include the partial replies. Should the partial replies be added to the sample size then the error margin will become smaller, not larger.

Based on the survey design, the survey sample and the precautions to be considered in the interpretation, the following Sect. 4 evaluates the survey data.

4 Discussion of Observations

This section outlines the third phase of the research. It is divided into two main parts. In line with the interview layout in Sect. 2 (Industry Interviews) the survey observations in this section are structured into the RFID industry's perspective of regulation by law (Sect. 4.1) and the RFID industry's perspective of the consumers (Sect. 4.2).

The discussion on the industry's experience with regulation by law (Sect. 4.1) is sub-divided into two parts: a discussion of the survey results with regard to law (radio spectrum, database and privacy law in Sect. 4.1.1.), and with regard to the regulators (legislator, lawyers and judges in Sect. 4.1.2). The discussion on the industry's experience with consumers (Sect. 4.2) is also divided into two parts: a discussion of the survey results with regard to the consumers' information knowledge

(knowledge level of consumers in Sect. 4.2.1.), and with regard to the industry's information policy (information policy to consumers in Sect. 4.2.2). Within each of these four discussions the survey results are grouped, on the one hand, by industry (telecommunication, retail, label/printing, logistics, information technologies, healthcare/life science, electronics, consulting, and miscellaneous industries); and, on the other hand, by geographical or political territory (USA, European Union, OECD, ITU regions 1–3).

Each of the four discussions includes two observations. One observation made by the authors and one survey feedback observation. Two observations are compared per discussion.

4.1 Industry's Experience with Regulation by Law

4.1.1 Radio Spectrum, Database and Privacy Law

An abstract search on the Emerald Internet database provides 16 journal articles relating to RFID and privacy over several years; in 2007 for instance: (Angeles 2007; Attaran 2007; Butters 2007; Hingley et al. 2007; Lee et al. 2007; Srivastava 2007); 1 journal article relating to RFID and database (Anonymous 2005); and zero journals relating to RFID and radio spectrum. This leads to a first observation: The RFID industry's problems with regulation range in descending order from privacy over database to radio spectrum regulation (O1).

O1 is discussed based on the survey observations of Figs. 1, 2, and 3 that formulate observation 2 (O2):

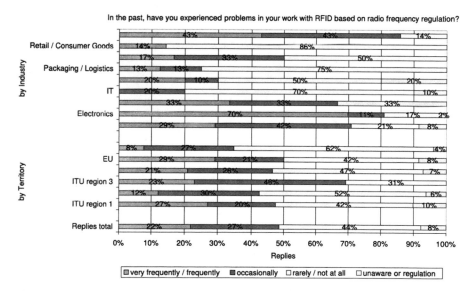

Fig. 1 Replies to survey question on radio spectrum regulation

398   D. Ronzani

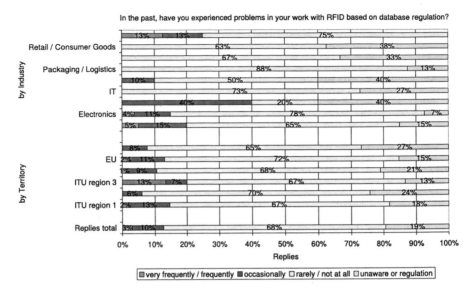

Fig. 2 Replies to survey question on database regulation

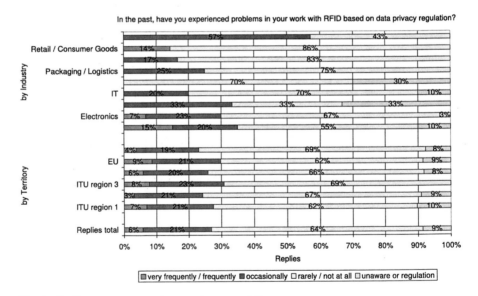

Fig. 3 Replies to survey question on privacy regulation

Spectrum regulation (Fig. 1): Overall by territory and industry only a small percentage of 8% is unaware of radio spectrum regulation. Of the remaining survey respondents aware of the regulation, 44% have no or rarely problems, 22% have occasionally problems, and slightly more than 20% indicate having frequently or very frequently problems with radio spectrum regulation.

By territory, between 0 and 10% of the responding companies and organisations indicate to be unaware of radio spectrum regulation. It is noticeable that survey respondents from ITU region 3 are all aware of spectrum regulation. Unsurprisingly they have the highest percentage of very frequent problems with this regulation. By industry, the knowledge or lack of knowledge of radio spectrum regulation varies. The telecom and ISV; retail and consumer goods; label, printing and paper; as well as the packaging and logistics industries are all aware of radio spectrum regulation. The remaining IT, healthcare and life science, electronics, consulting, and miscellaneous industries are all unaware of the radio spectrum regulation. Noticeable among the unaware industries is that the healthcare and life science industry shows considerable high lack of knowledge (33%).

Database regulation (Fig. 2): Overall by territory and industry, on the one hand, the unawareness about database regulation increases to 19% as compared to the results in Fig. 1. On the other hand, 68% indicate not having any or only rarely problems at all with database regulation. 10% of the survey respondents have occasionally problems with the regulation. Finally, only a small percentage encounters frequently or very frequently problems with database regulation.

Between 13 and 27% of all survey respondents grouped by territory claim to have no knowledge about database regulation. This lack of knowledge about database regulation pervades through the industries as well but increases significantly for certain industries. Hence, by industry, it is noticeable that the telecom and ISV industry is the only industry entirely aware of database regulation. In contrast, the healthcare and life science; retail and consumer goods; and the miscellaneous industries have a high percentage of lack of knowledge on database regulation (approximately 40%). The healthcare and life science industry also deviates from the overall pattern in that 40% encounter occasional problems with database regulation.

Privacy regulation (Fig. 3): Overall by territory and industry there is a small percentage of 9% that is unaware of privacy regulation and 64% do not have any or only rarely problems with data privacy regulation. About 20% have occasionally problems. Only a few indicate having frequently or very frequently issues with privacy regulations (6%).

By territory only a small amount of approximately 10% or less of the survey respondents indicate being unaware of data privacy regulation. It is noticeable that also in this question about privacy regulation the survey respondents from ITU region 3 are all aware of data privacy. By industry the telecom and ISV; retail and consumer goods; label printing and paper; as well as the packaging and logistics industries are all aware of data privacy regulation. It is noticeable and industry inherent – but not visible in Fig. 3 due to the allocation of frequent and very frequent problems into one category in all the graphs within the figures of this article – that the retail and consumer goods industry is the only industry that has very frequent problems. All the electronics and consulting industries in Fig. 3 actually only have frequent problems with privacy regulation.

A comparison of Figs. 1, 2, and 3 shows that of all three regulations the database regulation is the least known regulation. The awareness of radio spectrum and data privacy regulation is much higher. The telecom and ISV are the best informed industry. It is argued here that the telecommunication regulation of the 1990s has promoted the awareness of applicable regulation in this industry. Furthermore the healthcare and life science industry shows a higher trust in the RFID regulation. Here it is argued that the strong regulation in the healthcare and life science industry has promoted and strengthened that trust.

Figure 1 on frequency regulation shows the most frequent or very frequent problems with radio spectrum regulation. Figure 2 on database regulation shows a high lack of knowledge about database regulation. Figure 3 on privacy regulation shows the most regular response pattern. Based on the emphasis given in this research, from an industry perspective one would expect a different allocation and importance of the topics in the journal articles reviewed. First, one would expect most articles to research radio spectrum regulation because this is where the industry has the most problems in adoption diffusion of RFID technology (total of 50% of occasional, frequent and very frequent problems in Fig. 1). It has also been suggested that radio spectrum is fundamental to the Internet of Things, and its allocation is possibly the most important key issue for many regulators and government agencies (European Commission – Information Society and Media Directorate General 2006). Nonetheless, within the reviewed journal articles there is none on this topic.

Second, one would expect there to be the least articles about database regulation because a large part of the RFID industry is unaware of such regulation and because there are the least problems with database regulation (Fig. 2). Indeed there is only one article on RFID and databases.

Third, one would expect only a few articles about privacy regulation because, on the one hand, the overall awareness about privacy issues is good, and, on the other hand, 50% of the survey respondents indicate having some sort of problems with the regulation. But as the abstract search on the Emerald Internet database reveals, the contrary is true: privacy regulation is by far the most topical journal article subject.

O1 and O2 do not match: Privacy is the most contemporary research topic; database is the second most topical research subject, yet with only one article almost similarly insignificant as the inexistent articles about radio spectrum.

4.1.2 Legislator, Lawyers and Judges

Regulation by law involves the legislator, lawyers and judges. First, in one of the interviews an RFID industry partner alludes to the flawed impression by the regulator (legislator) influenced by pressure groups. Second, the RFID industry indicates in the survey that 50% of the survey respondents do not engage any legal experts for legal questions related to RFID. One quarter of the survey respondents uses internal legal experts and the remaining quarter outsources the legal issues to external legal experts. Third, to the knowledge of the authors there are not many court cases involving radio spectrum, database or privacy regulation relating to RFID technology. This leads to a third observation: From an RFID industry perspective the legal regulator's expertise on RFID technology is insufficient (O3).

O3 is discussed based the survey observation of Figs. 4, 5, and 6 that formulate observation 4 (O4):

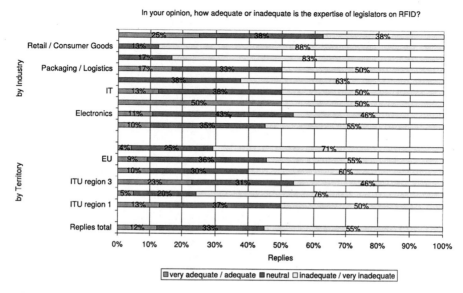

Fig. 4 Replies to survey question on legislator's RFID expertise

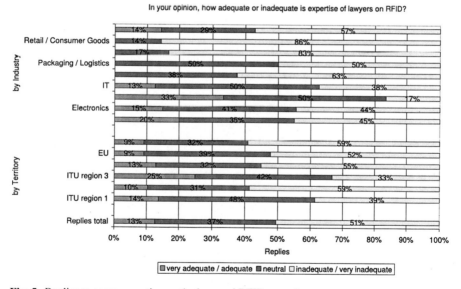

Fig. 5 Replies to survey question on the lawyers' RFID expertise

The survey respondents provide feedback on their experience with the legal regulator's expertise on RFID technology. The responses about the legislator's expertise (Fig. 4), the lawyers' expertise (Fig. 5), and the judges' expertise (Fig. 6) by territory are similar. Only a minority (approximately 10%) of the survey respondents in all territories indicate that the expertise of the legislator, lawyers and judges is

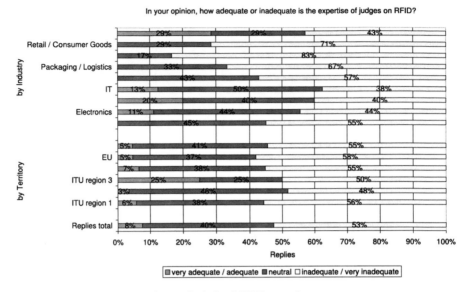

Fig. 6 Replies to survey question on the judges' RFID expertise

adequate or even very adequate. A constant number, of approximately 30–40%, rest neutral on the question. The remaining 50–70% of the responding companies and organisations think that the legislator, lawyers and judges have inadequate or very inadequate expertise. Two points are noticeable: First, of all bad marks given by the survey respondents, the legislator qualifies as having the best (of the bad) expertise (Fig. 4), followed by the lawyers' slightly worse expertise (Fig. 5), ending with the expertise of the judges being the worst (Fig. 6). Second, the ITU region 3 distributes high marks (approximately 25%) for adequate expertise to legislators, lawyers and judges.

Within the responses by industry there are five industries that trust more in the abilities of the legal regulator giving some adequate and very adequate marks: telecom and ISV; IT; healthcare and life sciences; and electronics (Figs. 4, 5 and 6). The retail and consumer goods, label and printing as well as the miscellaneous industries distrust more in the legal regulator and do not have adequate or very adequate opinion of the legal support (Figs. 4, 5 and 6). The packaging and logistics as well as the consulting industries do not show such a constant response pattern.

The discussion on the adequacy of the legal regulators RFID expertise shows that to a large part the RFID industry does not believe that legal support on RFID technology is in good legal hands. If one acknowledges that there have not been many court decisions on RFID technology, then it is understandable that the RFID industry would assess the judiciary's expertise as inadequate or very inadequate. However, of more concern is the fact that the legislator and lawyers get bad marks from the industry as well. The RFID industry does not seem to see itself represented satisfactorily by the legislator and lawyers.

O3 and O4 match: The survey respondents see themselves badly represented by the legal regulator. Within this bad assessment, the lawyers get the best (of the bad), the legislator the second worst, and the judges the worst marks.

4.2 Industry's Experience with Consumers (General Public)

4.2.1 Knowledge Level of Consumers

In 2006 the European Commission issued an online survey on the challenges, opportunities, and threats of RFID (Commission of the European Communities 2006). In that survey the availability of information for the consumer to make an informed decision about RFID is rated as insufficient by approximately 60% of the survey respondents, whereby nearly 65% of the survey respondents are citizens and not companies or organisations. As by definition of that survey all the survey respondents are aware of RFID technology, the Commission anticipates that there is a significant RFID knowledge gap by the general public (Commission of the European Communities 2007). If the consumer has insufficient information to make an informed decision then the consumer is obviously not knowledgeable about the topic. This leads to a fifth observation: From an RFID industry perspective the consumer (general public) is badly informed about RFID technology (O5).

O5 is discussed based the survey observations of Fig. 7 that formulates observation 6 (O6):

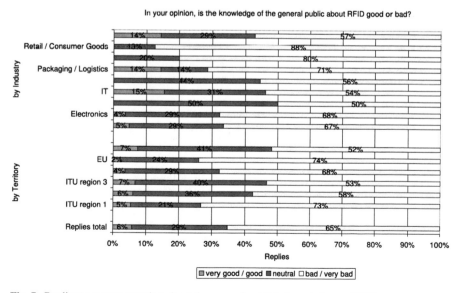

Fig. 7 Replies to survey question about the general public's knowledge on RFID

Overall, only a very small percentage of all survey respondents by territory view the knowledge of the general public with regard to RFID as good or even very good (6%). For approximately 10% of the companies and organisations participating in the survey the knowledge of the general public is very bad. The majority of 65% view the knowledge to be bad or even very bad. The remainder is neutral (Fig. 7).

The division by territory (as compared to the industries) shows all the respondents assess the consumers' knowledge good or even very good, even if only by a small percentage. The replies by industry, however, show that the retail and consumer goods, label and printing, miscellaneous, and healthcare and life science industries view the general public's knowledge on RFID to be bad or even very bad. Noticeable is the high percentage of bad or even very bad classification by the retail and consumer goods industry. The healthcare and life science industry stands out by only having a high (50%) neutral opinion of the general public's RFID understanding.

O5 and O6 match: The respondents of this online survey perceive the consumer (general public) to be badly informed about RFID technology.

4.2.2 Information Policy to Consumers

Confirmation of H 3 leads to the question where the communication to the consumer fails. Diffusion of innovation theory (Rogers 2003) explains the adoption and diffusion of different types of new information and communication technologies (Kim and Galliers 2004). Diffusion is a particular form of communication. Diffusion communicates a new idea from one party to another (or several other) party (ies). Among the many different characteristics of innovation (Damanpour and Wischnevsky 2006; Greenhalgh et al. 2004; Nutley and Davies 2000; Rogers 2003; Swanson and Ramiller 2004; Wolfe 1994) potential adopters engage in information seeking behaviours to learn about the expected consequences of using the innovation (Juban and Wyld 2004). This leads to a seventh observation: The RFID industry provides the best information about RFID technology to the consumer (O7).

O7 is discussed based on the survey observations of Figs. 8, 9, and 10 that formulate observation 8 (O8).

RFID Industry (Fig. 8): Overall 21% of the survey respondents believe the general public is informed well or very well by the RFID industry, whereas 37% believe the information by the industry is very bad or bad. The remainder of about 40% is neutral.

By territory it is noticeable half of the survey respondents in ITU region 3 believe the general public is well informed by the industry. Furthermore, on the European continent (EU and ITU 1) the information policy seems to be less good (13 and 14%) than in the other territories. Noticeable is also that in ITU region 3 there are 50% very good or good marks for the information policy.

The division by industry shows a similar response pattern as the responses by territory, with the following differences: The label, printing and paper industry does not believe the general public is well informed by the RFID industry itself. Noticeable is also that the telecom and ISV; retail and consumer goods; and label,

In your opinion, how good or bad is the general public informed about RFID by the RFID industry?

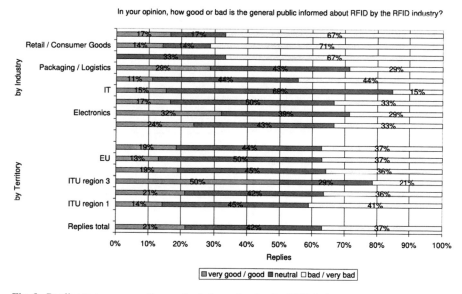

Fig. 8 Replies to survey question on the information by the RFID industry

In your opinion, how good or bad is the general public informed about RFID by the RFID regulatory?

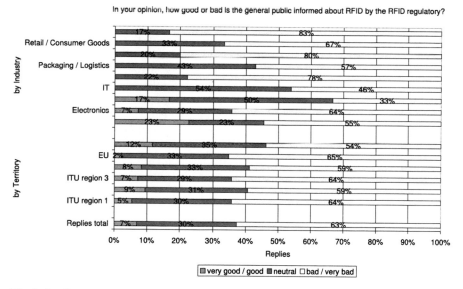

Fig. 9 Replies to survey question on the information by the regulator

printing and paper industries give high bad marks for the information policy by the RFID industry itself (approximately 70%). Furthermore, the label, printing and paper industry only gives bad marks altogether or remains neutral on the question.

Regulator (Fig. 9): Overall, the results get significantly worse in the assessment of RFID information policy by the RFID regulatory. Only 7% of the respondents

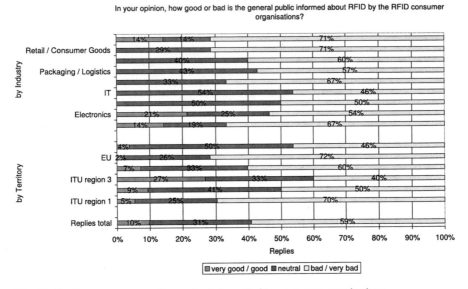

Fig. 10 Replies to survey question on the information by consumer organisations

believe the legal regulator has a good RFID information policy. 63% believe the information policy is bad or even very bad. Approximately a third remains neutral.

By territory the USA gives the best assessment on information policy to the legal regulator (12%), whereas in the EU the very good or good marks drop to 2%. The response allocation by industry shows that the telecom and ISV; retail and consumer goods; label, printing and paper; packaging and logistics; IT; and miscellaneous industries give only either bad or very bad marks to the information policy by the legal regulator, or remain neutral. There

As compared to the communication by the RFID industry (Fig. 8), the information policy by the RFID regulator (both by industry and territory) shows increased replies for the classifications bad and very bad (Fig. 9). Only the healthcare and life science; electronics; and consulting industries have certain good or very good opinion on the information policy by the RFID legal regulator.

Consumer organisations (Fig. 10): As compared to Figs. 8, 9, and 10 shows overall (by both industry and territory) a slightly more fragmented result. In direct comparison to the information policy by the legal regulator the information policy by consumer organisations improves slightly. 10% assess the information policy as good or very good, about 60% assess the information policy to be bad or even very bad, and the remainder of about 30% remains neutral on the question.

Within the division by territory the relatively high peak of good assessments by the survey respondents in ITU region 3 is noticeable. The allocation by industry shows furthermore that only the telecom and ISV; electronics; and consulting industry view the information by RFID consumer organisations as good, whereas the other six industries do not give any good points in this respect.

O7 and O8 match: The respondents of the survey perceive the consumer (general public) to be badly informed about RFID technology whereby the legal regulator provides the worst information, followed by consumer organisations. The RFID industry gives itself the best marks on information policy to the consumers.

5 Conclusion

This article provides insight into the RFID industry's experience with (i) regulation by law and (ii) the consumers. Eight observations have been discussed, yielding the following results in Table 1:

Overall, one can suggest that – from an RFID industry perspective – both the regulator and the consumer have insufficient knowledge and support with regard to the adoption and diffusion of RFID. The following three recommendations are proposed by the authors to the RFID industry to support the adoption and diffusion of RFID:

First, the RFID industry should engage in a better dialogue with the legal regulator, for instance by knowledge exchange between the RFID industry and in-house and external legal counsel to promote the constructive dialogue about RFID technology. It is acknowledged here that not everyone can and should become an expert on every topic. But this research has clearly shown that there is room for improvement in the exchange of professional expertise.

Second, the RFID industry should strengthen its knowledge on applicable regulation, for instance, by working closer together with the legal regulator, especially

Table 1 Summary of eight observations

Observation #	Authors' Observations 1, 3, 5 and 7	Survey Observations 2, 4, 6 and 8
O1 and O2	The RFID industry's problems with regulation range in descending order from privacy over database to radio spectrum regulation.	O2 does not match O1.
O3 and O4	From an RFID industry perspective the legal regulator's expertise on RFID technology is insufficient.	O4 matches O3.
O5 and O6	From an RFID industry perspective the consumer (general public) is badly informed about RFID technology.	O6 matches O5.
O7 and O8	The RFID industry provides the best information about RFID technology to the consumer.	O8 matches O7.

the in-house and external lawyers. This research shows that the RFID industry is unaware of certain regulations that could be used in its favour.

Third, the RFID industry should re-evaluate its information policy to the general public. Either the counter-information by the consumer organisations and the legal regulator are much better, or the RFID industry's information policy is not yet there where it should and could be to promote RFID technology to its advantage.

References

Anonymous (2005) The pros and cons of RFID: data analysis. Strategic Direction, 21(5):24–26

Albrecht K, McIntyre L (2005) Spychips: how major corporations and government plan to track your every move with RFID. Nelson Current, Nashville, TN

Angeles R (2007) An empirical study of the anticipated consumer response to RFID product item tagging. Ind Manage Data Syst 107(4):461 et seq

Attaran M (2007) RFID: an enabler of supply chain operations. Supply chain management: Int J 12(4):249 et seq

Butters A (2007) RFID systems, standards and privacy within libraries. Electron Libr 25(4):430 et seq

Commission of the European Communities (2006) The RFID revolution: your voice on the challenge, opportunities and threats. Commission of the European Communities, Brussels

Commission of the European Communities (2007) Results of the public online consultation on future radio frequency identification technology policy "The RFID Revolution: Your Voice on the Challenges, Opportunities and Threats". Commission of the European Communities, Brussels

Couper MP (2000) Web surveys. A review of issues and approaches. Public Opin Q 64:464 et seq

Damanpour F, Wischnevsky JD (2006) Research on innovation in organizations: distinguishing innovation-generating from innovation-adopting organizations. J Eng Technol Manage 23:269 et seq

EPCglobal Inc (2008). http://www.epcglobalinc.org. Accessed last 19 December 2008

European Commission (2006) 2006/804/EC, Commission decision of 23 November 2006 on harmonisation of the radio frequency identification (RFID) devices operating in the ultra high frequency (UHF) band

European Commission (2008) – Information Society and Media Directorate General (2006) From RFID to the internet of things – pervasive networked systems. Office for Publications of the European Commission. Luxembourg

Greenhalgh T et al (2004) Diffusion of innovations in service organizations: systematic review and recommendations. Milbank Q 82(4):581 et seq

Heckman J (1979) Sample Selection Bias as a Specification Error. Econometrica 47(1):153–61

Hingley M, Taylor S, Ellis C (2007) Radio frequency identification tagging: supplier attitudes to implementation in the grocery retail sector. Int J Retail Distrib Manage 35(10):803 et seq

ITU (2005) Radio regulations. http://www.itu.int/publ/R-REG-RR/en. Accessed 24 June 2009

Johnson JM (2001) In-depth interviewing. In: Gubrium JF, Holstein JA (ed) Handbook or interview research – context and method. Sage, London

Juban RL, Wyld DC (2004) Would you like chips with that?: consumer perspectives of RFID. Manage Res News 27(11/12):29 et seq

Kelly EP, Erickson GS (2005) RFID tags: commercial applications v. privacy rights. Ind Manage Data Syst 105(6):703 et seq

Kim C, Galliers RD (2004) Toward a diffusion model for internet systems. Internet Res 14(2):155 et seq

Lee SM, Park S, Yoon SN, Yeon S (2007) RFID Based ubiquitous commerce and consumer trust. Ind Manage Data Syst 107(5):605 et seq

Nardi PM (2003) Doing survey research – a guide to quantitative methods. Pearson Education, Boston, MA

Nutley S, Davies HTO (2000) Making a reality of evidence-based practice: some lessons from the diffusion of innovations. Public Money Manage 20(4):35 et seq

Rea LM, Parker RA (2005) Designing and conducting survey research – a comprehensive guide, 3rd edn. Jossey-Bass, San Fransisco, CA

Rogers EM (2003) Diffusion of innovation, 5th edn. Free Press, New York, NY

Rothensee M, Spiekerman S (2008) Between extreme rejection and cautious acceptance – consumers' reactions to RFID-based IS in retail. Soc Sci Comput Rev 26(1):75 et seq

Sapsford R (2007) Survey research, 2nd edn. Sage, London

Schaeffer NC, Presser S (2003) The science of asking questions. Annual Review of Sociology, 29:65–88.

Shank G, Cunningham DJ (1996) Modeling the six modes of peircean abduction for educational purposes. In: Annual meeting of the Midwest AI and cognitive science conference, MAICS 1996 Proceedings, Cincinnati, OH

Srivastava L (2007) Radio frequency identification: ubiquity for humanity. Info 9(1):4 et seq

Strickland LS, Hunt LE (2005) Technology, security, and individual privacy: new tools, new threats, and new public perceptions. J Am Soc Inf Sci Technol 56(3):221 et seq

Swanson EB, Ramiller NC (2004) Innovating mindfully with information technology. MIS Q 28(4):553 et seq

Vogt WP (2007) Quantitative Research Methods for Professionals in Education and Other Fields, Pearson Education, Boston

Wolfe RA (1994) Organizational innovation: review, critique and suggested research directions. J Manage Stud 31(3):405 et seq

Xiao Y, Yu S, Wu K et al (2007) Radio frequency identification: technologies, applications, and research issues. Wireless Commun Mobile Comput 7(4):457 et seq

Yin RK (2003) Case study research, 3rd edn. Sage, Thousand Oaks, CA

Investment Evaluation of RFID Technology Applications: An Evolution Perspective

Andriana Dimakopoulou, Katerina Pramatari, Angeliki Karagiannaki, George Papadopoulos, and Antonis Paraskevopoulos

Abstract Prior empirical research on the evaluation of RFID technology treats and assesses individual RFID applications independently and in isolation from each other. However, literature on investment evaluation of information technologies has recognised and utilised the significance of evaluating "interdependent" information systems (IS) projects with synergies. Moreover, previous studies when appraising the business value of an RFID investment ignore its opportunity to offer and evolve into additional follow-on investments in the future. Nevertheless, the importance of this notion has been acknowledged by the pertinent literature for the evaluation of other information technologies. This chapter proposes an approach for the investment evaluation of RFID applications considering them rather as a bundle of interdependent and sequential investments than as stand-alone ones. The results from a case study demonstrate how the proposed approach can be employed for the evaluation of RFID projects and offering an additional insight into evaluating investments in RFID applications.

1 Introduction

Radio Frequency Identification (RFID) is the generic name for technologies that use radio waves to automatically identify individual items that carry such identification tags (Cole and Ranasinghe 2008). This technology dramatically increases the ability of an organisation to obtain an enormous amount of data about the location movement and properties of any entity that can be physically tagged and wirelessly

A. Dimakopoulou (✉)
Department of Management Science and Technology, Athens University of Economics and Business, Athens, Greece
e-mail: andrianadima@aueb.gr

D.C. Ranasinghe et al. (eds.), *Unique Radio Innovation for the 21st Century*,
DOI 10.1007/978-3-642-03462-6_19, © Springer-Verlag Berlin Heidelberg 2010

scanned (Curtin and Kauffman 2007). Supply chain management, anti theft systems, asset tracking, airline baggage handling, electronic tolling, and facilities management (e.g. libraries) are examples of areas where RFID can be applied. Widespread adoption of this technology is observed in the retail industry. In this environment, RFID can support a range of applications from upstream warehouse and distribution management to retail-outlet operations including shelf management, promotions management and innovative consumer services, as well as applications for the whole supply chain such as product traceability (Pramatari et al. 2005).

Previous research estimates the business value of RFID applications considering them as independent projects. However, synergies among these applications exist as they share a common infrastructure. According to the literature, exploiting these synergies during the investment evaluation of technologies can result in sharing valuable resources, the decrease of total expenditures and the increase of benefits (Santhanam and Kyparisis 1996). Nevertheless, previous empirical studies have neglected the employment of these synergies for the investment evaluation of RFID.

In addition, prior research when evaluating an individual RFID application ignores its opportunity to offer and evolve to additional follow-on investments in the future. On the contrary, the pertinent literature on IT evaluation has acknowledged the importance of this notion. As Curtin and Kauffman (2007) argue, interesting opportunities for research on RFID emerge: "Researchers could test the notion that infrastructure technologies may not be the primary drivers of business value themselves but rather create real options for additional follow-on investments".

For the investment evaluation of RFID, this chapter proposes an approach which exploits the above ideas. This approach considers RFID applications rather as interdependent projects with synergies that can result in follow-on investments than as stand-alone applications. The results from a case study demonstrate how the proposed approach can be employed for evaluating RFID projects and enhancing business gains, offering an additional insight into evaluating RFID investments.

The chapter is organised as followed. The first section provides a brief literature review on RFID technology and its evaluation. The review continues with the analysis of the concepts of Information Systems (IS) interdependence and evolution to justify their application in the context of evaluating RFID applications. The gap of the existing literature and the aim of the chapter are discussed. The next section describes the proposed approach for the RFID investment evaluation, while the following section applies this approach to a case study. The chapter ends with conclusions, limitations of the study and further research avenues.

2 Background

2.1 Business Evaluation of RFID Technology

Applications of RFID technology based on a common infrastructure can operate simultaneously (Violino 2005) and support a variety of different business aims and processes. This trait constitutes one of the characteristics that distinguish RFID from other information technologies. Another distinctive trait of the technology is the need to invest in a new asset: "the tag". Although this type of cost is fixed per one unit (one tag); it can incur a variable cost which can be altered and highly increased according to several parameters such as the type of the application, the number of tagged objects and the level of tagging (item/case/pallet). For example, a company can invest in one RFID application during the first year by tagging one million of products, while investing further in subsequent applications in the future by tagging additional number of items (products). Consequently, the total cost of the tags can be varied year by year according to a company's investment.

As with all novel technologies such as RFID, projects and studies evaluate their potential. Although it is believed that RFID can offer a variety of business benefits, some initial adopters have stated that there will be no returns on investments in RFID in the end (Tajima 2007). Previous research anticipates the value of this technology in several areas related to "supplier facing" or "customer facing" activities (Lee et al. 2008).

The majority of the empirical studies on RFID evaluation have focused on the first category (Lee et al. 2008) which deals with back office operations such as inventory and warehouse activities. In particular, studies based on several tools such as mathematical or simulation models (Doerr et al. 2006; Fleisch and Tellkamp 2005; Kok et al. 2008; Rekik et al. 2008; Wang et al. 2008), assess the impact of the technology on inventory management and more specifically on inventory replenishment, inventory inaccuracies, product misplacement errors and shrinkage. In addition, research based on financial, simulation or hybrid approaches (Bottani and Rizzi 2008; Karagiannaki et al. 2007; Kim et al. 2008; Subirana et al. 2003; Wamba et al. 2008) has evaluated the use of RFID for warehouse and logistics operations such as shipping, orders receiving and put away processes, concluding that RFID can lead to labour, material and transportation cost savings.

Except for the above studies which deal with back office business activities, only limited empirical research studies (Lee et al. 2008; Tzeng et al. 2008) has anticipated the impact of RFID technology on customer facing activities such as customer service.

2.2 Application of Information Systems Interdependence to Evaluating RFID

Most of the mentioned empirical research on evaluating RFID appraises individual RFID applications, such as inventory replenishment and orders receiving, ignoring the assessment of extensive variety of the RFID applications and their opportunity to cooperate, through the presence of "interdependency". This effect results when specific sets of projects share a specific feature in common (Iniestra and Gutierrez 2009).

"General Systems Theory" (von Bertalanffy 1969; Yourdon 1989) and the literature regarding the evaluation and the selection of IS and other kinds of projects

(Iniestra and Gutierrez 2009; Lee and Kim 2000; Santhanam and Kyparisis 1996; Verma and Sinha 2002; Liesio et al. 2008) encompass the notion of interdependence between systems referring to the interaction that exists among them, as they rarely can work in isolation. According to the literature, the main types of interdependencies or synergies are:

1. "resource or cost", when IS or other kind of projects share the same resources and cost,
2. "benefit or value interdependencies", when the total organisational benefits can be increased if related projects are implemented together and
3. "technical" or "follow-on synergies", when the development of a project requires the development of a related one.

In other words, interdependent projects are interrelated projects which share common characteristics.

The above notion can be applied to the case of RFID. This technology based on one common infrastructure has the facility to interface and bring together the functionality of key systems (Curtin and Kauffman 2007) which can be viewed as interdependent.

The exploitation of these synergies among RFID projects is beneficial for an organisation. By evaluating and selecting interdependent projects valuable resources can be shared. Total resource expenditures (Santhanam and Kyparisis 1996) can be reduced and more benefits can be gained than those that would be derived if the IS projects are evaluated and implemented separately (Lee and Kim 2000). As a result, the investment evaluation of interdependent RFID projects can be more adequately justified than if these RFID projects are assessed in isolation. According to Panayi and Trigeorgis (1998), traditional valuation approaches often do not take into consideration the flexibility of investments treating them as 'stand-alone' investments. However, investment projects can be considered as a 'bundle' of interrelated investment opportunities, the earlier of which are prerequisites for others to follow (Panayi and Trigeorgis 1998).

2.3 Application of Information Systems Evolution to Evaluating RFID

Another issue which has been neglected by the studies mentioned in Sect. 2.1 on evaluating individual RFID applications is the evolution among these applications. That is the opportunity that occurs when an investment in one application can result in a second stage investment in another application. This idea is supported by and linked with the "follow-on synergy" interdependence (see Sect. 2.2), according to which a project depends on the execution of another.

According to this notion, technology evolves through several stages based on several factors such as its maturity level, the degree of the required business transformation, and the type of business benefits and goals that it underpins (Nolan 1973; Venkatraman 1990). The growth options that a technology may generate can be considered as a kind of evolution. For example, Taudes (1998) studied software growth options referring to software base configurations which can evolve and result in the introduction of additional IS functionalities.

Lee (2007) has applied this idea of IS growth to an RFID application for supply chain management. The author states that RFID applications such as inventory audit and checkout, which are part of the "substitution" stage as they mainly substitute barcodes, can evolve to the following "scale" level, which includes applications covering multiple points of the supply chain such as product origin tracking. The last evolution level, which is called "structural", consists of applications such as promotion management and reverse logistics which generate a higher business value, new business activities and processes.

The application of IS and RFID evolution is significant in an RFID investment evaluation context because RFID technology can serve many applications. It can, initially, for example support inventory business activities and ultimately customer service as a result of a follow-on investment. Thus, when evaluating an initial investment, the value of a potential second stage investment has to be considered as an added value to the first stage project. Otherwise, the value of the initial project can be underestimated. According to Dos Santos (1991), the value of a technology project is the sum of the value that a first stage project entails and the one derived from the second stage project. Due to the increasing number of RFID applications, several combinations and groups of RFID applications can be generated resulting in RFID 'evolution paths' comprised of initial and future projects, which should be evaluated jointly rather than in isolation. Consequently creating, as argued by Curtin and Kauffman (2007), interesting opportunities for research into evaluating RFID technology deployments: "Researchers could test the notion that infrastructure technologies may not be the primary drivers of business value themselves but rather create real options for additional follow-on investments".

The above issues of technology growth from initial to future investments and the interdependencies that these investments can encompass (Table 1) have not been implemented in empirical studies for evaluating RFID investments. This chapter proposes the application of these notions to the evaluation of RFID. For the above aim, this study seeks to address the following questions:

- What and how can RFID applications be considered for their evaluation as interdependent and implemented in sequence as initial and future investments?
- What is the business value of an investment in a group of initial and follow-on interdependent RFID applications compared to the one derived from an individual RFID application?

Table 1 The notions of IS interdependence and evolution

Notions	Definition	Business aim	Supporting literature	Application to RFID evaluation
IS interdependencies/ synergies	"Resource or cost synergies": IS or other kinds of projects share the same resources and cost "Benefit or value synergies": Total organisational benefits can be increased if related projects are implemented together "Technical" or "follow-on synergies", The development of a project requires the development of a related one	Reduction of the total resource expenditures More benefits can be gained	Iniestra and Gutierrez (2009); Lee and Kim (2000); Santhanam and Kyparisis (1996); Verma and Sinha (2002); Liesio et al. (2008), "General Systems Theory": von Bertalanffy (1969); Yourdon (1989)	RFID, based on one common infrastructure can interface and bring together many systems (Curtin and Kauffman, 2007) which have synergies
IS evolution/ growth options	Technology evolves through several stages Infrastructure technologies create real options for additional follow-on investments (Curtin and Kauffman, 2007)	Avoid value underestimations of an initial investment	Nolan (1973); Venkatraman (1990), Taudes (1998); Lee H. (2007); Curtin and Kauffman (2007)	RFID evolution (Lee H. 2007) Exploit the options for follow-on investments (Curtin and Kauffman 2007)

3 An Evolution Approach for RFID Investment Evaluation

In order to address the above questions regarding the evaluation of investments in RFID technology, the following approach is proposed, comprised of the two following phases:

- Identification of groups of interdependent and sequential investments in RFID applications ("RFID evolution paths") and
- Financial assessment of these "RFID evolution paths".

3.1 "RFID Evolution Paths"

The aim of creating groups of interdependent RFID applications is to understand the kinds of applications that can be implemented together in order to exploit their synergies towards business benefit. The goal of creating a sequence within this group is to help a firm identify RFID applications that can be seen as prerequisites or initial investments for the implementation of other follow-on applications.

A required step for identifying the groups is to form a list with candidate RFID applications which can be implemented in an organisation. Towards this goal, the current basic business processes of an organisation can be analysed and modelled in order to identify possible inefficiencies and opportunities for improvement. Based on the identified inefficiencies, the basic aim and the needs of an organisation, as well as the variety of the applications from the literature, a list with the candidate RFID applications is created. This list can cover the whole supply chain of an organisation including logistics, sales, operations, and service. These applications can be differentiated based on their aim, the business process they support, and the application user.

The grouping of the listed RFID applications can be based on Porter's (1985) business activities classification. Applications which support the same business process or activity can be grouped together. As a following step, the "follow-on or technical interdependence" can be exploited in order to identify and position first the applications which are prerequisites for implementing the remaining application in the group. In this manner, "RFID evolution paths" are created as groups comprised of initial and follow-on interdependent RFID applications.

3.2 Financial Assessment of "RFID Evolution Paths"

The aim of the assessment is to compare the business value which is created by the joint appraisal of the grouped RFID applications to that which can be derived from individual assessment of each application separately. This comparison serves to provide evidence to support the approach that better justifies the investment in RFID.

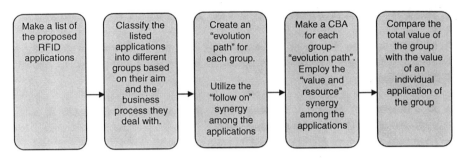

Fig. 1 An evolution approach for RFID investment evaluation

As a first step, the implementation cost of RFID applications in the same group is estimated and classified into individual applications and into a common application which can be shared among the grouped applications ("resource or cost interdependence"). Cost categories are capital costs consisting of the cost of hardware and software that the applications require (RFID equipment such as the readers, antennas, servers and the PCs, customization of the software, integration with the legacy systems, installation cost, software commercial licenses) and the operational costs comprised of the tagging cost, the maintenance and the updates of the software and hardware, and training. The quantification of this cost can be based on market prices and the literature.

Monetary benefit categories taken from the literature such as operational cost decreases can be considered to estimate the gains derived from the introduction of RFID. The previous estimations can be based on the anticipated impact of this technology on current business processes of a firm by evaluating related performance metrics. An example metric is to estimate the decrease in the time required to complete a specific business process as a result of utilising the RFID enabled solution.

Adding the estimated costs and the benefits of each of the applications (utilizing the "benefit or value synergy") within the examined group, a cost benefit analysis (CBA) referring to the total value of the examined group of applications is developed. A comparative analysis among the latter total value from the group and those derived from individual appraisals of each application in the group is made. Fig. 1 depicts the steps of the proposed evaluation approach.

4 Application of the Proposed Approach to a Retailing Organisation

The approach described above is applied to a retailing organisation which merchandises telecommunication services and products. The organisation is interested in improving customer service and considers RFID technology as a new technology that can support this objective. In order to justify investments in RFID applications the firm needs to understand the variety and the aim of these applications and anticipate their value.

4.1 Identification of RFID Group-Evolution Paths

Interviews with the managers and the employees in the retail stores of the enterprise, observations, and business process modelling were employed as tools to analyze the current business processes of the organisation with regards to sales, service and warehouse operations. Based on the previous, specific inefficiencies were revealed, such as: duplication of work, difficulties and delays for the location of the required documents and products in the inventory, lack of information on the products received and sold, customer delays, incomplete and inadequate product and promotion information during sales service.

Given the identified flaws, the aim of the organisation to improve customer service and the variety of the RFID applications derived from the literature, a list of potential RFID applications (presented in Table 2) for the specific organisation emerged. Porter's (1985) categorization of the primary business activities of an organisation into logistics, sales, operations and after sales service is utilized as one form of interdependency among the listed RFID applications in order to classify them into groups.

The following step is to arrange the grouped RFID applications in a sequence in order to identify the applications which are required for the operation of others, based on the "follow-on or technical synergy" (Liesio et al. 2008; Santhanam and Kyparisis 1996). The Delphi approach was utilised as a means to gather information from IT and business experts for this purpose.

An example of an application grouping included those that dealt with store inventory. In order for the "inventory replenishment" application to operate and notify that stock levels have reached safety stock levels require the "inventory audit" application monitoring inventory level in real-time. Similarly, in order for the "inventory audit" to operate, the "orders receiving" and the "check-out" applications are required. Collectively, these applications automatically update inventory levels when products enter and leave a retail store. These updates play an important role for accurately auditing inventory. As a result of the grouping and the sequential arrangement an evolution path for the inventory is generated, as depicted in Fig. 2. According to this "evolution path", a firm can invest initially in RFID for supporting the "orders receiving" activities. Then invest further in other related applications (such as the audit or the inventory replenishment application). These follow-on applications build on the technological capabilities yielded by the initial investment in the "orders receiving" application.

As shown in Fig. 3, following an identical process yields other "RFID evolution paths" consisting of initial and interdependent applications in different of activity, such as sales and promotions management. In particular, an organisation can initially invest in an RFID application which involves the installation of

Table 2 The list of candidate RFID applications

Category	RFID applications	References	Description
Logistics	Order receiving	Bottani and Rizzi (2008); Karagiannaki et al. (2007)	Ordered products entering the inventory room are automatically identified by the readers placed at the entrance to the inventory room. The number of products in the inventory is updated.
	Inventory audit/count	Fleisch and Tellkamp (2005), Rekik et al. (2008)	Products stored in the inventory room are automatically identified and counted on a real time basis.
	Inventory replenishment	Wang et al. (2008), Kok et al. (2008)	System provides alerts to the inventory managers notifying that the stock in the inventory has reached a critical level and needs replenishment
Sales	Check out	Moon and Ngai (2008), Uhrich et al. (2008)	Products sold at the check-out point are automatically identified. The number of the products in the inventory is updated
	Product information to the consumer	Roussos et al. (2002); Moon and Ngai (2008)	Consumers are able to obtain product related information using an information terminal or a display screens attached to shopping carts
	Product and promotion information to the consumer	Uhrich et al. (2008), Jones et al. (2004)	Consumers get product information and information on specific promotions, on an in-store information terminal or on a display screens attached to shopping carts

Table 2 (continued)

Category	RFID applications	References	Description
	Personalised product and promotion information to the consumer	Roussos et al. (2002);Ngai et al. (2008), Moon and Ngai (2008)	Consumer gets personalised product information and promotional offers on an in-store information terminal or display screens attached a shopping carts
Operations	Anti-theft	Jones et al. (2005), Moon and Ngai (2008)	Products leaving the store are automatically identified. Automated alerts in case of stolen goods
After sales service	Automated service	Lee et al. (2008), Tzeng et al. (2008)	Products brought for service by a customer are automatically identified
	Service traceability	Lee et al. (2008), Tzeng et al. (2008)	The products which are sent to the service companies can be tracked and traced during their transportation

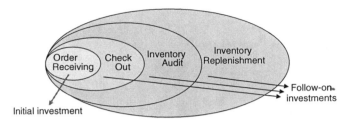

Fig. 2 "RFID evolution path" of applications for inventory management

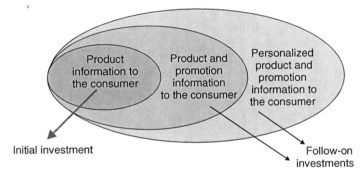

Fig. 3 "RFID evolution path" of applications for sales and promotion management

RFID-enabled information terminals around a store. These terminals can inform shoppers about the characteristics of products or promote specific products on display screens (Jones, et al. 2004). Every time a customer passes a tagged product through an information terminal, he can receive information on the product on the display screen and related promotional information. Alternatively, product information can be displayed on a screen located inside a shopping cart which is equipped with an RFID reader to detect the tagged goods within the cart (Ngai et al. 2008).

Furthermore, the organisation can invest in relevant subsequent RFID applications. For example, the company can add additional required features in the initial RFID infrastructure and support personalised offers and promotions to the consumers. As Ngai et al. (2008) present in their article, a consumer can use an RFID-enabled shopping cart which is networked with system databases such as CRM. As a result, consumers using their loyalty cards in their shopping carts can receive personal recommendations for products on their display screens within the carts. Antennas within the cart can detect the in and out movements of the items and the networked application systems can present shoppers other associated brands or products on their display screens. Marks and Spencer stores, Prada Epicenter stores, Gap, Benetton and DHL fashion are some examples of retailers who have implemented RFID pilots of the above applications (Moon and Ngai 2008). These applications, based on the initial RFID application, can be considered as follow-on investments.

It is expected that the retailing enterprise can increase their benefits and decrease their total expenditures, if they invest once in a central and common infrastructure and invest further in other related RFID applications, rather than investing in only one application. The anticipated result is tested through the following financial assessment.

4.2 Financial Assessment

The aim of this section is to financially assess the investment of the examined organisation in the two applications: "orders receiving" and "check out" derived from the "RFID evolution path" for the inventory. The specific group and the two applications are chosen because of the data availability and the opportunity to estimate and quantify the impact of the technology on the specific retailing enterprise. The ultimate goal is to compare the evaluation outcome coming from the individual appraisal of these two separate applications to the one derived from the combined assessment of the two applications together to conclude whether or not the exploitation of interdependencies among related applications enhances the economic results.

The financial approach of the discounted cash flow, net present value, is used as the value estimation method due to the fact that the specific organisation is interested in an economic and monetary approach for the specific investment. The joint net present value of the two applications is estimated and then compared to the individual net present value of each of the two applications separately. The value of RFID is estimated by considering the annual cash flows based on a cost benefit analysis. The estimations are calculated for 153 retail stores of the examined organisation.

4.2.1 Quantification of RFID Implementation Costs

Market prices and the Delphi approach with IT experts were used to quantify the RFID investment cost for the two applications. This cost encompasses capital (see Table 3) and operational expenses (see Table 4).

Estimation of the capital cost: The capital costs include the cost of the RFID equipment, installation and required software. Based on "resource-cost interdependence", the cost is estimated and divided into a common cost which can be shared among the two applications and the separate costs which result from implementing each individual application. Table 3 contains the details of the division of the capital cost of the investment. This is a prerequisite for the comparative analysis between the value resulted from both of the applications and the value obtained from each individual application.

Estimation of the operational cost: The operational expenditures include the following categories:

- RFID tags cost, the cost of the tagging process and training which remain the same for both applications regardless of the type of application implemented and
- The cost of HW maintenance and SW upgrades is calculated separately for each application and aggregated if the applications are implemented together.

Table 4 depicts the calculated operational cost, if both applications are implemented.

Table 3 Division of the capital cost (in €) into the shared (common cost) and the individual cost for each application based on the "resource-cost interdependence" among the two applications examined

Capital cost	Shared cost	Application 1 cost	Application 2 cost	Total cost (Applic.1 +2)
RFID equipment (store type 1)	145,000	46,400	69,600	261,000
RIFD equipment (store type 2)	237,500	76,000	76,000	389,500
Equipment for the central warehouse	13,800			13,800
Installation Cost	77,800			77,800
Servers cost	1,500			1,500
Central infrastructure cost (pcs)	50,000			50,000
S/W Commercial licences	61,200			61,200
S/W customisation for each application		64,000	40,000	104,000
ERP Systems Adjustment Cost		18,000	16,000	34,000
Total cost	586,800	204,400	201,600	992,800

Note: For the assumptions and calculations behind the capital cost estimation, see the Appendix

Table 4 Estimation of the operational cost (in €) of both RFID applications

Operational expenses (OPEX) for both applications	Year 1	Year 2	Year 3	Year 4	Year 5
RFID tags cost	144,000	129,600	116,640	104,976	94,478
Equipment maintenance	–	85,896	85,896	85,896	85,896
SW updates and maintenance	–	12,480	12,480	12,480	12,480
Training	73,440	11,016	11,016	11,016	11,016
Tagging cost	100,00	100,00	100,00	100,000	100,000
Total cost	317,440	338,992	326,032	314,68	303,870

Note: For the assumptions and calculation behind the operation cost estimation see the Appendix

Quantification of the cost savings due to the RFID implementation: To quantify the estimated benefits of the two RFID applications a questionnaire survey was carried out. The questionnaire was distributed to 39 retail stores of the organisation (out of the 153 stores which were the basis for the calculations). The aim of this survey was to collect information about the performance of the current business processes for orders receiving and checking out. The questionnaire consists of specific questions mainly regarding the average time it currently takes for orders receiving and checking out processes, without the use of RFID technology. The average time is a key performance indicator, as shown in Table 5. An estimation of the decrease in processing time for current business processes due to the introduction of RFID technology was made using the Delphi approach with interviews with IT and business people. This estimation is exploited for the assessment of the operational cost reduction as a result of the introduction of RFID technology. While the main benefit considered for the two applications is in terms of operational cost reductions,

Table 5 Estimation of the effect of RFID on key performance indicators utilised for the costs saving calculation

RFID application	Key performance indicators	As -Is (Without RFID)	To-Be (RFID enabled)	Unit
Orders receiving	Average time needed for processing a received order	70	15	Minutes
	Average number of employees occupied with the receipt and the processing of one received order	2	1	Units
Check out	Average time needed for checking-out one sales receipt (pricing of the products purchased)	5,00	2,00	Minutes
	Average time needed for finding the product code ("KAY") in the case of it not being included on the product package	5,00	0,00	Minutes
	Frequency of the above incidents (within one year)	104,00	0,00	Units

the RFID implementation is not only about cost saving but it can also create additional types of benefits such as sales or competitiveness increase which are however difficult to adequately anticipate and quantify.

For the cost saving estimation the following three types of savings are considered (see Table 6):

1. Annual operational cost saving due to the decrease in the time needed for processing orders, due to the RFID implementation. This process includes the identification and documentation of the received products. It also contains the examination of the received products to reconcile the goods ordered with those received.

2. Annual operational cost saving due to the decrease in the time needed for the check out process, specifically for the production of a sales receipt, as a result of the RFID implementation.

3. Annual operational cost savings derived from the elimination of the time needed from the employees to locate a missing product code during the check-out process.

Table 6 Cost saving estimation (in €) due to the introduction of the RFID applications

RFID application	Cost saving Category	Store type 1	Store type	Cost saving per year	Calculation Formula
Orders receiving	Annual operational cost savings due to the decrease in the time needed for order processing	409, 688[1]	186, 875	596, 563	[(Time (as is) × number of employees (as-is) occupied)- (time (to-be) × No of employees (to-be) occupied)]/60 × man-hour cost × Total annual number of orders
Check out	Annual operational cost savings due to the decrease in time needed for the check out process (product pricing)	427, 500[2]	195, 000	622, 500	Time reduction per sales receipt/60 × man-hour cost × Annual total number of sales receipts
	Annual operational cost savings due to the decrease in time needed for locating the product code when it is not included on the product package or not read by the barcode reader	5, 027[3]	8, 233	13, 260	Time reduction per incident × total annual number of incidents/60 × man-hour cost × Number of stores
Both RFID applications				1, 232, 323	

For the detailed calculations regarding (1), (2) and (3) and the exploitation of the formulas see the Appendix

The cost savings are estimated based on the following method. The time decrease is multiplied with the average man-hour rate which is 15 €/h for the employee who is responsible for the orders receiving and 10 €/h for the employee at the check-out points. The figures are multiplied either by the number of the stores or the number of the received orders or the sales receipts, according to each separate application. These numbers are taken from the accounting department of the retailer and they are included in the appendix (see "assumptions for the cost saving estimation"). Table 6 depicts the calculated cost saving estimates.

NPV estimation: After identifying the costs and the benefits for the two applications, the NPV method is used for the financial evaluation of the investment (see Table 7). The joint NPV of the RFID application for the "order receiving "and its interdependent follow-on investment "check out" proved to be positive (886,585 €) in contrast to the negative separate NPV (–466,789 €) of the RFID application "order receiving" and the NPV (–356,173 €) for the "check out" application, if these are considered as stand-alone and independent applications (see Table 4). This occurs because of the fact that these RFID applications when implemented as a group, share common resources as their capital and operational costs, based on the "resource interdependence" among them (Santhanam and Kyparisis 1996, Verma and Sinha 2002, Liesio et al. 2008). Because of the cost sharing, the total

Table 7 Estimation of the Joint NPV for the two applications and the NPV for each application separately (where all amount are expressed to the nearest €)

	Year					
	0	1	2	3	4	5
NPV estimation for the RFID application: "Order Receiving"						
TS		596,563	596,563	596,563	596,563	596,563
TC	791,200	317,440	316,720	303,760	292,096	281,598
GM	−791,200	41,762	42,482	55,442	225,346	314,964
NPV	−466,789					
NPV estimation for the RFID application "Check Out"						
TS		635,760	635,760	635,760	635,760	635,760
TC	788,400	317,440	316,624	303,664	292,000	281,502
GM	−788,400	81,800	82,616	95,576	264,920	354,257
NPV	−356,173					
Joint NPV estimation (both applications)						
TS		1,232,323	1,232,323	1,232,323	1,232,323	1,232,323
TC	992,800	317,440	338,992	326,032	314,368	303,870
GM	−992,800	617,043	595,491	608,451	818,675	928,452
NPV	886,585					

The NPV is estimated based on the discounted value of the Net Cash Flows of the investment (*Total savings - Total cost - Amortisation- Taxes*).
Here, TS: Total Savings, TC: Total Cost = CAPEX+OPEX, GM: Gross Margin = *Total savings –Cost - Amortisation*)

expenditures decrease (Santhanam and Kyparisis 1996). In addition, these two applications create a "value synergy" (Liesio et al. 2008) and a "benefit interdependence" (Santhanam and Kyparisis 1996) because when implemented together they create an extra benefit for the organisation as opposed to when the applications are implemented separately. In particular, when these two applications are considered as interdependent investments and evaluated jointly, the savings per year (1,232,323 €) in the organisation (as a result of the decrease in operational costs) is higher than that which is conceived when the two applications are evaluated separately. This synergetic effect could not be exploited if these two applications were developed and evaluated as stand-alone applications.

As a result, it is worthwhile for the organisation to invest in the RFID application for the orders receiving taking into consideration the value that is offered by the other related and follow-on application for automated check out. However, if the organisation had evaluated only one of the two applications in isolation from the other, then based on the negative NPV, it would have rejected the investment.

5 Discussion

The approach outlined in this chapter for evaluating investments in RFID applications differentiates the evaluation outcome with regards to that one derived from a standard NPV approach. If the two examined RFID applications are evaluated in isolation from each other based on the standard NPV approach, then the investment is rejected, as the NPV is negative for the two individual applications. On the contrary, when these two applications are evaluated jointly taking into consideration their interdependencies, the investment is adequately justified and not rejected, as the joint NPV is positive. Thus, an RFID application which is difficult to be justified because of a negative net present value estimate can be accepted if the value of an embedded growth opportunity for a follow-on investment in another related application is considered. This is as a result of the cost sharing among the two applications and their value synergies.

Both of the evaluation approaches are based on a cost-benefit analysis and the estimation of the NPV. However, the standard approach does not take into consideration and exploit the synergies among RFID applications such as cost and value interdependencies as done by the approach in this study. Considering RFID applications as a related group of investments, which can share costs and accumulate business value, can help an organisation assess their investment and decide whether or not to deploy RFID technology.

The notion of the interdependence (Lee and Kim 2000; Santhanam and Kyparisis 1996) among IS and other kinds of projects has been exploited and supported in this study for the case of RFID technology. It is concluded that by evaluating and selecting interdependent and interrelated RFID projects, cost savings and business benefits other than those that are derived if the RFID projects are evaluated and implemented separately can be generated. In addition, the notion of evolution has been applied to the case of RFID evaluation based on Lee's model (2007). It is concluded that when evaluating an initial investment in RFID, the value of a potential follow-on investment has to be considered as an added value to the first stage project. Otherwise, the value of the initial project can be underestimated.

Due to the fact that the proposed approach is based on issues from the IS literature which have been exploited by different kinds of IS investments, it is believed that this approach is applicable beyond the context of investments into RFID technology and the examined organisation. Every other organisation can take advantage of the common features of its candidate investment projects and their potential to offer follow-on investments utilizing the approach outlined in this chapter.

However, organizations may not have the option to invest in a set of RFID applications and they may only be interested in one limited application. In such a scenario our demonstrated approach is not useful. In addition, due to the fact that the results of the approach are based only on one case study and two RFID business applications, the opportunity to generalise the outcome is limited.

The followed financial approach is differentiated from standard techniques due to the interdependence and the evolutionary notion that it employs. However, it still utilises the discounted cash flow method (DCF) for evaluating RFID investments. This method has been criticised for its inability to value investments that are strategic in nature (Dos Santos 1991). Thus, a limitation of this study is the fact that the value of the examined RFID applications is anticipated based on financial only measures and not strategic. Nevertheless, value derived from an RFID implementation is not only about saving costs. Consumer service can be enriched as a result of the RFID implementation. For example through the streamlining of stock operations to improve shelf availability of products or the prevention of long payment queues (Jones et al. 2005). Thus provide an organisation with a competitive advantage. Furthermore, value added services and enhanced customer relationships management can improve the competitiveness of a company and consequently increase sales. Future studies are required to study these anticipated benefits.

In addition, the exploitation of the NPV method for the proposed approach ignores the flexibility and the future opportunities that an organisation can have regarding an investment. The followed approach of the study assumes that an organisation invests simultaneously in initial and follow-on investments. But what if the opportunity of an organisation to invest further in a follow-on investment occurs in the future after the deployment of the initial investment? This opportunity cannot be taken into account through our approach, however, it may be evaluate by the alternative methodology of Real Options. Further research is needed to use Real Options approach for evaluation investments in RFID technologies (Yeo and Qiu 2003). Based on Benaroch's (2001, 2006) options typology, infrastructure based projects with multiple product generations is one kind of growth option. RFID projects can be viewed as such and evaluated through this method, as they are based on a basic infrastructure which generates multiple business applications and opportunities for an organisation.

6 Conclusions

In this chapter we have explored the application of interdependence property and the evolution theory to the RFID investment evaluation. It is concluded that an investment evaluation approach which exploits interdependence and considers RFID applications as a group of related and sequential applications than stand-alone applications can more appropriately assess investments in RFID technology. This approach results in decreasing the total expenditure and increasing the organisational benefits through cost sharing and the benefits derived from multiple applications.

From a theoretical perspective, this study seeks to fill an identified gap in the literature concerning the impact of the "interdependence property" and the notion of "evolution" on the assessment of investments in RFID technology. From a practical perspective, it is argued that in order to justify investments in RFID technology companies need to consider investments in initial RFID applications as the basis for further, follow-on investments in related applications. Organisations need to realise the variety of possible RFID applications that they can implement while taking advantage of their synergies to maximise the benefits and reduce investment costs.

Further research is needed to evaluate investments in RFID technology using alternative approaches, such as real options analysis, to exploit the option that an organisation can have to invest in related follow-on RFID applications in a staged investment project pipeline. In addition, further research is needed to evaluate and compare different sets of RFID projects to evaluate the most beneficial groupings.

Acknowledgments The author Andriana Dimakopoulou would like to thank the Alexander S. Onassis Public Benefit Foundation for supporting this research through their scholarship fund for her PhD studies.

Appendix

Calculations and Assumptions for the Capital Cost Estimation (Table 3)

For the estimation of the capital cost, several assumptions have been made. Due to space limitations, some examples of these assumptions and the pertinent calculations are mentioned below. Regarding the RFID equipment cost, two different types of stores have been taken into consideration. For example for the store type 1, it is assumed that 1 reader per store is shared by the two applications. Furthermore, two antennas are needed for application 1 and three antennas are required for application 2 per store of type 1. The prices of one reader, antenna and printer are assumed to be 2500 €, 400 € and 4000 € respectively. The calculated figures are multiplied by the number of the stores (58 stores for the type 1 and 95 stores for type 2). It is assumed that RFID equipment will be also be required for the central warehouse of the retailer, apart from the store. In particular, 2 readers with 2 antennas and 2 printers are required resulting in a total shared cost of 13,800 €. In addition, for the installation cost it is assumed that 3 h per store are needed with a man-hour rate of 200 €/h. For example, for stores of type (1), 34,800 € (200 € × 3 h × 58 stores) is the total installation cost.

Calculations and Assumptions for the Operational Cost Estimation (Table 4)

- *RFID tag cost*: For the RFID tag cost estimation, it is assumed that the tagging is made at an item level. The number of the tagged items (products) is estimated to be 1,200,000 €. The price of one tag is assumed to decrease each year by 10% starting from 0.12 € in year 1 and reaching 0.08 € in year 5.
- *Tagging cost*: It is assumed that 15 s are required per item to be tagged. Thus, the total tagging cost is estimated to be 100,000 € based on the man-hour rate of 20 €/h and 1,200,000 tagged items. This cost is assumed to be fixed per year.
- *Training cost*: It is assumed that for the first year, 2 employees are required to be trained per store for 8 h each with a cost of 30 € /h. Thus, the total cost for the first year for all the 153 stores is 73,440 €. For the following years, it is estimated that 15% of this cost will be spent for additional training per year resulting in a cost of 11,016 €.
- *Equipment & SW update and maintenance cost*: Finally, the HW maintenance and update cost is estimated as a percentage (12%) of the capital cost (Table 3) for the equipment (store equipment, central warehouse equipment, servers, PCs). Likewise, the SW updates and maintenance cost is calculated as a percentage (12%) of the software customisation capital cost. The types of the capital cost which are considered are highlighted in Table 3 (in bold). The rest of the capital cost categories (e.g. Installation cost) are not considered to result in operational expenses as they take place only in the beginning of the investment. In particular, this kind of cost is calculated separately for the two applications and it is the same for each year. For example, if both applications are implemented, the estimated cost of the equipment maintenance is 85,896 € (Table 4) per year, whereas if the orders receiving process is implemented without the check out process, this cost will be 68,424 €.

Assumptions for the Cost Saving Estimation

Table 8   Assumptions for the cost saving estimation

Store type	No of stores	Total no. of received orders for all the stores per year	No. of sales receipts for all the stores per year
1	58	13, 110	855,000
2	95	5, 980	390,000

Calculations for the Cost Savings Estimation (Table 6)

(1) (Time (as is) × number of employees (as is) occupied)- time (to be) × No of employees (to be) occupied)/60 × man-hour cost × Total annual number of orders => (75 min to process one order × 2 occupied employees –15 min to process one order × 1 occupied employee)/60 = 2.08 € cost saving per hour per order × 15 € /h (man-hour rate)= 31.25 € is the cost saving per order × 13,110 orders per year = 409,687.50 € is the total cost saving for processing all the orders in all the stores of type 1.

(2) *Time reduction per sales receipt/60 × man-hour cost × Annual total number of sales receipts* => 3 min less time per sales receipt/60 = 0.05 hours per sales receipt × 10 € /h (man-hour rate) = 0.5 € per sales receipt × 855,000 sales receipts (for the 58 stores) = 427,500 € is the total cost saving for the checking out process in all the stores of type 1.

(3) *Time reduction per incident × Number of stores × man-hour cost × total annual number of incidents* => 5 min less time per incident × 104 incidents per year per store= 520 min/ year/60= 8, 67 hours per year per store × 10 €/h (man hour rate) = 86.67 € per store × 58 stores= 5,027 € is the total cost saving for saving time on searching for the product code per year for all the stores in type 1.

References

Benaroch M (2001) Option-based management of technology investment risk. IEEE Trans Eng Manage 48(4):428–444

Benaroch M (2006) On the valuation of multistage information technology investments embedding nested real options. J Manage Inf Syst 23(1):239–261

Bottani E, Rizzi A (2008) Economical assessment of the impact of RFID technology and the EPC system on the fast-moving consumer goods supply chain. Int J Prod Econ 112: 548–569

Cole PH, Ranasinghe DC (2008) Networked RFID system: raising barriers against counterfeiting, Springer, Germany

Curtin J, Kauffman RJ (2007) Making the most out of RFID technology- a research agenda for the study of the adoption, usage and impact of RFID. Inf Tech Manage, 8:87–110

Doerr KH, Gates WR, Mutty JE (2006) A hybrid approach to the valuation of RFID/MEMS technology applied to ordnance inventory. Int J Prod Econ 103:726–741

Dos Santos B (1991) Justifying investments in new information technologies. J Manage Inf Syst 7(4):71–90

Fleisch E, Tellkamp C (2005) Inventory inaccuracy and supply chain performance: a simulation study of a retail supply chain. Int J Prod Econ 95:373–385

Iniestra JG, Gutierrez JG (2009) Multi-criteria decisions on interdependent infrastructure transportation projects using an evolutionary-based framework. Appl Soft Comput 9(2): 512–526

Jones P, Clarke-Hill C, Shears P, Comfort D, Hillier D (2004) Radio frequency identification in the UK: opportunities and challenges. Int J Retail Distrib Manage 32(3):164–171

Jones MA, Wyld DC, Totten JW (2005) The adoption of RFID technology in the retail supply chain. Coastal Bus J 4:1

Karagiannaki A, Mourtos I, Pramatari K (2007) Simulating and evaluating the impact of RFID on warehousing operations: a case study. In summer computer simulation conference, July 15–18, San Diego, CA

Kim J, Tang K, Kumara S, Yee S, Tew J (2008) Value analysis of location-enabled radio-frequency identification information on delivery chain performance. Int J Prod Econ 112(1): 403–415

Kok AG, van Donselaar KH, van Woensel T (2008) A break-even analysis of RFID next term technology for inventory sensitive to shrinkage. Int J Prod Econ 112(2):521–531

Lee H (2007) Peering through a glass darkly. ICR 7:60–68

Lee JW, Kim SH (2000) Using analytic process and goal programming for independent information system project selection. Comput Operat Res 27:367–382

Lee LS, Fiedler KD, Smith JS (2008) Radio Frequency identification (RFID) implementation in the service sector: a customer-facing diffusion model. Int J Prod Econ 112:587–600

Liesiö J, Mild P, Salo A (2008) Robust portfolio modeling with incomplete cost information and project interdependencies. Eur J Oper Res 190:679–695

Moon KL, Ngai EWT (2008) The adoption of RFID in fashion retailing: a business value-added framework. Ind Manage Data Syst 108 (5):596–612

Ngai EWT, Moon KKL, Liu James NK, Tsang KF, Law R, Suk FFC, Wong ICL (2008) Extending CRM in the retail industry: an RFID-based personal shopping assistant system. Commun Assoc Inf Syst 23(16):277–294

Nolan RL (1973) Managing the computer resource: a stage hypothesis. Communications of the ACM 16(7):399–405

Panayi S, Trigeorgis L (1998) Multi-stage real options: the cases of information technology infrastructure and international bank expansion. Q Rev Econ Finance 38:675–692

Porter ME (1985) Competitive advantage: creating and sustaining superior performance., The Free Press, New York, NY.

Pramatari KC, Doukidis GI, Kourouthanassis P (2005) Towards 'smarter' supply and demand-chain collaboration practices enabled by RFID technology In: Vervest P, van Heck E, Louis-François P, Kenneth P (eds) Smart business networks, Springer, Heidelberg

Rekik Y, Sahin, Dallery Y (2008) Analysis of the impact of the RFIDnext term technology on reducing product misplacement errors at retail stores. Int j Prod Econ 112(1):264–278

Roussos G, Koukara L, Kourouthanasis P, Tuominen J, Seppala O, Frissaer J (2002) A case study in pervasive retail. Proceedings of the 2nd international workshop on mobile commerce, September 28, Atlanta, GA, USA. doi:10.1145/570705.570722

Santhanam R, Kyparisis G (1996) A decision model for interdependent information system project selection. Eur J Oper Res 89(2):380–399

Subirana B, Eckes C, Herman G, Sarma S, Barret, M (2003) Measuring the impact of information technology in value and productivity using a process-based approach: the case for RFID technologies. MIT sloan working paper No 4450–03

Tajima M (2007) Strategic value of RFID in supply chain management. J Purch Supply Manage 13:261–273

Taudes A (1998) Software growth options. J Manage Inf Syst 15(1):165–185

Tzeng SF, Chen WH, Pai FY (2008) Evaluating the business value of RFID: evidence from five case studies. Int j Prod Econ 112(2):601–613

Uhrich F, Sandner U, Resatsch F, Leimeister JM, Krcmar H (2008) RFID in retailing and customer relationship management. Commun Assoc Inf Syst 23(13):219–234

Venkatraman N (1990) IT-induced business reconfiguration In: Scott-Morton, M (ed) The corporation of the 1990s. Oxford University Press, New York, NY

Verma D, Sinha K (2002) Toward a theory of project interdependencies in high tech R&D environments. J Operat Manage 20:451–468

Violino B (2005) Leveraging the internet of things. RFID J. November/December:1–2.

Von Bertalanffy L (1969) General systems theory: foundations, development, applications. G.Braziller, New York, NY

Wamba SF, Lefebvre L, Bendavid Y, Lefebvre E (2008) Exploring the impact of RFID technology and the EPC network on mobile B2B eCommerce: a case study in the retail industry. Int j Prod Econ 112:614–629

Wang SJ, Liu SF, Wang WL (2008) The simulated impact of RFID next term-enabled supply chain on pull-based inventory replenishment in TFT-LCD industry. Int j Prod Econ 112:570–586

Yeo KT, Qiu F (2003) The value of management flexibility-a real option approach to investment evaluation. Int J Proj Manage 21(4):243–250

Yourdon E (1989) Modern structured analysis. Yourdon Press, Prentice-Hall International, Englewood Cliffs, NJ

An Analysis of the Impact of RFID Technology on Inventory Systems

Yacine Rekik

Abstract Nowadays, most enterprises undertake large investments in order to implement information systems that support decision making for managing inventories. Nevertheless, if data collected from the physical processes used to feed these systems are not correct, there will be severe impacts on business performance. Inventory inaccuracy occurs when the inventory level in the Information System is not in agreement with the physically available inventory. In this chapter, we first describe the major factors generating inventory inaccuracy. Then, we provide situations permitting to manage an inventory system subject to errors. We provide a framework to model the inventory inaccuracy issue and focus on the impact of advanced identification systems, such as that provided by RFID technology, in improving the performance of a supply chain subject to inventory inaccuracies.

1 Introduction

The standard literature on inventory models has rarely differentiated between the inventory record and the physical inventory. The two have always been considered to be the same and the main concern was on how, having observed demand and the resulting inventory levels, an inventory manager should determine when and how much to replenish. Based on empirical observations this implicit assumption has proven to be wrong. In fact, based on a study done with a leading retailer, Raman et al. (2001) reports that, out of close to 370,000 SKUs (Stock-Keeping Unit) investigated, more than 65% of the inventory records did not match the physical inventory at the store-SKU level. Moreover, 20% of the inventory records differed from the physical stock by six or more items.

Y. Rekik (✉)
EMLYON Business School, 23 avenue Guy de Collongues, 69134 Ecully, France
e-mail: rekik@em-lyon.com

D.C. Ranasinghe et al. (eds.), *Unique Radio Innovation for the 21st Century*,
DOI 10.1007/978-3-642-03462-6_20, © Springer-Verlag Berlin Heidelberg 2010

A general definition of accuracy includes obtaining the correct value for a measurement at the correct time (Schuster et al. 2004). According to Iglehart and Morey (1972) and DeHorativs and Roman (2004), inventory inaccuracy occurs when the system inventory, i.e., what is available according to the information system does not match the physical inventory (what is actually available). Various other definitions provide the same sense and measures of inventory accuracy are presented in Ernst et al. (1984); Buker (1984); Bernard (1985); Chopra (1986); Young (1986) and Martin and Goodrich (1987). For example, Ernst et al. (1984) proposes using a control chart to monitor the changes in the inventory accuracy. Another definition provided by Bernard (1985) considers the percentage (and not the difference) error in the inventory records. Martin and Goodrich (1987) define accuracy as the total dollar deviation between the actual dollar value of inventory and recorded dollar value of the inventory. As a conclusion of the definitions provided, we say that an inventory stock is inaccurate when the record stock is not in agreement with the physical stock.

The aim of this article is to provide a comprehensive study on the issue of inventory inaccuracy and to show the manner in which RFID technology can improve the performance of supply chains subject to such inaccuracies. The remaining part of this section will be concerned with the main sources causing inventory inaccuracy: we focus on the impact on the performances and we briefly detail main investigations performed for each type of source. Then, in Sect. 2, we provide different ways to cope with the inventory inaccuracy issue. For this purpose, we show that RFID technology could be an interesting alternative as it is stated in many recent investigations. The aim of Sect. 3 is to present a framework to model an inventory system subject to inaccuracies and to show how we can measure the impact of RFID on such a system. Finally Sect. 4 concludes the chapter.

1.1 State of the Art in Inventory Inaccuracy

Inventory inaccuracy can be a major obstacle to improvements in firms' performance (Kök and Shang 2007). While companies have undertaken large investments to automate and improve their inventory management processes, inventory information system and physical inventory are rarely aligned (Raman et al. 2001). Inventory inaccuracy might result from several factors. The aim of this section is to present a comprehensive analysis of the factors generating inventory inaccuracy. Based on empirical and qualitative investigations, we try to focus on the order of magnitude of these errors. We also provide main quantitative investigations for each error type.

1.1.1 Transaction Errors

Transaction errors are unintentional errors occurring during inventory transactions. Some of these transactions happen when counting the inventory, receiving an order or checking out at the cash register.

Errors at check outs occur if the cash register scans one item twice, rather than each item separately, when a customer is buying two similar (but not identical) items with the same price. This innocuous action by the cash register ensures that the customer pays the correct amount and may even save the customer time as the cash register avoids handling the additional item. However, it generates a discrepancy in the inventory information system. Similarly, errors when picking impacts inventory records. A warehouse employee can unintentionally ship the wrong quantity of a particular item to a store or even send the wrong item altogether. According to DeHoyatius and Raman (2004), in an apparel warehouse, it is quite easy for an employee to mistakenly pick a "medium" instead of a "large" garment. Stores typically do not scan merchandise on receipt allowing for such errors to remain invisible. In her PhD dissertation, Sahin (2004) provides a comprehensive analysis of inventory systems subject to perturbations in nominal flows. According to the author, the major defects resulting in transaction errors are:

- The technical limitations of the bar code system.
- The potential failures stemming from the interaction between inventory operators and the bar code system. Those errors result in (*i*) errors made when identifying entities, (*ii*) errors made when counting and (*iii*) errors made when keypunching data.

1.1.2 Misplacement Errors

Misplacement errors occur when a fraction of the inventory is misplaced, it is not available to meet a customer demand until it is found. According to Chappell et al. (2003), there are several sources generating misplacement errors such as: (i) Consumers picking up products and then putting them down in another location, (ii) Clerks not storing products on the correct shelf at the right time and (iii) Clerks losing products in the backroom. From a Ton and Raman (2004)-year longitudinal study of 333 stores of a large retailer, 4 show that an increase in product variety and inventory level per product are associated with an increase in misplaced products. The authors also show that increasing misplaced products is associated with a decrease in store sales. According to Çamdereli and Swaminathan (2009), misplaced inventory is also present in warehouse operations. G.T Interactive, the creator of computer games like Doom II and Driver, suffered from low productivity due to inventory misplacement in the warehouse. Fundamentally what happens in these settings is that the product is misplaced in the supply chain and is unavailable during the sales period but can be retrieved when a cleanup is performed.

Misplaced inventory can be quite large and have a significant impact on the inventory performance. It is reported in Raman et al. (2001) that customers of a "leading retailer" cannot find 16% of the items in the store because of misplacement errors. The consequence is that misplacement errors reduced profit by 25% at this retailer.

The authors in Rekik et al. (2008) model the consequences of random misplacement errors on the retail supply chain by comparing three approaches. In the first approach, the retailer is unaware of the misplacement and places his orders as if inventory information is perfect. In the second case, the retailer is aware of the errors in his inventory status information, and adjusts his ordering policy accordingly. In the final approach, perfect inventory status information based on RFID technology is assumed. They also provide a critical value for the cost of RFID technology that makes its deployment cost effective. As an extension of the last investigation, the authors in Rekik et al. (2007), consider the impact of misplacement type-errors and the effect of RFID technology on a decentralised supply chain. In particular, they compare two possible strategies where the first one deals with the deployment of RFID technology and the second one consider the coordination of the channel by using a modified buy-back contract. Gaukler et al. (2007) investigate the effects of RFID technology within the context of the retail supply chain. They build a newsvendor model that takes into account the inefficiency of the replenishment process from the backroom to the shelf in the retail store. Then, based on this model, they examine how the cost of the RFID implementation should be shared among supply chain actors, and determine coordinating contracts for the RFID-enabled supply chain within a Newsvendor framework. Çamdereli and Swaminathan (2009) also study misplacement type errors in a decentralised supply chain for uniform distributions of demand and error.

1.1.3 Damage and Spoilage

For supermarkets, perishables are the driving force behind the industry's profitability and represent a significant opportunity for improvement, accounting for up to $200 billion in U.S. sales a year but subjecting firms to losses of up to 15% due to damage and spoilage (Ketzenberg and Ferguson 2006). Examples with limited lifetime products are drugs and food products. In retail stores, customers can cause damages to products and as a consequence making them unavailable for sales. Some examples are tearing of a package to try on the contained cloth item, wearing down a shoes by trying it on and walking, erasing software on computers on demonstration, spilling food on clothes, and scratching a car during a test drive (Bensoussan et al. 2005). Those damage may not be detected by the inventory manager and as a consequence would cause inventory inaccuracy.

According to Sahin 2004, items reaching their lifetime limit during storage is due to:

- Errors when forecasting customer demand which may lead to overestimating demand and as consequence an important quantity of products are not sold.
- The inability to accurately track the location, condition and age of products stored within a facility.

An industry survey performed by the Joint Industry Unsaleables Steering Committee (Lightburn 2002) provides data on the level of unsaleable products in the US retail industry. According to the survey results, which are based on responses from over 60 manufacturers and retailers, the cost of unsaleable food and grocery products amount to 1% of sales in the US. Damage is the biggest cause accounting for 63% of all unsaleables, followed by expired (16%) and discontinued items (12%). More recently, Ilic et al. (2009) provides a simulation study on the effect of sensor information in supply chains of perishable goods.

1.1.4 Theft

Inventory theft is defined as a combination of employee theft, shoplifting, internal and external theft, vendor fraud and administrative error. The ECR[1] (Efficient Consumer Response) Europe project on shrinkage subsequently analysed the causes of stock loss and proposed a systematic and collaborative approach to reduce the phenomenon throughout the supply chain. ECR defines "Shrinkage" as being due to process errors, deceptions and internal and external thefts). according to Beck (2003), some specific types of internal theft include:

- Staff stealing goods by either hiding them in their bags or intentionally placing them outside the building for later collection.
- Collusion occurring when a staff member collaborates with a customer to steal products. During such an incident, the staff member may not scan the item or the security personnel may intentionally ignores the offense as it occurs.
- Grazing occurs when items stored in the warehouse are consumed by the warehouse staff

The results from the research carried by ECR Europe have shown that the scale of shrinkage in the fast moving consumer goods sector is estimated to 24 billion euros in 2003 (weekly, 465 million euros is lost irreparably within the fast moving consumer goods turnover), which is 2.41% of the whole turnover value of the sector. The process errors present 27% of the whole shrinkage value, 7% deceptions, 28% internal thefts and 38 external thefts.

Based on survey data, internal and external theft, administrative errors and vendor fraud accounted for an estimated 1.8% of sales in the US retail industry in 2001, costing US retailers USD 33 billion (Supermarket 2001). For US supermarkets, the NSRG (National Supermarket Research Group) survey (Hollinger and Davis 2001) estimates that internal and external theft, receiving errors, damage, accounting errors and retail pricing errors amount to 2.3% of sales. These figures only take the value of the item into account, but not any process-related costs (e.g. for handling of damaged items). Kang and Gershwin (2004) use simulation to analyze the consequences of inventory inaccuracy. They show that even small undetected losses can lead to important stock-outs. They also propose several ways to tackle this problem

---

[1] http://www.ecr.org/

including the deployment of RFID technology. The authors in Fleisch and Tellkamp (2005) also use simulations to show the consequences of inventory inaccuracy in a three stage supply chain. Considering *misplaced type errors*, the authors in Rekik et al. (2008) show, analytically, the impact of errors on the inventory decision strategies and derive a threshold cost value, at which an RFID deployment would become cost effective and consider the case of a decentralized supply chain with one manufacturer and one retailer whose inventory is subject to inaccuracies. They develop the optimal ordering strategies under centralised and the decentralised scenarios. They also study the impact of RFID deployments on such supply chains. Heese (2007) studies a supply chain consisting of a Stackelberg *leader* manufacturer and a retailer with inventory inaccuracy and random demand. Using specific distribution functions, he derives the threshold value of tag cost in order for a firm to adopt RFID. More recently, Rekik et al. (2009) consider a finite horizon, single product periodic review store inventory in which inventory records are inaccurate due to theft errors. They analyze the problem of having theft in store by optimizing the holding costs under a service level constraint and they analyze the impact of theft errors and the value of RFID technology to managing inventory.

1.1.5 Supply Errors

When the product quality is low or a production process has a low yield or a supply process is unreliable, the physical inventory may not be known. As a consequence may be different from the inventory in the information system (Inderfurth 2004; Rekik et al. 2007; Yano and Lee 1995). Products that are not conforming to quality standards can also make the inventory inaccurate. According to Bensoussan et al. (2005) receipts are usually added to the inventory without a full inspection process. The consequence is that the information system may consist of both non defective products and defective products which are not available for sale.

2 An Inventory Model Subject to Inaccuracies: The Impact of RFID

2.1 The Framework

We consider a supply chain composed of three actors: the manufacturer, the warehouse and retailers. The manufacturer produces products, the retailers are the actors selling products to the final customers. Between the manufacturer and the retailers, there may exist a warehouse who is an intermediate actor that buys products from the manufacturer and resells them to the retailers.

From an inventory control point of view, satisfying customer demands may be different within the retail store and the warehouse. For this reason two structures may exist:

- *Structure A*: this structure focuses on the retail store context. The end customers are physically present in the retail store and their demand is confronted by the physical on shelf inventory. The Information System (IS) does not play a major role in this structure.
- *Structure B*: this structure focuses on the warehouse context. Customers are not physically present in the warehouse and demand satisfaction is based on the inventory level shown in the Information System. Based on the demand received, the warehouse observes its IS stock and makes a commitment. The commitment is later satisfied based on the physical available stock.

In the presence of inaccuracies, the two structures do not behave in the same way from an inventory control point of view. We argue that the main difference between the two structures concerned is the commitment by the warehouse under *Structure B*. In *Structure A*, no commitment is given by the retailer since the end customers are physically present in the store.

Remark 1: *Note that an Internet retailer's inventory system can be modeled as Structure B since decisions concerning demand satisfaction are based on the IS inventory.*

Remark 2: *From an inventory control point of view, Structure A can be considered as a particular case of Structure B where, the* Available for Sales Quantity *in Structure A corresponds to Q_{IS} of Structure B with assuming that $Q_{IS} \leq Q_{PH}$. In fact, under the retail context, the available for sales quantity is always lower than the total physical stock in the presence of errors.*

Under *Structure B*, we can consider the following sequence of events. Before the beginning of the selling season, the warehouse manager orders a quantity from the manufacturer. This quantity is established based on forecast information available to the inventory manager regarding future demand. The warehouse receives the goods and stores them. Just before the beginning of the selling season, the inventory manager receives orders from the customers. He compares the cumulative orders from all the customers to the quantity observed in the Information System. If the cumulative orders are less than the Information System quantity Q_{IS}, he accepts the orders. Otherwise, he only accepts orders summing up to Q_{IS}. Later on, the products are shipped from the warehouse and delivered to the customers. All the orders that the inventory manager has committed himself to should be satisfied, except in the case where the physical inventory is not able to satisfy the committed quantity. Unsatisfied demand during the commitment and unsatisfied commitments are lost since there is no opportunity for replenishment during the selling season.

The inventory manager's decision is to determine the best quantity to order from the supply system before the selling season to satisfy the aggregate demand from customers. He faces three risks: (*i*) risk of having unsold products at the end of the selling season; (*ii*) risk of shortage and (*iii*) risk of not being able to deliver the quantity that he has made a commitment for. Within this framework, the inventory manager faces three types of costs:

- h: the unit overage cost due to products unsold at the end of the selling season.
- u_1: the first type of unit underage cost due to orders rejected by the inventory manager during the commitment.
- u_2: the second type of unit underage cost due to orders initially accepted by the inventory manager but finally not delivered in total to the customers.

We note that the second type underage cost is the parameter that characterizes the warehouse context. In a retail context, this cost does not exist since the customers are physically present at the retail store: if the product is not available, the demand is lost otherwise it is satisfied immediately. The second type underage cost occurs when the inventory level maintained by the Information System confronts customer demand.

To model the impact of inaccuracy errors on the performance of the inventory system, it is important to characterise the error type. In fact, in a general setting, if we let Q the quantity ordered from the supply process, the physical and the IS inventory can respectively be written as the following: $Q_{PH} = \gamma_{PH}Q + \varepsilon_{PH}$ and $Q_{IS} = \gamma_{IS}Q + \varepsilon_{IS}$ where the random variables $(\gamma_{PH}, \varepsilon_{PH})$ and $(\gamma_{IS}, \varepsilon_{IS})$ characterizes errors on the physical inventory level (IS inventory level). From this general setting which is called the *mixed error setting*, one can distinguish two particular cases:

- The additive error setting: in this case $Q_i = Q + \varepsilon_i$ where $i \in \{PH, IS\}$
- The multiplicative error setting: in this case $Q_i = \gamma_i Q$ where $i \in \{PH, IS\}$

Second, it is also important to discuss the manner in which an inventory system subject to inaccuracy problems can be managed. In the presence of errors, one should distinguish between two situations depending on whether the inventory manager is aware of or not of the existence of errors. We will refer to the following two situations to differentiate the previous cases throughout the chapter.

- *Situation 1*: It is where the inventory manager is unaware or simply ignores errors.
- *Situation 2*: It is where the inventory manager is aware of errors occurring in the inventory system.

For Situation 2, we can also distinguish between two cases based on the information the inventory manager has about the error parameters. The first case occurs when the inventory manager has statistical information about the error parameter (such as the mean or the distribution of the error). The second case occurs when exact information on the occurrence of the error is known. The difference between the two cases makes sense especially in a multi-period framework.

Concerning Situation 2, we notice that an estimation of the error parameters can be realised based on statistical sampling methods as reported by Pergamalis (2002) who proposes a methodology for measuring a store's inventory accuracy.

In contrast to Situation 2, Situation 1 is easier to model and optimize. In fact, in Situation 1, the inventory manager acts as if there were no errors. So, his ordering decisions or his replenishment policy coincides simply with the ordering quantity or the replenishment policy of the model without errors.

2.2 Impact of RFID Technology: Quantitative Analysis

RFID (Radio Frequency IDentification) technology is a wireless technology including RFID tags and readers often linked to an information system through networked computers. An RFID tag may be active or passive. An active tag is powered through a battery, while a passive tags scavenge energy. This affects performance and price. Due to its characteristics, an RFID tag may be read without visual or physical contact. To facilitate physical distribution and inventory management, tags are applied to packaging or unit loads, for instance, transport packaging, pallets or containers.

RFID technology is developing at a rapid pace and offering companies the opportunity to improve supply chain visibility. In fact, the number of RFID applications in supply chains is increasing steadily, which is indicated by the escalating level of investment in this technology. The Aberdeen Group Klein (2007) surveyed over 600 companies using RFID in the last 12 months. According to this study, the average manufacturer's annual RFID budget is growing rapidly, from $50,000–$75,000 in 2006 to $100,000–$200,000 in 2007. The study shows that the investments already made greatly improved cycle time, on-time delivery, safety stock and changeover time.

Several authors highlight the opportunities presented by RFID technology. In the physical distribution field, one set of authors Jim Wu and Chen (2007) use a simulation study to predict operational efficiency in a distribution centre. Thanks to a perfect RFID system (100% read rate), the authors show an inventory reduction of 15%, lowering of space usage by 15% and elimination of out-of-stock items. The authors in Holmquist and Stefansson (2006) study a mobile RFID system designed to manage arrival inspection and loading of containers. They find both time savings in arrival inspection and lead-time reduction in loading, except when there are only a small number of consignments per container. A third set of authors, Jargumilli and Grasman (2007), use invented data to model a vendor-managed inventory setting with RFID tags applied to goods. In Choy et al. (2007), the authors propose a logistics information system to simplify the distribution process with the help of RFID technology. A case study is discussed in applying the proposed information system to solve distribution management problems of a medium-sized logistics service company. By using this RFID-enabled system, the authors show that the overall distribution performance of the company is greatly improved.

In the inventory control side, as it is stated in Sect. 1, many recent investigations concerned with inventory inaccuracy issues are motivated by RFID technology and its impact on supply chain performance. Indeed, RFID deployments contribute to the elimination of sources of inaccuracies throughout the supply chain (Hardgrave et al. 2008):

- Errors resulting from the unreliability of the supply system can be detected thanks to the automatic counting performed during receiving orders. RFID can reduce the errors in receiving via the RFID-enabled process referred to as electronic proof-of-delivery (or ePod) (Mason et al. 2006). However, with barcode, mistakes are made by misidentifying the quantity and type of product.
- For food and perishable products, RFID can:
  - Facilitate FIFO (First In First Out) and pricing management.
  - Facilitate tracking and maintenance of temperature conditions supporting the integrity of the cold chain for certain food products.
  - Facilitate more accurate information about demand and supply forecasts. Accurate and timely information about stock levels, sales throughput and production data allows re-engineering the overall business to minimise buffer and obsolesce stock. Thus, leading to lower inventory. Consequently, products can be brought closer to the market in form of mini stocks once visibility of products can be guaranteed. Also combining product flows from different sources on their way to the receiver becomes faster and more efficient. Overall this will lead to increased product availability and thereby increased sales. Sales can be further accelerated by retail applications like automatic cross selling or product information displays .
- Inaccuracies resulting from shrinkage errors can be reduced by discouraging thieves and by facilitating the location of quantities and timings of theft losses across the supply chain. Loss prevention during transport between sending and receiving parties or even within sites can be achieved by increased transparency and real-time visibility over the goods flow. Basically the level of tagging (pallet, case or item) and the frequency of read events determine the degree of loss prevention.
- Products subject to misplacement errors can be detected by deploying RFID readers within shelves and in the backroom.
- Improving shipping accuracy. Distribution centers and manufacturers often make mistakes by loading product on the wrong truck. With RFID, the system can send an alert (visible, audible, etc.) to the person loading the truck that a mistake was made. This alert could save a company money from shipping and reshipping the same items as well as enhance inventory control.

The decision to invest in RFID technology will depend on the trade-off between the cost associated with the deployment of RFID technology and savings resulting from the investment. Taking into account variable[2] costs associated with RFID, i.e., the RFID cost of the RFID tag embedded in each product, we are able to model

---

[2]The introduction of fixed cost can be done by an additional ROI analysis

the RFID enabled process. We note that RFID can bring two significant benefits to managing inventory at a warehouse:

- *Increased visibility*. For example, RFID enables tracking and tracing of items in stock and in the pipeline, thus, creating complete inventory visibility, leading to an accurate account of inventory discrepancy. In such a case, the inventory problem under study is nothing other than the basic random yield problem.
- *Prevent or reduce the magnitude of some sources of inventory inaccuracy*. For example, being able to distinguish customer demand from other sources mimicking demand, such as inventory errors due to theft, can allow the inventory manager to take action to prevent or discourage the sources of theft.

Now we can consider a third situation.

- *Situation 3*: The inventory inaccuracy problem after RFID is deployed. This is also a basic inventory problem where the unit purchase (production) cost of the the product includes the RFID tag cost.

To summarise, Fig. 1 presents the three different situations to model the impact of inventory inaccuracies and the benefits pertaining to the deployment of RFID technology.

We can obtain significant insights into the benefit of gathering information on errors and establishing a better inventory policy, taking into account the inaccuracy issues, by comparing Situations 1 and 2. Comparing Situations 2 and 3 permits us to to obtain insights into the real benefit of RFID, in particular, to answer the following key questions.

- Is RFID deployment an economically feasible solution?
- If yes, under what tag price is RFID a cost effective solution?

The aim of the next section is to answer these questions from an analytical and numerical point of view under the supposition that errors are additive.

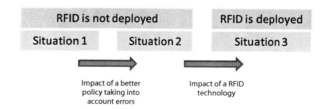

Fig. 1 Modelled scenarios

2.3 The Impact of RFID Technology: Analytical Analysis

To analyse the impact of RFID technology on inventory performance, we should compare Situations 2 and 3. For this purpose, consider a supply chain modelled as *Structure B* (refer to Sect. 2.1 for the sequence of events facing the inventory manager). Considering the case of the additive error setting mentioned previously, the following results state the optimal ordering strategies under Situations 2 & 3.

Under an additive error setting, $Q_{IS} = Q + \varepsilon_{IS}$ and $Q_{PH} = Q + \varepsilon_{PH}$ where Q is the ordered quantity and ε_{IS} and ε_{PH} are the random variables describing the errors on the IS and the PH inventory, respectively. Let's also define two additional random variables. $D_m = D + \varepsilon_{IS}$ is a random variable with the probability density function f_m and the cumulative distribution function F_m. Then, $e = \varepsilon_{IS} - \varepsilon_{PH}$ is the random variable with the probability density function g and the cumulative distribution function G. Additional technical details and proofs of the following results can be found in Rekik (2006).

Result 1 *Optimal ordering strategy under Situation 2: Under Conditions 1 and 2 (see below), there exists a unique optimal ordering quantity Q^* that minimises the expected cost function of the inventory manager in Situation 2. Q^* solves the following equation:*

$$(u_1 + h)F_m(Q^*) + (u_2 + h)\int_{e=0}^{+\infty} g(e)\big[F_m(Q^* - e) - F_m(Q^*)\big]de = u_1 \qquad (1)$$

where

- *Condition 1: The u_1 cost and the h cost are such that $u_1 \geq \frac{F_m(0)}{2 + F_m(0)}h$*

- *Condition 2: The modified demand distribution is such that: $\frac{f_m(x)}{f_m(x)}$ is an increasing function of x.*

The optimal expected cost[3] incurred by the inventory manager is given by:

$$C(Q^*) = u_1 E[D_m] - (u_1 + h)\int_{D_m=0}^{Q^*} D_m f_m(D_m)dD_m$$

$$+ (u_2 + h)\int_{e=0}^{+\infty}\left[e - eF(Q^* - e) + \int_{D_m=Q^*-e}^{Q^*} D_m f(D_m)dD_m\right]g(e)de - hE[e]$$

---

[3]The sign $E[.]$ in the results is used to express the expected value

Taking into account only variable[4] costs associated with RFID technology, i.e., the RFID tag cost embedded in each product, the aim of the next result is to provide the optimal ordering strategy under the RFID-enabled scenario. Contrary to past investigations that assume RFID technology to be a 100% perfect solution eliminating all errors, we assume in this section that RFID permits only to align the IS and the PH inventories. We agree that once product visibility is achieved with the aid of RFID technology, the inventory problem is nothing else but a classical random yield problem with an additional tag price t. Instead of purchasing the product by a its original unit cost, the inventory manager is purchasing a product with its price increased by the unit cost, t, of an RFID tag. Such an increase in the purchase cost has a direct impact on the holding cost (increased by t) and the first shortage cost (decreased by t). Since IS and PH inventory levels are aligned, there is no longer a second shortage cost in the RFID-enabled situation.

Result 2 *Optimal ordering strategy under Situation 3:* *The optimal ordering quantity in the RFID enabled situation is given by:*

$$Q^*_{RFID} = F_m^{-1}\left[\frac{u_1 - t}{u_1 + h}\right]$$ (2)

The associated optimal expected cost is obtained as the following:

$$C^*_{RFID} = u_1 E[D_m] - (u_1 + h)\int_{D_m=0}^{Q^*_{RFID}} D_m f(D_m)dD_m$$ (3)

By comparing Situations 2 and 3, the next result state the critical RFID tag cost to ensure that the technology is cost effective:

Result 3 *We can identify a critical tag cost t_c such that for $t \leq t_c$, the implementation of the RFID technology yields a positive benefit. t_c solves:*

$$\int_{F_m^{-1}[\frac{u_1-t_c}{u_1+h}]}^{F_m^{-1}[\frac{u_1}{u_1+h}]} D_m f_m(D_m)dD_m = \frac{u_2 + h}{u_1 + h}(1 - G(0))$$ (4)

We end the analytical analysis by a numerical example where we assume that the inventory manager faces a normally distributed demand $N(10, 3)$. Both IS and PH errors are normally distributed with a mean equal to zero (standard deviation of the IS error, σ_{IS}, is set to 3). We set $h=1$, $u_1 = 0.4$ in Fig. 2 and $h=1$, $u_1 = 4$ in Fig. 3. We analyze the behavior of t_c with the standard deviation of PH error, σ_{PH}, for different values of P where P measures the impact of the u_2 penalty ($u_2 = u_1 + P$).

Note that the evolution pertaining to Fig. 3 is more intuitive, i.e., if the inventory system is subject to more inaccuracies, the RFID deployment is easier and the

---

[4]The introduction of fixed cost can be done by an additional ROI analysis

Fig. 2 Behavior 1 of t_c

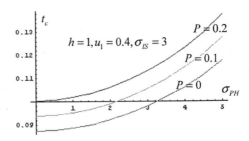

Fig. 3 Behavior 2 of t_c

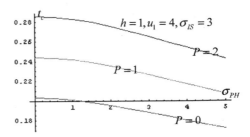

critical tag cost decreases with σ_{PH}. For the case where the sum $u_1 + u_2$ is less important than h (see Fig. 2), the evolution can be explained by the fact that the RFID solution also suffers form inaccuracies (such as missed reads, hardware failures, etc.) since we assumed that RFID does not remove errors but it only provides visibility into the discrepancy between the PH and IS levels. The result show that C_{RFID}^* increases more than $C(Q^*)$ when σ_{PH} increases which explain the decrease in the critical RFID tag cost. As expected, t_c increases with the penalty P for both evolutions.

3 Conclusion

Our major contribution in this chapter was to provide a comprehensive study around inventory inaccuracy and the way the RFID technology can tackle this issue. We listed the main sources of errors with a focus on their impacts on the supply chain performances and briefly detailed main research investigations in this field. We proposed a conceptual framework permitting us to model the inaccuracy errors and the way RFID can contribute to improving such errors. In fact, three modeling situations permit us to show the impact of a better inventory policy taking into account error distribution and the deployment of RFID technology. We concluded the chapter with an analytical study to quantify the impact of RFID technology on inventory systems subject to inaccuracies. The study permits us to derive a critical RFID tag cost that will make the technology deployment cost effective. For further research, we are extending the proposed framework for the multiplicative error setting as well as the case of a multi-period ordering strategy.

References

Beck A. 2003 How can rfid help to reduce shrinkage. Technical report, University of Leicester

Bensoussan A., Cakanyildirim M., Sethi S.P. 2005 Partially observed inventory systems: the case of zero balance walk. Technical report, School of Management, University of Texas a Dallas

Bernard A. 1985 Cycle counting: the missing link. Prod inventory manage, 25(4):27–40

Buker D.W. 1984 Inventory accuracy. Inventory management reprints, The American Production and Inventory Control Society, pp 221–223

Çamdereli A. Z. Swaminathan J.M. (2009) Coordination of a supply chain under misplaced inventory. Technical report, Kenan-Flagler Business School

Chappell G., Durdan D., Gilbert G., Ginsburg L., Smith J., Tobolski J. 2003 Auto-id in the box: the value of auto-id technology in retail stores. Auto-ID Center. http://www.autoidlabs.org/single-view/dir/article/6/143/page.html. Accessed 24 June 2009

Chopra V. 1986 How to set up an effective inventory accuracy program. 29th Annual APICS International Conference Proceedings, Alexandria, VA, USA

Choy K.L., Henry W.L.C, Kwok S.K. , Stuart C.K.S., Chow, K.H., Lee, W.B. 2007 Using radio frequency identification technology in distribution management: a case study on third-party logistics. Int J Manuf Technol Manage, 10:19–40

DeHoratius N. Raman A, 2004 Inventory record inaccuracy: an empirical analysis. Technical report, Graduate School of Business, University of Chicago, Working paper, Graduate School of Business, University of Chicago

Ernst R., Guerrero J.L., Roshwalb A. 1984 A quality control approach for monitoring inventory stock levels. J. Opl Res. Soc., 44:1115–1127

Fleisch E. Tellkamp C. 2005 Inventory inaccuracy and supply chain performance: a simulation study of a retail supply chain. Int J Prod Econ 95:373–385

Gaukler G., Seifert R.W., Hausman W.H. 2007 Item-level rfid in the retail supply chain. Prod Oper Manage 16:65–76, Working Paper, Stanford University, Stanford, USA

Hardgrave B., Riemenschneider, C.K., Armstrong D.J. 2008 Making the business case for RFID. In: Haasis H-D, Kreowski H-J, Scholz-Reiter B (eds) Dynamics in logistics Springer, Berlin Heidelber

Heese, H. 2007 Inventory record inaccuracy, double marginalization and rfid adoption. Prod Oper Manage 16:542–553

Hollinger R.C. Davis J.L. 2001 National Retail Security Survey. Department of sociology and the center for studies in criminology and law, University of Florida. http://diogenesllc.com/NRSS_2001.pdf. Accessed 24 June 2009

Holmqvist M. Stefansson G. 2006 Smart goods' and mobile rfid: a case with innovation from volvo. J Bus Logist, 27:251–272

Iglehart D., Morey R.C. 1972 Inventory systems with imperfect asset information. Manage Sci, 18(8):B388–B394

Ilic A., Schuster E.W., Staake T., Fleisch E. 2009 Sensors in retailing: reducing the quality uncertainty of perishable goods. Technical report, ETH Zurich

Inderfurth K. 2004 Analytical solution for a single-period production-inventory problem with uniformly distributed yield and demand. Central Eur J Oper Res, 12(2):117–127

Jarugumilli S., Grasman S.E. 2007 Rfid-enabled inventory routing problems. Int j Manuf Technol Manage, 10:92–105

Jim Wu, Y.-C. Chen J.-X. 2007 Rfid application in a cvs distribution centre in taiwan: a simulation study.' Int J Manuf Technol Manage, 1:121–136

Kök A. G. Shang K.H. 2007 Replenishment and inspection policies for systems with inventory record inaccuracy. Manuf Serv Operat Manage, 9:185–205

Kang Y. Gershwin S.B. 2004 Information inaccuracy in inventory systems - stock loss and stockout. IIE Trans, 37:843–859

Ketzenberg M.E. M. Ferguson (2006) Managing slow moving perishables in the grocery industry. Technical report, College of Business Colorado State University and The College of Management Georgia Institute of Technology, June

Klein R. 2007 Can RFID deliver the goods?- the manufacturer's visibility into supply and demand. Technical report, Anderson Group. http://www.aberdeen.com/Aberdeen-Library/3758/RA_RFIDMFG_3758.aspx. Accessed 24 June 2009

Lightburn A. 2002 Unsaleables Benchmark Report. Joint industry unsaleables steering committee, food distributors international, food marketing institute and grocery manufacturers of America

Martin W., Goodrich S. 1987 Minimizing sample size for given accuracy in cycle counting. Prod inventory manage, 28

Mason M., Langford S., Supple J., Spears M., Lee R., Dubash J., Roth L., Subirana B., Sarma S., Ferguson C. 2006 Electronic proof of delivery. Technical report, EPCglobal. http://www.epcglobalinc.org/home/. Accessed 24 June 2009

Pergamalis D. 2002 Measurement and checking of the stock accuracy. Article in www.optimum.gr (www.optimum.gr/Knowledge_Center/articles/)

Raman A., DeHoratius N., Ton Z 2001 Execution : the missing link in retail operations. Calif Manage Rev, 43:136–152.

Rekik Y. 2006 The Impact of the RFID Technology in Improving Performance of Inventory Systems subject to Inaccuracies. PhD thesis, Ecole Centrale Paris

Rekik Y., Jemai Z., Sahin E., Dallery Y. 2007 Inventory inaccuracy in the retail supply chain: Coordination versus rfid technology. OR Spectr, 29:597–626

Rekik Y., Sahin E., Dallery Y. 2007 A comprehensive analysis of the newsvendor model with unreliable supply. OR Spect, 29:207–233

Rekik Y., Sahin E., Dallery Y. 2009 Inventory inaccuracy in retail stores due to theft: an analysis of the benefits of rfid. Int J Prod Econ, 118:189–198

Rekik Y., Sahin E., Dallery Y. 2008 Analysis of the impact of the rfid technology on reducing product misplacement errors at retail stores. Int J of Prod Econ, 112:264–278

Sahin E. 2004 A qualitative and quantitative analysis of the impact of Auto ID technology on the performance of supply chains. PhD thesis, Ecole Centrale Paris

Supermarket. 2001 Supermarket Shrink Survey. National Supermarket Research Group. http://www.docstoc.com/docs/19781149/Supermarket-Shrink-Report. Accessed 24 June

Schuster E.W., Scharfeld T.A., Kar P., Brock D.L., Allen S.J. 2004 The prospects for improving erp data quality using auto-id. Cutter IT Journal: In the Pursuit of Quality. http://www.cutter.com/itjournal/fulltext/2004/09/index.html. Accessed 24 June 2009

Ton Z. Raman A. 2004 The effect of product variety and inventory levels on misplaced products at retail stores: a longitudinal study. Technical report, Harvard Business School, June

Yano C.A. Lee H.L. 1995 Lot sizing with random yields: a review. Oper Res, 43:311–334

Young J.B. 1986 The limits of cycle counting. 29th Annual APICS international conference proceedings

Index

D.C. Ranasinghe et al. (eds.), *Unique Radio Innovation for the 21st Century*,
DOI 10.1007/978-3-642-03462-6, © Springer-Verlag Berlin Heidelberg 2010

Breinigsville, PA USA
11 October 2010
247163BV00002B/25/P